WAVES IN DUSTY SPACE PLASMAS

T0214155

ASTROPHYSICS AND SPACE SCIENCE LIBRARY

VOLUME 245

WAVES IN
DUSTY SPACE PLASMAS

by

FRANK VERHEEST

Sterrenkundig Observatorium,
Universiteit Gent,
Gent, Belgium

KLUWER ACADEMIC PUBLISHERS

DORDRECHT / BOSTON / LONDON

A C.I.P. Catalogue record for this book is available from the Library of Congress.

ISBN 0-7923-6232-2

Published by Kluwer Academic Publishers,
P.O. Box 17, 3300 AA Dordrecht, The Netherlands.

Sold and distributed in North, Central and South America
by Kluwer Academic Publishers,
101 Philip Drive, Norwell, MA 02061, U.S.A.

In all other countries, sold and distributed
by Kluwer Academic Publishers,
P.O. Box 322, 3300 AH Dordrecht, The Netherlands.

Printed on acid-free paper

CONTENTS

PREFACE

Dusty plasmas

Plasmas and dust are both ubiquitous ingredients of the universe. The interplay between both has opened up a new and fascinating research domain, that of dusty plasmas, containing charged dust grains besides the usual plasma constituents. The original impetus was given by the Voyager observations in the early 1980s, that showed phenomena in the rings of Saturn which could not really be explained on purely gravitational grounds alone. Telltale were the spokes in the B ring and the braids in the F ring, the latter ring itself being discovered by these missions. Other examples in the solar system are circumsolar dust rings, noctilucent clouds in the arctic troposphere as the closest natural dusty plasmas, or even the flame of a humble candle. Other dusty plasmas occur in the asteroid belt, in cometary comae and tails, in the rings of all the Jovian planets and in interstellar dust clouds, to name but a few.

Charging of typical micron-sized grains can lead to several thousand electron charges, for masses of million to billion proton masses. The ensuing, extremely small charge-to-mass ratios lead to new plasma eigenmodes at the very low-frequency end of the spectrum, the full implications of which have not been completely digested yet. A truly self-consistent study of various types of modes in dusty plasmas is, however, still in its infancy, the many papers published during the last decade notwithstanding. Many obstacles prevent analogies with standard plasmas to be fully exploited, since grain charges are determined by the plasma potentials and can fluctuate, and dust comes in all sizes, in an almost continuous range from macromolecules to rock fragments.

The daunting task to describe such a variety of sizes, masses and of course also charges by any form of tractable distribution has so far deterred all but preliminary efforts. Naturally, most efforts have gone in modelling the dust grains as if they were an additional, heavy ion species. Even within these severe limitations, however, the study of dusty plasmas, and of the wave modes possible therein, have yielded a great many interesting and new results.

The scarcity of adequate space missions and observations in the planetary realm had caused theory to surge far ahead of observations. Luckily, the thread has been picked up by experimentalists, who have produced an impressive amount of verifications of dusty plasma phenomena during the present decade, and moreover given us the real pleasure of seeing these plasmas with the naked eye! In addition, the very recent upswing in solar system missions, both launched and considered, now holds the promise that the three strands of dusty plasma research, namely space observations, theoretical modelling and experimental verifications, will become nicely woven together again in the next decade.

About this book

From the little which has been said about dusty plasmas so far, it is clear that this monograph can only try to give a flavour of such a new and fascinating subject.

There have been authoritative reviews, dealing with space applications, like those of Goertz [1989], Mendis and Rosenberg [1994] and Horányi [1996], in addition to conference reports and proceedings. At the time of writing, however, only one monograph on dusty and self-gravitational plasmas in space had been published [Bliokh *et al.* 1995], which is rather narrowly focused on self-gravitational aspects. Hence, after summarizing the state of the art about waves and instabilities in space plasmas some years ago [Verheest 1996], I felt the time was ripe to give a more detailed coverage, incorporating also many of the newer results. Very recently Bouchoule [1999] edited a book on the more technological aspects of dusty plasma research, in the laboratory and in industry, which is the ideal complement of what I am trying to cover.

However, when writing this book, I was quickly confronting old but ever vexing questions: who would be reading it, what could I assume to be known, from what level should I start? Is it that only experienced researchers, already well versed in dusty plasma physics, will turn to this book for reference, or can I also help newcomers to this exciting field? Of course, space does not permit me to develop the subject totally *ab initio* like one can do in real textbooks, but on the other hand it would have to be more than a mere compendium about dusty plasma waves. Hence my attempt to steer a middle course, by assuming that the reader is not totally innocent of some basic notions about plasma physics.

In many ways, dusty plasmas can be viewed as the multispecies plasma *par excellence*. Nevertheless, multispecies plasmas are not really familiar to many plasma physicists, more used to think in terms of electrons and protons. This has led me to try and give a fairly self-contained and logical expose, where some of the familiar waves in plasmas are briefly recalled, before we venture out to newer applications. In this way, I feel, new dusty plasma modes can be linked to known waves, in a structured progression. Of course, the latter reflects my own taste and inclination, and if this explicit approach is too pedantic for some, I hope it will serve others well.

Acknowledgments

It really is a great pleasure to thank many people! But let me begin by the beginning.

I always have been interested in multispecies treatments of plasma waves and instabilities, not only because of their intrinsic potential, but also due to the attraction of a neat, systematic description. However, this strand of research has in the past long been regarded as of rather academic, not to say limited interest, and it is my good friends G.S. Lakhina and P.K. Shukla who rekindled my efforts in this direction, first for cometary, and then for dusty plasmas. Especially P.K. Shukla involved me, and I guess many others, in dusty plasma research, for which he has been an enthusiastic and tireless advocate and, equally important, an efficient organizer. Needless to say, the field has rapidly progressed.

Many of my other co-authors have also been lured to this new field and all the discussions with them have greatly contributed to my understanding of the subject. In particular (and in alphabetical order to be on the safe side!) I would like to mention here, as a small token of my gratitude: B. Buti, N.F. Cramer, P. Meuris, and N.N. Rao. Furthermore, great mentors and friends like P.K. Kaw, J.F. Lemaire, D.A. Mendis, F.W. Sluijter and L. Stenflo have influenced and supported me over the years.

In addition, I would especially like to thank V.M. Čadež and P. Meuris, not only for the stimulating way in which we interacted, but also for their patience in reading through earlier drafts of the manuscript and suggesting many clarifications and improvements. Of course, in spite of their efforts, all remaining idiosyncrasies are fully mine! They have also helped with the figures, as did G. Jacobs and H. Vermis.

For many years my research has been materially supported by the Fund for Scientific Research – Flanders. More recently new avenues have opened. The Flemish Government (Department of Science and Technology) and the (South African) Foundation for Research Development have, in the framework of the Flemish-South African Bilateral Scientific and Technological Cooperation on the "Physics of Waves in Dusty, Solar and Space Plasmas", allowed me far greater contacts and deeper interaction with R. Bharuthram, M. Goossens, M.A. Hellberg, R.L. Mace, and S.R. Pillay than were possible before. Then the International Space Science Institute (Bern, Switzerland) helped me in two ways, once as a member of the International Team on "Dust Plasma Interaction in Space", and on another occasion, in offering me as a Visiting Senior Scientist a gracious and efficient haven, where part of this book was finalized. It is a pleasure to express here my gratitude to the ISSI Directors, Professors J. Geiss, B. Hultqvist and R. Von Steiger. Finally, the Bijzonder Onderzoeksfonds of the Universiteit Gent is now my main and generous sponsor.

A special word of gratitude also goes to E. De Geus, then at Kluwer

Academic Publishers, who first suggested I work on this book, and to his successor at Kluwer's, H. Blom, who helped with the material preparations, and provided for a quick publication schedule.

Dedication

Last but very much not least, I wish to thank my wife Francesca and dedicate this book to her. Her love has provided me with the peace of mind so essential to continue, nay persevere with this work, and with the strength to see it through to the end, to what is now in your hands.

CHAPTER 1

PLASMAS AND DUST

1.1 Plasmas as the fourth state of matter

Life on the surface of the Earth has made us familiar with three states of matter, the solid, liquid and gas phases. The transition is easily described in terms of increasing temperatures, whereby the constituent atoms and molecules have ever more freedom of movement. When the atoms in a gas are heated further and accumulate enough energy, they decompose into ions and electrons and the gas becomes fully or partially ionized. We have thus reached the plasma state, in this picture the fourth state of matter. Ancient man recognized four holy elements: Fire, Earth, Water and Air, and the latter three readily symbolize the solid, liquid and gas states of matter. What about Fire? The flame of a burning candle is ionized, as we now know, and thus a plasma. Even better, it is contaminated with soot and other remnants of the burning material, and as such, a perfect example of a dusty plasma.

However, before we go on with the multiple facets of dusty plasmas as a modern-day incarnation of Fire, let us come back to plasmas in general. Since a plasma is made up of electrically charged particles, it is strongly influenced by electromagnetic phenomena, in sharp contrast to neutral gases. Plasma physics hence needs to draw on many different branches of physics, and is rightly seen as a fascinating but demanding subject, the frontier of classical physics. Of course, at terrestrial temperatures and shielded by the mostly neutral atmosphere above our heads, most of us think of plasmas as very exotic, needing rather special circumstances to create and sustain them, even though many modern contraptions in daily life, like fluorescent tubes, contain plasmas when operating.

This myopic view is shattered when we leave the Earth's surface, and encounter an ionized atmosphere, the ionosphere, already at altitudes above 80 km. Further exploration of the whole environment of the Earth shows that the neutral atmosphere occupies only a tiny fraction of the magnetosphere, defined as the region where the Earth's magnetic field dominates all plasma behaviour. Then we reach interplanetary space, which is not empty as previously imagined but pervaded by electromagnetic fields and ionized media like the solar wind. Our star, the Sun, is so huge and hot that it is a plasma rather than a gas sphere, and so are the other stars. The solar magnetic field dominates what is called the heliosphere, the plasma environment encompassing the solar system. Beyond that, even interstellar space is now known to contain plasmas, albeit very tenuous ones. We can go on like this, and truly empty space probably does not exist, because some form of very dilute plasma is always present.

In the end, plasmas are so prevalent in astrophysics that Hannes Alfvén (Nobel Laureate 1970) coined the expression *Plasma Universe* to really express what is dominant. In this sense, it is often said that 99% of the observable universe is in the plasma state. It is through the electromagnetic radiation, originating from these astrophysical plasmas and reaching us here on Earth, that we have learned most of what is known about the Universe. Far from being exotic, plasmas are the normal state of affairs, and it is we who are living in an essentially non-plasma corner! This realization has given a tremendous impetus to the development of plasma physics, coupled to and spurred also by the quest for less polluting, less dangerous and more lasting sources of energy in the form of fusion rather than fission reactors or fossil fuel burning.

This interest and research in various aspects of plasma physics and its applications has been recorded in countless research papers and a whole host of textbooks and research monographs. Some of these will be referred to later, in particular the ones specifically dealing with space and astrophysical plasmas.

1.2   Dust

If the claim is made that more than 99% of the observable universe is in the plasma state, then it can be jokingly asserted that the remainder is dust. In this context dust means many things to different people, and we need to be more specific. Dust particles are of macroscopic dimensions, compared to atoms and ionized nuclei, and are typically of the order of a micron. The composition greatly varies with the astrophysical environment, and it can be carbonaceous, silicate, or ferrite in nature, or an alloy of these. Furthermore, macromolecules and frozen ices are also part of the family of dust grains. Indeed, closer inspection reveals that the transition from gas to large dust particles in astrophysics is almost continuous, from macromolecules, clusters of molecules, small sub-micron sized grains, micron-sized grains, larger grains and boulders to asteroid remnants, etc.

The presence of dust in astrophysical environments has been known for a long time, from different types of remote observations, as for the dust around and between stars. There are beautiful examples of dust, like the molecular clouds seen in the Orion, Coalsack, Horsehead and Eagle nebulae. We observe these because of the attenuation and extinction of the light coming from more distant stars.

There is also plenty of dust in the heliosphere, associated with planetary rings, cometary comae and tails, meteoric impacts, zodiacal light, to name only some of the more conspicuous occurrences. For the solar system, of course, we do not only rely on the traditional remote observations that served us well, like for the more distant astrophysical objects, but have also collected a wealth of information from various space missions. These *in situ* measurements have been a resounding success, tremendously expanding our knowledge about space plasmas and the conditions in which they occur. Nevertheless, by far not enough of these have been carried out to solve all our riddles, to answer all our questions, and great store is set in future space missions. The reverse of the coin is that we have been forced to vastly increase the sophistication of our models, so that some of our too simple ideas and naive pictures had to be radically altered or even abandoned.

1.3   Dusty plasmas

And thus we come to the interplay, in space and astrophysics, of the all pervading plasmas and the ubiquitous dust. As dust grains are more often than not immersed in ambient plasmas and radiative environments, they will become electrically charged by various processes, the most simple one being capture of electrons and ions from the plasma. Other mechanisms will be discussed in the next chapter. Although the idea of dust grain charging was put forward more than half a century ago, by Spitzer [1941], it only came back on the agenda of space physicists when the very successful missions to planets and comets brought compelling evidence for the existence, indeed necessity of dusty plasmas. Otherwise, if all the dust were strictly neutral, some of the observed phenomena could not properly be explained.

Dusty plasmas is a somewhat ambiguous term for the mixture of charged dust grains with the electrons and ions we find in normal plasmas. Two of the new and unexpected observations by the Voyager missions in the 1980s were in the rings of Saturn [Smith *et al.* 1981,1982]. Most intriguing were deviations in the rings from what we could expect from a classical, purely gravitational point of view. One issue concerns the observations of radial spokes in the B ring [Smith *et al.* 1982] as illustrated in Figure 1.1. These are not permanent structures, yet live long enough so that their rotation and dynamics can be observed. Various explanations have been put forward, either in terms of combinations of gravitational and electromagnetic forces on individual charged grains, or else through the generation of wave modes. More about these competing ideas will be given at later stages, first when discussing space observations in Chapter 3, and after that, in subsequent chapters, when reviewing the manifold wave types possible in dusty plasmas.

Then there is the braiding observed in the F ring [Smith *et al.* 1981] as visible in Figure 1.2. This even more transient and elusive phenomenon has defied generally agreed explanations. As the Voyager missions revealed that all the Jovian planets have a more or less well developed ring system,

Figure 1.1. Images of spokes in the B ring of Saturn. *Voyager 2 photograph, reprinted with permission of JPL and NASA*

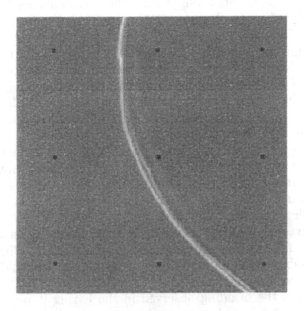

Figure 1.2. Braiding of the F ring of Saturn. *Voyager 1 photograph, reprinted with permission of JPL and NASA*

further enigmas arose.

It is fair to say that these pictures, in particular those coming from Saturn, provided the main impetus for the study of dusty space plasmas in earnest. Ever since, the field has grown rapidly, fueled by a mixture of space observations, theoretical interest in challenging and unforeseen aspects of pure plasma physics problems, and last but not least, compelling experimental evidence. The excitement of being able to see dusty plasmas in the laboratory with the naked eye has been allied to the beauty of doing fundamental but yet rather simple experiments, simple at least compared to their fusion counterparts with all the inherent complications. Experimentally one has also been able to investigate strongly coupled dusty plasma regimes and dusty plasma crystals. Although not really expected to be of direct importance in space, strongly coupled and crystal-like dusty plasmas have opened up new avenues in our understanding of how system of macroscopic charges can interact and influence each other.

1.4 Basic properties

Once a dust grain is charged, it starts to react to electromagnetic forces, in addition to the ordinary gravitational ones. Hence there is a coupling to the ambient plasma, through electromagnetic fields, whether these are pre-existent or self-induced. Such a coupling tends to get stronger as the grain size decreases. Boulders in orbit around Saturn, even when charged, would still have their Keplerian motions determined exclusively by gravitation, as electromagnetic forces are then far too weak. On the other hand, electrons and ions are usually affected by electromagnetic rather than gravitational forces. The intermediate range is most interesting, with charged dust where most of the forces come into play. It turns out that there is enough, albeit circumstantial evidence, that micron-sized or even smaller grains dominate, in the solar system at least. It is this average size that will be kept in mind when discussing wave modes in dusty space plasmas.

One can think of three characteristic length scales for such a combined dust and plasma mixture. First of all, there is a typical dust grain size a, then comes the plasma Debye length λ_D, and finally an average inter-grain distance d, roughly related to the dust density n_d by $n_d d^3 \sim 1$. The definition of a proper Debye length is not trivial, but will be discussed later.

For cosmic plasmas containing charged dust there are essentially two regimes, depending on the concentration of the dust grains. In both cases the size of the dust grains is the smallest of the three lengths. One can treat the dust from a particle dynamics point of view, provided $a \ll \lambda_D < d$, and in that case we speak of a plasma containing isolated screened dust grains, or of dust-in-plasma. On the other hand, collective effects of the charged dust become important when $a \ll d < \lambda_D$, and it is for this latter case that I would prefer to reserve the name of dusty plasmas, as now the dust is an essential ingredient of the total plasma mixture.

1.5   Analogies and differences

When we turn to collective phenomena, observations and charging models suggest that typical micron-sized dust grains can have very high negative charges, up to 10^4 e, and in proportion to electrons and ordinary ions even higher masses, some $10^6 \sim 10^{18}$ m_p [De Angelis 1992]. Here e is the unit charge and m_p the proton mass. I will, in what follows, use averages for dust grains with similar characteristics, which is of course far from true in general, but serves to fix the ideas. In Chapter 9 I will address possible treatments of charge and mass distributions of grains with different sizes.

There are several immediate consequences of such typical dust grain charges and masses. One is that the frequencies characteristic for the dust components are considerably smaller than the corresponding electron or ion quantities, as seen below. First comes the dust plasma frequency ω_{pd}, defined through

$$\omega_{pd}^2 = \frac{N_d Q_d^2}{\varepsilon_0 m_d} = \frac{Z_d(N_i - N_e)e^2}{\varepsilon_0 m_d} = \frac{Z_d m_i}{m_d}\omega_{pi}^2 - \frac{Z_d m_e}{m_d}\omega_{pe}^2, \quad (1.1)$$

with $Q_d = Z_d e$ and N_d referring to the equilibrium charges, charge numbers and number density of the dust. The electron and ion plasma frequencies, ω_{pe} and ω_{pi}, respectively, are given through $\omega_{pe}^2 = N_e e^2/\varepsilon_0 m_e$ and $\omega_{pi}^2 = N_i e^2/\varepsilon_0 m_i$. Here N_e and N_i are the electron and ion equilibrium densities, m_e and m_i the corresponding masses, and ions are taken to be singly charged, which is not a serious restriction.

I have assumed that the grains are preferentially charged negatively, which seems to be true in many solar system dusty plasmas, and that in equilibrium there is complete charge neutrality, expressed by

$$N_i = N_e + N_d Z_d. \quad (1.2)$$

We will later have to look at the variations in the charges themselves due to fluctuations of different kinds in the plasma as a whole.

Similarly, there is the dust gyrofrequency in absolute value

$$\Omega_d = \frac{|Q_d|B_0}{m_d} = \frac{Z_d e B_0}{m_d} = \frac{Z_d m_i}{m_d}\Omega_i, \quad (1.3)$$

the ion gyrofrequency being given by $\Omega_i = eB_0/m_i$. The latter itself is much smaller than the absolute value of the electron gyrofrequency, $|\Omega_e| = eB_0/m_e$.

Due to the unusually low charge-to-mass ratio of the dust, the characteristic dust frequencies are well below those typical for an ordinary hydrogen plasma, so abundant in stellar and cosmic plasmas, and in laboratory experiments. As a consequence,

$$Z_d m_i \ll m_d, \tag{1.4}$$

in other words, the dust mass per unit charge is very large. Small characteristic frequencies will give rise to new low-frequency eigenmodes of the combined plasma, clearly separated from the usual plasma modes. The latter, of course, can also occur, but sometimes appreciably modified by the charged dust.

A second important change, compared to ordinary plasmas, comes from charge neutrality in equilibrium, which dictate that for negatively charged dust there are more free protons than electrons. This will also affect wave phenomena, sometimes to a considerable extent. At strong enough electron depletion, the electron and ion plasma frequencies could even become of the same order of magnitude, a situation not encountered in ordinary plasmas.

On top of all that, one of the really outstanding distinctions with ordinary plasmas is that the dust charges are not fixed at all like for elementary particles or ionized nuclei, but can fluctuate with perturbations in the plasma potentials responsible for the charging, as will be discussed in the next chapter. This has proved to be one of the main stumbling blocks for the theoreticians to arrive at a proper self-consistent treatment. In addition, the scientific community is far from unanimous about how the charged dust grains interact between themselves!

Hence many of the efforts have gone towards describing the charged dust in first instance as one or more negative (or positive) ion fluids, sometimes with variable charges, besides the classic electron and proton fluids. This assumes a pointlike, heavy-ion character for the dust grains, and the main distinction with the usual multispecies or multi-ion treatments then lies in the widely separated frequencies and spatial scales associated with the dust component of the plasma mixture. This simple minded picture has served us well, at least to get a feeling for typical dust modes and modifications to existing plasma modes.

Of course, fluctuating dust charges mean that electrons and protons are given up or captured by the dust, so that source and/or sink terms have to be incorporated in the different electron and proton equations. Plasma densities, momenta and energies are then no longer conserved per species. Otherwise the system has to viewed as an open one, with new but still largely uncharted possibilities. The precise forms of these source/sink terms are highly nontrivial, but imply new electrostatic and electromagnetic instabilities, both at the linear and the nonlinear level.

Nevertheless, at the end of the day, charged dust grains are not simply heavy ions, but the special, complicated character of their charging and interaction properties still awaits proper self-consistent descriptions. Little analytical progress has been made here, but is has to be admitted that the task is daunting. In addition, as alluded already to in the beginning of this section, the dust grains come in a range of masses and sizes, and also this aspect calls for an appropriate treatment. Because of all the complexities sketched here, dusty plasma research is still a wide open field.

1.6   Reviews and books

Besides many well-known introductions to plasma physics in general, there have recently been a spate of books more concerned with the astrophysical aspects of plasmas. Among the more readable and valuable additions to this field I can cite here the textbooks by Cravens [1997], Choudhuri [1998], Kallenrode [1998] and Krishan [1999]. Of course, dusty plasmas are too young a specialty to be treated in these books in any but the most cursory way, if at all.

The vast output of scientific papers in the research literature will be dealt with in the next chapters, when we delve further into the subject. There have also been a string of authoritative reviews, which helped the domain along by providing the necessary overviews and summaries. In chronological order and restricting myself to the ones mostly dealing with space applications, I have greatly benefited from the valuable reviews by Goertz [1989], Mendis and Rosenberg [1994] and Horányi [1996]. More specifically for waves and instabilities in space plasmas I summarized the state of the art some years ago [Verheest 1996].

Attention is also drawn to several conference reports and proceedings. Among these, there is the trail-blazing topical section of *Physica Scripta* [45, 465–544, 1992], followed by the proceedings of several meetings exclusively devoted to dusty plasmas, like the ones in San Diego 1995 [Shukla *et al.* 1996], Goa 1996 [Shukla *et al.* 1997a], Boulder 1998 [Horányi *et al.* 1998] and Hakone 1999 [Nakamura *et al.* 2000]. Finally, at the time of writing, only one monograph on dusty and self-gravitational plasmas in space had been published, the one by Bliokh *et al.* [1995], to which I will return in Chapter 8 when dealing with self-gravitational effects. The very recent book by Bouchoule [1999] and co-workers is heavily oriented towards the technological aspects of dusty plasmas in the laboratory and in industry. It thus promises to be a valuable source of information for those aspects that are totally outside the scope of the present monograph.

1.7   Structure of the book

After these preliminaries, I will start in Chapter 2 with a discussion of different dust charging mechanisms, and give some of the single-particle behaviour that can be derived from that. I also include there a brief description of more recent experimental results. Next I detail in Chapter 3 some of the observational results from space missions, before waves and instabilities are dealt with in later chapters. I wanted to include these observational results, because too many papers on wave phenomena in dusty plasmas describe really interesting phenomena, that are, alas, only very qualitatively related to what we see. It is great time, especially for dusty plasmas in space, that the real connections between theory and observations are made.

I have long been in two minds about the order of these two chapters. It would have been very natural to start with the space observations, since that is what set everything in motion. However, a proper understanding of these observations requires some notions about the charging mechanisms, and there is no simple way out of this chicken-and-egg dilemma. In the end I have opted to first deal with the charging mechanisms, supported by experimental evidence, and only then go the observations. Moreover, these chapters are not strictly needed to understand the theoretical developments included in the following chapters, although they reflect the underlying theme.

The remaining chapters then follow a more or less logical order. Chapter 4 addresses the general framework in which waves and instabilities will be treated, by going from a general description in phase space to multi-species fluid treatments. I found it necessary to include sufficient details, so as to better delineate the different approximations made further on and also to make the treatment reasonably self-contained. However, discussions on how a dusty plasma should really be described are nowhere rounded off, or more precisely, generally agreed upon kinetic theories are still in the process of being developed. This chapter thus is certainly far from definitive, even though its, maybe naive, logical progression might lead you to think otherwise!

In Chapters 5 and 6 I review electrostatic and electromagnetic waves, in a mixture of rather detailed theoretical steps combined with summaries of what different experts have contributed. The precise degree of how much detail to include is of course open for debate, but has a double aim. One is to facilitate the introduction of new modes and concepts by linking them to what is known from standard analysis, and exploit the analogies also for beginning plasma physicists, without having to read the present monograph in parallel with an existing treatise. In this way I hope that the literature is put in a systematic perspective. Another goal is to clear up some of the misunderstandings that have cropped up in the literature, that otherwise might propagate unchecked.

Whereas in these two fairly lengthy exposes charged dust grains are dealt with as if they were purely heavy ions, it is then time to add on specific but complicating features. One is that dust charges can fluctuate.

Because we have no real theory here worth the name to describe this, I will use the phenomenological approach though specific attachment frequencies, as it has found its way in the literature. That is the way how Chapter 7 is structured, by going back and enlarging upon the material presented in Chapters 5 and 6.

In Chapters 8 and 9 I abord waves and instabilities which involve possible self-gravitational effects and mass and size distributions. Here the discussion centers on the two faces of the Jeans instabilities that are so characteristic for problems involving self-gravitation. Mass and size distributions have not been studied extensively, and also here the paucity of analytical models necessitates a simple minded approach for the time being. For the sake of completeness, I have included in Chapter 10 some results on other modes, without going in too much theoretical detail, however, otherwise the monograph would become too bulky.

Finally, I conclude in Chapter 11 with a summing up of what has been achieved so far, and try once again to relate theoretical models to real dusty space plasma waves. Of course, in such a rapidly expanding domain of space plasma physics, nothing is set in stone and the real conclusions are rather an outlook, to future work and to a whole range of open problems.

CHAPTER 2

CHARGING MECHANISMS AND EXPERIMENTS

2.1 Grain charging

Before we can embark on a review of waves in dusty plasmas, a closer look at grain charging processes is in order. Much of the material in this chapter has been gleaned and adapted from what Peter Meuris reviewed in his PhD thesis [1997b]. As stated in the previous chapter, I prefer to first discuss different charging mechanisms, before going to the observational evidence from space missions in the next chapter.

Furthermore, towards the end of this chapter I will briefly describe some of the experimental results and indicate verifications of dusty plasma modes. In addition, I will include here possible explanations of typical planetary ring phenomena in terms of single particle dynamics, rather than through waves and instabilities. These two topics are really outside the scope of the present monograph, but have to referred to at some stage, and I felt they did not belong in the later chapters, where the treatment of wave modes flows in a logical order from the simple to the more complicated.

The charging of a dust grain embedded in a plasma is a very elementary, and easily posed problem: we put a dust grain in a plasma, and ask ourselves what charge it will get, and how this will depend on the local plasma parameters? And what really happens when there is not one, but a collection of dust grains? Thus formulated, this old problem looks deceptively simple, but has turned out to be very intricate and the fascination with it has never really ended.

It was first tackled by Mott-Smith and Langmuir [1926] in the framework of probe theory. Indeed, the analysis of the floating potential of a plasma probe closely mimics the computation of the equilibrium potential of a dust grain in a plasma. Such a dust grain can draw no net current and has to be considered as a probe at the floating potential. Solving such a charging problem can become quite difficult, aspects of which are even today far from clear. To fix the ideas in most of what comes, I will usually think of micron-sized grains, the ones that have been documented from space missions as prominent in heliospheric plasmas, albeit indirectly.

Let us consider a single dust grain immersed in a plasma environment. Its surface will continually be bombarded by incident charged particles and photons. Some of the particles can be captured by or will stick onto the dust grain, resulting in a change of the grain charge at random intervals, with probabilities that depend on the grain potential. This mechanism is called primary charging and will be discussed in more detail below, as will be other mechanisms.

Initially, the electrons being far more mobile than the other plasma particles, they will impact more onto the grain surface, and the grain will become negatively charged, acquire a negative potential with respect to the ambient plasma. Immediately the random motions of the ions and electrons in the neighbourhood of the charged grain are disturbed. Electrons are repelled and ions are attracted. Of course, as the grain gets more negatively charged, the probability of another electron hit decreases, due to electrostatic repulsion. Similarly, the chance to be hit by or interact with a positively charged ion will increase. Ultimately, the electron flux is reduced by repulsion just enough to balance the ion flux. On a totally different scale, similar charging occurs on spacecraft when returning in the denser atmosphere, as friction creates a plasma around the vessel which then charges negatively.

In reality, however, this sketch of what happens is far too simple. Some of the incident electrons, especially those with a high enough energy, can pass through the grain, before being stopped (tunnelling). Moreover, dust particles can reflect and scatter (elastic and inelastic electron reflection) or liberate secondary electrons (true secondary electron emission), while incident energetic photons can cause the emission of photoelectrons (photo-emission). All these processes result in a charge transfer between the plasma and the grain, which as a consequence might even become positive.

When the dust grain acquires a high potential, the electrostatic repulsion of the surface charges produces a tension in the grain. If the grain stability and internal coherence are insufficient, this tension can cause electrostatic disruption. Furthermore, when a critical surface electric field is exceeded, electrons can be emitted by a negatively charged grain (electron field emission), or similarly ions by a positively charged grain (ion field emission). For a spinning grain the centrifugal stress can also cause disruption. You can probably think of other physical phenomena that might be important.

The rate at which charge transfer occurs depends not only on grain properties (like charge, geometry and composition) but also on the plasma parameters (like densities and temperatures) and on other characteristics (like relative grain-plasma velocities and ambient magnetic fields). The rates for these different charging mechanisms are I_α, I_t, I_s, I_r and I_p, respectively for the primary currents due to capture of species α ($\alpha = e$ for electrons and $\alpha = i$ for ions), the electron tunnelling current, the secondary electron current, the reflected electron current and the photo-emission current.

Throughout this chapter and indeed most of this book, I shall assume that a steady state exists for these charging processes. Whether this as-

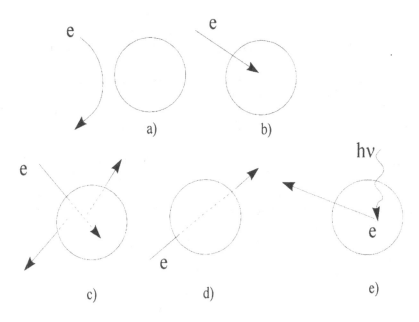

Figure 2.1. Different charging mechanisms: (a) reflection, (b) primary charging, (c) true secondary electron emission, (d) tunnelling, and (e) photo-emission

sumption is reasonable cannot be answered, unfortunately. Since the exact nature of the interactions between the charged grains in a plasma is still hotly debated, together with the open or closed nature of the combined system, definite answers have to wait. Unless I close the book here, I have to proceed with the imperfect picture we have formed so far, consoled by the idea that some of the experimental results seem to tally very well with the theoretical models used up to now. This will be further discussed towards the end of this chapter.

A steady state is reached when the sum of all currents to the grain surface is zero, and determined by a nonlinear equation in V_0

$$\frac{dQ_d}{dt} = \sum_\beta I_\beta(V_0, ...) = 0. \tag{2.1}$$

Here V_0 is the equilibrium value of the grain surface potential V. In all what follows, equilibrium values will be denoted, either by a subscript 0 or, when possible, by using the corresponding capital letters to avoid too many subscripts, like Q_d for the equilibrium value of the grain charge q_d. It needs to be stressed that, depending on the specific form of the different charging currents, this equation might not have a unique solution. Especially when secondary charging becomes important, hysteresis effects can occur. The equilibrium potential of a grain then depends on its history, producing both positively and negatively charged particles in the same plasma environment [Meyer-Vernet 1982]. These findings were confirmed experimentally [Walch et al. 1995].

An interesting parameter is the characteristic charging time t_{ch}, defined more or less arbitrarily as the time needed for a neutral grain to reach 90% of its equilibrium value. This parameter may be regarded as a time constant for the charging mechanism, even though the charging process is highly nonlinear.

Furthermore, in most of the cases the charging process is treated as if it were a continuous process, whereas the real charging is obviously discontinuous. Plasma particles are absorbed at the grain surface at random times and in a random sequence, resulting in equilibrium charge fluctuations. The discontinuous character of the charging process was studied by Cui and Goree [1994] for the primary charging currents, and later by Khrapak et al. [1999]. They find that the fractional rms fluctuation level is given by the simple law

$$\frac{< (q_d - Q_d)^2 >^{1/2}}{|Q_d|} = \frac{1}{2}\sqrt{\frac{e}{|Q_d|}}, \qquad (2.2)$$

where $< ... >$ stands for the average over the different charge levels. For a large object (carrying a large equilibrium charge) in a plasma (think of a spacecraft here) the fluctuations are totally negligible. On the other hand, for a tiny dust grain with an average charge of only a few electrons, the fluctuations are enormous, and the grain can have a positive charge momentarily, even in the absence of electron emission. In the remainder, I will assume, just for simplicity and tractability, that the charging process is continuous, as done in all papers reviewed in this and the other chapters.

When there is a unique equilibrium potential, but many grains are injected into or formed in an initially neutral plasma, they become negatively (positively) charged, leaving excess ions (electrons) in the plasma. Therefore the ion (electron) flux decreases and the grain does not have to become as negative (positive), in order to equalize the ion and electron currents to its surface. The resulting equilibrium charge hence is lower in absolute magnitude, than if the grain were completely alone, isolated [Goertz 1989; De Angelis 1992; Northrop 1992]. It is then necessary to examine the interaction of the different grains on the charging processes.

In addition, the resulting charging current has to be adapted when a strong external magnetic field is present, for then we would expect that the geometric cross section of the grain becomes $2 \times \pi a^2$ rather than $4\pi a^2$, because the charging will take place only by grains moving parallel to this magnetic field. A discussion of this will be given in a further section.

There are related problems, when the charged grains tend to shield part of the electron and ion fluxes from each other [Tsytovich *et al.* 1997], leading to possibilities of shadowing attraction rather than purely electrostatic repulsion between grains with similar charges. This novel and rather controversial shadowing is still in its infancy, as is attraction between identical grains when dipolar effects are fully taken into account, besides the usual monopole terms. Until there is some consensus among the scientists about the true quantitative implications of these ideas, I have to leave it at this, unfortunately.

Although the present chapter with a discussion of many charging mechanisms is needed for completeness, I shall mostly use a simple, phenomenological picture, with corresponding parameters, when reviewing different wave theories in the later chapters.

2.2 Charging mechanisms for isolated grains

In the following discussion the different charging mechanisms are treated in decreasing order of importance, at least for the astrophysical applications I have in mind. In view of its analytical simplicity, I start with and devote most of the attention to primary charging. An additional motivation for doing so is that the treatment of waves and instabilities in space plasmas, as far as the papers quoted further on are concerned, usually restricts the dust charging to primary electron and ion currents. This has emerged as the commonly used paradigm. First isolated grains will be considered, whereas grain ensembles are discussed in the next section.

2.2.1 Standard model: primary charging

The primary charging mechanism, described by the standard model, was first described by Mott-Smith and Langmuir [1926]. Suppose that all other charging mechanisms can be neglected for a single neutral grain, at rest in a Maxwellian two-component plasma with $c_{si} \ll c_{se}$. Here $c_{s\alpha}$ stands for the different thermal velocities, in this simple case given through $c_{s\alpha}^2 = \kappa T_\alpha / m_\alpha$, with κ being Boltzmann's constant, and T_α referring to the respective temperatures. The initial ion flux hence is smaller than the initial electron flux and it is mostly electrons that will hit the grain, the difference in cross sections notwithstanding. As the grain becomes more and more negatively charged, the ion flux increases but the electron flux decreases, until an equilibrium value is reached.

At this point the frequency of negative charges hitting the grain equals the frequency of positive hits, a dynamic equilibrium is reached and the floating potential of the grain is smaller than the plasma potential (V_p). Of course, only potential differences have a physical meaning and hence the difference $V - V_p$ is considered rather then the grain potential V proper. For isolated grains, the plasma potential equals the potential at infinity, and the latter can be taken to be zero ($V_p = 0$) for the remainder of this section. The charging process is driven by the difference between the instantaneous surface potential and the equilibrium potential ($\Delta V = V - V_0$).

Although the standard charging theory is probably too simple for some applications, it is a good starting point to see what the consequences are of various, commonly made assumptions. These include

- A steady state exists for the charging process.
- There is no external magnetic field.
- Only primary charging mechanisms are taken into account.
- Dust grains are identical spheres, perfectly sticky and conducting.
- Plasma particles have Maxwellian velocity distributions at infinity, ideally, far from the dust grain considered.
- Currents are orbital motion limited, based on the assumption that some particles of every energy range can graze the grain surface. Implicitly this excludes trapped orbits for the plasma particles.

When $f_\alpha(\mathbf{v})$ represents the velocity distribution at infinity, the charging current I_α due to plasma species α is given by

$$I_\alpha = n_\alpha q_\alpha \int_{v_0}^{|\mathbf{v}|=\infty} v\sigma_\alpha f_\alpha(\mathbf{v})d^3\mathbf{v}. \tag{2.3}$$

Here σ_α is the charging cross section and v_0 is the smallest particle velocity required to hit the grain. Furthermore, the capacitance C for a spherical grain equals [Goertz and Ip 1984; Whipple et al. 1985; Houpis and Whipple 1987]

$$C = 4\pi\varepsilon_0 a, \tag{2.4}$$

provided that $a \ll \lambda_D$, so that the dust charge becomes

$$q_d = 4\pi\varepsilon_0 aV. \tag{2.5}$$

The charging cross section can be found as follows. A plasma particle starts with a velocity v outside the Debye sphere of a dust grain, with impact parameter b_c. When it enters the Debye sphere, the particle feels the influence of the grain, and its path changes due to electrostatic forces. Consider a plasma particle grazing the dust grain surface, with a velocity v_g. When the impact parameter decreases, for a given velocity, the plasma particle will hit the grain, and therefore the collision cross section is given by πb_c^2. Momentum and energy conservation require that

$$m_\alpha v_g a = m_\alpha v b_c,$$
$$\frac{1}{2}m_\alpha v^2 = \frac{1}{2}m_\alpha v_g^2 + \frac{q_d q_\alpha}{4\pi\varepsilon_0 a}. \tag{2.6}$$

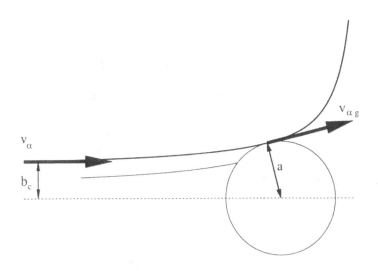

Figure 2.2. Grazing collision between an electron and a charged particle

Eliminate v_g from the above equations, use (2.4) and find that the cross section is given by

$$\sigma_\alpha = \pi b_c^2 = \pi a^2 \left[1 - \frac{q_\alpha q_d}{2 m_\alpha \pi \varepsilon_0 a v^2} \right]$$
$$= \pi a^2 \left[1 - \frac{2 q_\alpha V}{m_\alpha v^2} \right]. \qquad (2.7)$$

When $q_\alpha V < 0$, the particle and the grain attract each other. The integration in (2.3) is performed over the complete v domain and thus $v_0 = 0$. On the other hand, for $q_\alpha V > 0$, plasma particles with too small velocities are repelled by the electrostatic force and cannot reach the grain. The limiting orbit, which just reaches the grain surface, has to start with a velocity v_0, found from energy conservation as

$$E = \frac{1}{2} m_\alpha v_0^2 + q_\alpha V = 0. \qquad (2.8)$$

Furthermore, the velocity distribution is taken to be Maxwellian at sufficiently large distances

$$f_\alpha(v) = n_\alpha \left(\frac{m_\alpha}{2 \pi \kappa T_\alpha} \right)^{3/2} \exp \left[-\frac{m v^2}{2 \kappa T_\alpha} \right]. \qquad (2.9)$$

Substituting (2.9) and (2.7) in (2.3) and using spherical coordinates, it follows for an attractive potential ($q_\alpha V < 0$) that [Whipple 1981]

$$I_\alpha = \pi a^2 n_\alpha q_\alpha \sqrt{\frac{8\kappa T_\alpha}{\pi m_\alpha}} \left(1 - \frac{q_\alpha V}{\kappa T_\alpha}\right),$$ (2.10)

whereas for a repulsive potential ($q_\alpha V > 0$) this becomes

$$I_\alpha = \pi a^2 n_\alpha q_\alpha \sqrt{\frac{8\kappa T_\alpha}{\pi m_\alpha}} \exp\left[-\frac{q_\alpha V}{\kappa T_\alpha}\right].$$ (2.11)

Isothermal electron-ion plasmas

When the plasma is isothermal ($T_e = T_i = T$), and consists only of two plasma species ($n_e = n_i = n$), the grain will be negatively charged, and I can rewrite the expressions for the currents as

$$I_i = \pi a^2 ne \sqrt{\frac{8\kappa T}{\pi m_i}} \left(1 - \frac{eV}{\kappa T}\right)$$ (2.12)

and

$$I_e = -\pi a^2 ne \sqrt{\frac{8\kappa T}{\pi m_e}} \exp\left[\frac{eV}{\kappa T}\right].$$ (2.13)

In this case the evolution of the normalized surface potential $\chi = eV/\kappa T$ is obtained from (2.1) and (2.5) as

$$\frac{d\chi}{dt} = \frac{a\omega_{pe}^2}{c_{se}\sqrt{2\pi}} \left\{\sqrt{\frac{m_e}{m_i}}(1 - \chi) - \exp[\chi]\right\}.$$ (2.14)

The immediate conclusions are that

- The equilibrium solution of (2.14) is independent of the plasma density N and the grain size a, as it is determined only by the variables between the curly brackets.

- The natural time scale for the charging is given by

$$t_0 = \frac{c_{se}\sqrt{2\pi}}{a\omega_{pe}^2}.$$ (2.15)

The charging time for a neutral grain is proportional to $\sqrt{T}/(Na)$, and therefore larger grains, or grains embedded in a denser or colder plasma, will reach equilibrium faster. This can be readily explained [Choi and Kushner 1994], seeing that small dust particles charge slowly because they are small targets, and less plasma particles can hit them. This is also the case in very hot plasmas, because the final equilibrium potential is large. Quite to the contrary, dust particles in dense plasmas charge quickly because of the high plasma fluxes.

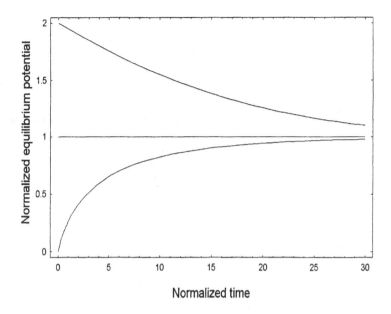

Figure 2.3. Charge evolution for a single dust grain embedded in a proton-electron plasma described by (2.14). Time is normalized by t_0 and potential by its equilibrium value

- From the numerical solution for a proton-electron plasma (Figure 2.3), I see that the equilibrium charge is obtained faster for grains that obey $|V| < |V_0|$. When a grain is "not negative enough", the charging will be faster than for a grain that is "too negative", due to the higher electron mobility.

Drifting electron-ion plasmas

In many cases, however, the different plasma species can have different equilibrium drifts (U_α) in the reference frame of the body to be charged [Whipple 1981]. The velocity distribution is then taken to be a Doppler-shifted Maxwellian

$$f_\alpha(\mathbf{v}) = \left(\frac{m_\alpha}{2\pi\kappa T_\alpha}\right)^{3/2} \exp\left[-\frac{m(\mathbf{v} - \mathbf{U}_\alpha)^2}{2\kappa T_\alpha}\right]. \qquad (2.16)$$

Substituting this in (2.3) and using spherical coordinates, one obtains for an attractive potential $(q_\alpha V < 0)$ that [Whipple 1981]

$$I_{\alpha,att} = \pi a^2 n_\alpha q_\alpha c_{s\alpha} \mathcal{F}_\alpha \left[\frac{U_\alpha}{\sqrt{2}c_{s\alpha}}\right], \qquad (2.17)$$

with

$$\mathcal{F}_\alpha(x) = \sqrt{2}x \left\{ \left(1 + \frac{1}{2x^2} - \frac{q_\alpha V}{\kappa T_\alpha x^2} \right) \text{erf}[x] \right.$$
$$\left. + \frac{1}{x\sqrt{\pi}} \exp\left[-x^2 \right] \right\}. \tag{2.18}$$

On the other hand, for a repulsive potential ($q_\alpha V > 0$) the current is given by

$$I_{\alpha,rep} = \pi a^2 n_\alpha q_\alpha c_{s\alpha} \mathcal{G}_\alpha \left[\frac{U_\alpha}{\sqrt{2}c_{s\alpha}}, \sqrt{\frac{q_\alpha V}{\kappa T_\alpha}} \right], \tag{2.19}$$

where

$$\mathcal{G}_\alpha(x, y) = \frac{x}{\sqrt{2}} \left\{ \left(1 + \frac{1}{2x^2} - \frac{y^2}{x^2} \right) (\text{erf}[x + y] + \text{erf}[x - y]) \right\}$$
$$+ \frac{1}{\sqrt{2\pi}} \left\{ \left(\frac{y}{x} + 1 \right) \exp\left[-(x - y)^2 \right] \right.$$
$$\left. - \left(\frac{y}{x} - 1 \right) \exp\left[-(x + y)^2 \right] \right\}. \tag{2.20}$$

As can be seen, the influence of the drift velocity is expressed as a function of $U_\alpha/c_{s\alpha}$. In most applications, this ratio becomes negligibly small for the electrons and hence (2.11) can be used for the electron charging current. For the ion charging current, however, the thermal velocity is much smaller, and it is more prudent to use (2.17) when drifts are present.

In Figure 2.4, we can see the variation of the equilibrium grain potential as a function of the drift velocity relative to the ion thermal velocity. One would expect the grain to sweep up more ions per time unit when moving than at rest, and therefore to be less negatively charged. This is indeed the case for high values of U_i/c_{si}. However, for low values of this parameter the grain becomes more negative, because the increase in ion flux to the front side, due to the motion, is more than offset by reduction in the flux to the backside.

Discussion of the assumptions

- The grains are modelled as spherical, for reasons of mathematical tractability. As will be shown later, nonspherical grains show a tendency to become more spherical by electrostatic disruption, and hence the simplifying assumption of sphericity is maybe not bad after all.

- For poorly conducting grains, the charges cannot easily redistribute themselves over the grain surface, itself no longer an equipotential

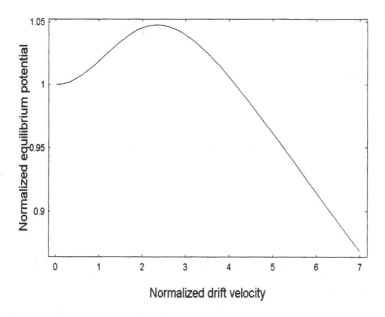

Figure 2.4. The equilibrium potential of a single dust grain embedded in a proton-electron plasma. The potential is normalized by its equilibrium value, while the drift velocity is normalized by the ion thermal velocity c_{si}

surface. Hence, the electric field near the surface has to be known in order to obtain the currents. This is a very complicated problem.

- Using the orbital motion assumption, particles are assumed to effectively graze the grain. This implicitly excludes trapped orbits, although in reality some of the plasma particles can be trapped by the grain. This is often included in probe theory by assuming that an absorption radius exists outside the probe, which in a sense replaces the probe radius proper. Electrons and ions which cross this absorption radius are bound to be collected by the probe [Allen 1992].

The discussion of the validity of the orbit-limited motion (OML) approach is still going on. Ideally, the equations of motion and Poisson's equation should be solved together for a large amount of plasma particles, which is far easier said than done, although some numerical simulations have been performed by Sonmor and Laframboise [1991].

To examine whether or not trapped orbits can exist, a brief analysis is given, using spherical coordinates with origin in the center of the grain. For the total energy E_α of a plasma particle of species α (attracted or repelled) one gets

$$E_\alpha = \frac{1}{2}m_\alpha \left(v_r^2 + v_\theta^2 \right) + q_\alpha V(r). \tag{2.21}$$

The canonical angular momentum component J_α about any z-axis is conserved and given by

$$J_\alpha = m_\alpha r v_\theta. \tag{2.22}$$

Combining these equations yields that

$$E_\alpha = \tfrac{1}{2} m_\alpha v_r^2 + U_\alpha(r), \tag{2.23}$$

where

$$U_\alpha(r) = q_\alpha V(r) + \frac{J_\alpha^2}{2m_\alpha r^2}. \tag{2.24}$$

These equations describe particle motion in one dimension with an effective potential $U_\alpha(r)$ and the analysis goes as follows.

In principle, it might be possible that the effective potential has a local minimum, causing trapped orbits. However, particles will not populate such orbits when there are no collisions. Indeed, particles coming from $r = \infty$ with a positive energy will be collected or reflected, but keep their initial energy, and therefore trapped orbits will be empty.

On the other hand, collisions will scatter particles, and a possibility of trapped particles exists if the effective potential has a local minimum. It is easily verified that if $V(r)$ decays faster with r than r^{-2}, such a minimum can exist. Trapped orbits will be populated due to collisions, and the OML-theory will not hold in that particular case.

2.2.2 Secondary electron emission

When an electron is absorbed by a dust grain, it may be stopped immediately (primary charging), pass through the grain (tunnelling) or release secondary electrons (true secondary emission). On the other hand, electrons may also get scattered (elastic and inelastic scattering).

Tunnelling

Tunnelling is distinguished experimentally from true secondary emission primarily by the energy of the emitted electrons. It has only recently been recognized that this phenomenon might be important for the charging of very small dust grains [Chow *et al.* 1993]. Assuming that the primary electron current density is conserved within the grain, and neglecting drift velocities, the electron tunnelling current can be computed as follows

$$I_t = \frac{8\pi^2 a^2 e}{m_e^2} \int_{\max[E_{min}, E_{min} - eV]}^\infty E f_e(E + eV) dE. \tag{2.25}$$

E_{min} denotes the minimum energy required for an electron to tunnel through the grain and f_e is the electron velocity distribution. This energy is given by $E_{min} = \sqrt{K_W a}$, with K_W the constant of Whiddington [Chow *et al.* 1993]. This constant is of the order of 10^{14} eV^2 m^{-1}, both for insulators and for conductors, and not really dependent on the composition of the grains.

Evaluating the integral (2.25) over the usual Maxwellian distribution yields for negatively charged grains $(V \leq 0)$

$$I_t = -\pi a^2 n_e e \sqrt{\frac{8\kappa T}{\pi m_e}} \exp\left[\frac{eV}{\kappa T}\right] \exp\left[-\frac{E_{min}}{\kappa T_e}\right]\left(1 + \frac{E_{min}}{\kappa T_e}\right), \qquad (2.26)$$

whereas for positively charged grains $(0 \leq V)$ this becomes

$$I_t = -\pi a^2 n_e e \sqrt{\frac{8\kappa T}{\pi m_e}}\left(1 + \frac{E_{min} + eV}{\kappa T_e}\right)\exp\left[-\frac{E_{min}}{\kappa T_e}\right]. \qquad (2.27)$$

This tunnelling effect is only of importance for very small grains, as the number of electrons with energy $E \geq 10^4 eV \sqrt{a_{\{\mu m\}}}$ is usually negligible.

True secondary emission

If sufficiently energetic particles are present, true secondary emission becomes important. Electrons are usually the most energetic plasma particles, and therefore it is mostly electron driven secondary emission that has been taken into account.

The secondary yield $\delta(E)$ is defined as the ratio of emitted to incident electrons with energy E. The secondary electron emission charging current can be written [Meyer-Vernet 1982; Chow *et al.* 1993] for negative grains $(V \leq 0)$ as

$$I_s = \frac{8\pi^2 a^2 e}{m_e^2} \int_0^\infty E\delta(E)f_e(E - eV)dE, \qquad (2.28)$$

and for positive grains $(0 \leq V)$ as

$$I_s = \frac{8\pi^2 a^2 e}{m_e^2} \exp\left[-\frac{eV}{\kappa T_s}\right]\left(1 + \frac{eV}{\kappa T_s}\right)\int_{eV}^\infty E\delta(E)f_e(E - eV)dE. \qquad (2.29)$$

Now T_s stands for the temperature of the emitted secondary electrons, which can adequately be described by a Maxwellian velocity distribution [Meyer-Vernet 1982], with T_s in the range $1 \sim 5$ eV. This tacitly assumes that the escaping electron flux is independent of the incident electron energy. For plasma temperatures of the order of T_s a (significant) fraction of the secondary electrons can escape with energies greater than the incident energy which is clearly unphysical, and the results need to be modified. A better approximation is obtained by describing the emitted secondary electrons by the same Maxwellian, but with a cutoff at the incident electron energy [Jurac *et al.* 1995].

For explicit expressions of $\delta(E)$ I refer to the papers mentioned and the references therein. In general $\delta(E)$ exhibits a maximum at an optimum incident energy E_m, indicating that low-energy primary electrons will not produce secondaries, because of their lack of energy, while energetic primaries penetrate deep in the grain and the produced secondaries cannot escape. Laboratory data are available, but the yield for dust grains can differ appreciably from these values, due to the known strong dependence of the yield on the physical and chemical structure of the surface. Whipple [1981] gives expressions for this.

Also, secondary emission starts at some threshold energy E_{th} for the impacting electrons and not at zero energy, as almost always assumed in derivations of $\delta(E)$. The surface barrier for insulators is determined by the electron affinity E_A, which is the energy difference between the vacuum level and the bottom of the conduction band. Only those electrons for which the component of kinetic energy perpendicular to the surface is greater than E_A will escape from the material. The resulting function for $\delta(E)$ corresponds to replacing E with E_{th} and E_m with $E_m - E_{th}$, and the threshold energy is usually between 5 and 10 eV for insulators. Therefore, especially in plasmas where the temperature is in the same range, significant mistakes can be made by ignoring this effect [Jurac *et al.* 1995].

It has also been shown [Chow *et al.* 1993] that two grains with the same history can have charges of the opposite sign, depending on their sizes. This is the case when besides the primary currents, the secondary electron and tunnelling currents are taken into account. The effects of tunnelling and secondary electron current are therefore strongly related to the grain size distribution.

Elastic and inelastic reflection

When discussing the total electron fluxes, we expect a peak due to true secondary emission electrons, and a second peak at the incident electron energy, corresponding to reflected electrons. The maximum corresponds to elastic reflection, while inelastically reflected electrons lose some of their energy in the interaction with the grain. The distinction between inelastically scattered and secondary electrons is somewhat arbitrary, but when the true secondaries are modelled as a (truncated) Maxwellian, the remainder of the flux can be considered as reflected electrons.

For negative grain potentials all emitted electrons can escape, regardless of their energy that is, and the reflected current can hence be computed using

$$I_r = \frac{8\pi^2 a^2 e}{m_e^2} \int_0^\infty E R(E) f_e(E - eV) dE. \qquad (2.30)$$

The reflection coefficient $R(E)$ denotes the mean number of ejected electrons per emitted electron by inelastically, as well as by elastically backscattered primaries.

For positive grains only electrons with sufficient energies can escape. Therefore, a model is needed for the reflected current as a function of the escaping electron energy. A semi-Gaussian velocity distribution (with a spread measured by T_{ref}) up to the elastic peak at incident electron energy can be used. The resulting current is [Jurac *et al.* 1995]

$$I_r = \frac{8\pi^2 a^2 e}{m_e^2} \int_{eV}^\infty E f_e(E - eV) R(E) \times$$
$$\times \left[1 - \frac{E}{\kappa T_{ref}} + \left(\frac{eV}{\kappa T_{ref}} - 1 \right) \exp\left(\frac{eV - E}{\kappa T_{ref}} \right) \right] \times$$
$$\times \left[1 - \frac{E}{\kappa T_{ref}} - \exp\left(-\frac{E}{\kappa T_{ref}} \right) \right]^{-1} dE. \qquad (2.31)$$

The reflection coefficient approaches zero as $E \rightarrow 0$, reaches a maximum below ~ 20 eV and decreases slowly at higher energies. Unfortunately, the number of available measurements of the reflection coefficient is very limited, especially at low energies.

Ion induced secondary electron emission

The impact of highly energetic ions on a target (sputtering) has been extensively examined in laboratory simulation, due to its use in the production of integrated circuits. Sputtering causes erosion of the target, and releases secondary electrons. Secondary emission due to ion impact is important for ion energies above several keV, but therefore less relevant for space applications [Whipple 1981].

2.2.3   Photo-emission

The absorption of photons can release photoelectrons and contribute to a positive charging current. The absorption characteristics of electromagnetic radiation are strongly dependent on grain size and type, and on the radiation wavelength [Havnes 1984]. For regularly shaped grains, the absorption characteristics may be computed by Mie theory, which is, unfortunately, a poor representation in the case of irregularly shaped grains.

The general expression for the photo-emission current of a negatively charged grain is given by

$$I_p = e\pi a^2 \int \xi(E_\nu) S(E_\nu) dE_\nu, \qquad (2.32)$$

where $\xi(E_\nu)$ is the photo-electric efficiency and $S(E_\nu)$ the flux of photons of energy E_ν onto the dust particle. High values of I_p can be found as a result of rather large photo-electric yields for many materials in the extreme ultra-violet wavelength range, together with significantly energetic solar photons. The integration could be carried out using typical solar spectra and photo-efficiencies. The spectrum of the photo-electrons released is often assumed to be a Maxwellian with a temperature T_p in the range $1 \sim 3$ eV. The grain surface lit by the Sun emits photo-electrons, all of which escape in the plasma when the grain potential is negative. This leads to a constant current

$$I_p = e\pi a^2 \Gamma, \qquad (2.33)$$

where Γ denotes the number of photo-electrons per square meter, per second. When the potential is positive, a fraction of these electrons return to the surface, and only the most energetic ones overcome the retarding potential and escape, contributing to a net current

$$I_p = e\pi a^2 \exp\left[-\frac{eV}{\kappa T_p}\right] \Gamma. \qquad (2.34)$$

The importance of the photo-emission for the charging can be quantified [Havnes *et al.* 1990] by the ratio between the photo-electron flux and the electron flux

$$R_{p/e} = \eta \frac{40 \times 10^9}{r_{h\{A.U.\}}^2 N_{e\{m^{-3}\}} \sqrt{T_{e\{K\}}}}. \qquad (2.35)$$

2.2.4   Electrostatic disruption

When dust grains acquire a very high potential, electrostatic repulsion of its surface charges produces an electrostatic tension in the grain. Öpik [1956] was the first to compute this tension for a charged spherical and conducting grain, and showed that the charged grain would blow apart if its tensile strength F_t is exceeded by the electrostatic repulsive force at a surface potential V_0,

$$F_t \leq \varepsilon_0 \frac{V_0^2}{a^2}. \qquad (2.36)$$

It is important to note that this expression underestimates the importance of the process, since irregularities in the grain geometry or surface conditions are not taken into account.

Hill and Mendis [1981] examined the disruption of conducting ellipsoidal grains, and showed that the electrostatic pressure increases monotonically from a minimum at the center to maxima at the ends of the main axis of the ellipsoid. As long as the tensile strength of the grain is larger everywhere than the electrostatic pressure, the grain remains intact. However, as a grain with a given tensile strength is charged more and more, it will chip off at its oblong ends whenever the electrostatic pressure exceeds the tensile strength, and this mechanism produces a more spherical grain. This is a nice conclusion for the modellers, because grains are very often assumed, just for simplicity, to be spherical.

It is also clear that when the grain disrupts, its size becomes smaller, so that the value of F_t required to prevent grain disruption increases rapidly. This implies that when a grain begins to disrupt electrostatically, it will continue to do so until small enough fragments are produced, for which the macroscopic condition (2.36) is no longer fulfilled.

2.2.5   Field emission

For micron- and submicron-sized particles, the tensile strength can be much larger, because they may consist of mono crystals. For these particles the maximum electric field attainable at the surface is limited by ion field emission for positively charged grains, and electron field emission for negatively charged grains. The critical surface electric field for ion and electron field emission is $E_{ec} = 5 \times 10^{10}$ V/m and $E_{ic} = 10^9$ V/m, respectively [Draine and Salpeter 1979].

When a highly negatively charged grain breaks up, the grain radius decreases, and the surface electric field increases to the critical value E_{ec}. At this value, electron emission occurs and the absolute value of the grain potential decreases to a value given by the size alone, $V_0 = E_{ec}a$. With the help of (2.36), it is seen that materials with a tensile strength larger than $\varepsilon_0 E_{ce}^2/2 \simeq 4 \times 10^6$ Pa are stabilized against electrostatic destruction by electron field emission. Silicates, glasses and metals are in this case.

2.2.6   Centrifugal disruption

For spherical particles in an isothermal plasma, and assuming that the maximum tensile stress is independent of the grain size, a spinning grain will be disrupted [Meyer-Vernet 1984] if

$$F_t \leq \frac{\pi}{8}\rho a^2 \omega^2, \qquad (2.37)$$

with ρ the grain mass density and ω the angular rotation frequency. If equipartition holds, the *rms* angular speed, due to random collisions with

plasma particles, satisfies $I\omega^2 = 3\kappa T$, where $I = 8\pi\rho a^5/15$. This yields the
condition

$$F_t \leq \frac{45\kappa T}{64a^3}. \tag{2.38}$$

Due to its cubic dependence on the grain size, this mechanism could become
relevant for submicron sized grains.

2.3 Charging model for grain ensembles

In the previous section I discussed several charging mechanisms, always
assuming that a single grain is placed in a plasma. If the dust density
is increased, with an inverse effect on the average distance d between two
grains, the equilibrium charge on the grains decreases dramatically. Two
key effects will play a role here, in opposite directions [Goertz 1989].

- Increasing the dust density also increases the capacitance of the grains
  [Houpis and Whipple 1987]. Indeed, the grain together with its Debye sheath is qualitatively like a spherical capacitor, the outer shell of
  which is replaced by the sheath. As grain spacing becomes comparable to or less than the Debye length, the positive sheath of each grain
  is forced closer to the grain surface, thus decreasing the capacitor gap
  and increasing its capacitance.

- There is, however, a strong counter effect. When there is a unique
  equilibrium potential and grains are injected into or formed in an
  initially neutral plasma, they become negatively (positively) charged,
  leaving excess ions (electrons) in the plasma. Hence each grain does
  not have to become so negative (positive) in order to equalize the
  ion and electron currents to its surface. The absolute value of the
  equilibrium charge obtained for a grain ensemble is smaller than the
  single grain value [Havnes et al. 1987; Goertz et al. 1988]. This effect was experimentally investigated by Barkan et al. [1994]. If the
  equilibrium potential of the grain is not unique, however, the role of
  ensemble effects is not obvious and needs further investigation.

An increase in dust density means that the grains together have a larger
appetite for electrons, but the number of available electrons per grain decreases. The latter effect quickly overtakes the former, and the mean charge
for each grain decreases, compared to the equilibrium charge of a single
grain. These conclusions might have great influence on the still open problem of strong versus weak coupling in many dusty plasma experiments. In
this context, De Angelis and Forlani [1998] discuss a simple charging model
that can be used to determine the grain charge in experiments, and conclude that the coupling scales with the dust charge squared. Hence reducing
the grain charges in an ensemble very rapidly leads to weak coupling.

2.3.1 Havnes model

For a plasma where only primary currents and photo-emission are relevant for the dust grain charging, the influence of high dust densities has been quantified by Havnes *et al.* [1987,1990]. Let us consider a dusty plasma cloud with one mono-sized dust population, in which only primary charging takes place, with currents given by (2.10) and (2.11). Furthermore, the system is charge neutral in equilibrium,

$$q_i N_i + Q_d N_d - e N_e = 0, \tag{2.39}$$

and then the total current on the grain vanishes

$$I_e(Q_d) + I_i(Q_d) = 0. \tag{2.40}$$

The plasma species are considered as thermalized, with number densities given by Boltzmann relations,

$$n_\alpha = N_0 \exp\left[-\frac{q_\alpha U}{\kappa T_\alpha}\right], \tag{2.41}$$

with $N_0 = N_e = N_i$ the undisturbed plasma density outside the dust cloud. There are thus two equations involving Q_d, to be solved for U (the cloud potential) and $V - V_p$. The results are usually expressed for a proton-electron plasma with a common temperature $T = T_e = T_i$, as a function of the Havnes parameter

$$P = \frac{4\pi\varepsilon_0 a N_d \kappa T}{N_0 e^2} = 60 \times 10^3 \frac{N_d a T}{N_0}. \tag{2.42}$$

When $P \ll 1$, the values obtained for Q_d indicate that results from the previous section remain valid and the grains can be considered as isolated. However, when P increases, the cloud potential increases because of the imbalance between electron and ion charges in the surrounding plasma. As a result, the surface potential decreases because there are less free electrons left to charge the grain, which has also been verified experimentally [Xu *et al.* 1993].

A possible drawback of the Havnes model was given by Wilson [1991], saying that the effect of absorption of charged particles is ignored. Indeed, both the ions and electrons were assumed to be and remain Boltzmann distributed, and eventual sink/source terms, occurring in continuity and momentum equations due to the charging process, are not included in the model. When the particle densities in the cloud are not too high and the loss of plasma particles due to the charging process is rather small, the neglect of these sink/source terms is reasonable.

It is also important to note that the number of dust particles in a Debye cube can be written as

$$N_d \lambda_D^3 = \frac{P \lambda_D}{8 \pi a}. \tag{2.43}$$

As in later chapters, the electron and ion Debye lengths, are defined as in terms of thermal velocities and plasma frequencies as $\lambda_{De} = c_{se}/\omega_{pe}$ and $\lambda_{Di} = c_{si}/\omega_{pi}$, respectively. For those regimes where both electrons and ions participate in the shielding of a dust grain, the global Debye length λ_D is given through

$$\frac{1}{\lambda_D^2} = \frac{1}{\lambda_{De}^2} + \frac{1}{\lambda_{Di}^2} = \frac{\omega_{pe}^2}{c_{se}^2} + \frac{\omega_{pi}^2}{c_{si}^2}, \tag{2.44}$$

and could be heavily influenced by the ion screening effects.

As can be readily verified, even when $P \ll 1$ and grains can be considered as isolated as far as their charges are concerned, the number of grains in a Debye cube can be large, provided that $\lambda_D \gg a$, which is usually the case in dusty plasma environments. Ensemble effects are often explained by overlapping Debye spheres [Goertz 1989]. This picture, however, is clearly not always valid, because ensemble effects can be neglected when $P \ll 1$, even when the number of grains in a Debye cube is large.

Continuing with denser grain packing, Ikezi [1986] predicted that a lattice can be formed by the charged dust grains, provided Q_d^2/Td exceeds a critical value, where d is the average intergrain distance. This paper was the start of amazingly successful dusty plasma crystal experiments, making it possible to examine wave phenomena, the structure of dust crystals and the charging of grains in strongly coupled systems. Some of these will briefly be discussed in the section dealing with experimental results, but most of it is beyond the scope of a treatment of dusty plasma waves in space.

2.4 Charging in magnetized plasmas

The presence of an external magnetic field will significantly alter the charging currents, as the paths of the plasma particles are affected. A review on the charging of a spherical conducting probe in a laboratory magneto-plasma is given by Laframboise and Sonmor [1993], from which some salient elements are recalled here.

The relevant length scales for this problem are the grain radius a, the length scale of the electric potential $V(\mathbf{r})$ variation (denoted as λ_D^* because for a Debye shielded potential this equals the traditional Debye length), the mean gyroradius ρ_α and the mean free path length r_{mfp}. For a collisionless plasma it is assumed that $r_{mfp} \to \infty$, while for all relevant dusty plasma applications $a \ll \lambda_D^*$.

The computation for a spherical conducting grain in a collisionless magnetoplasma, at infinitely large Debye lengths, was performed by Sonmor and Laframboise [1991], while upper and lower bounds for the charging currents were obtained by Rubinstein and Laframboise [1982]. In this approach the charging current is a function of the normalized magnetic field and electric potential, denoted respectively by $\beta_\alpha = a/\rho_\alpha$ and $\chi_\alpha = q_\alpha V/\kappa T_\alpha$.

2.4.1   Collisionless theory for infinite Debye lengths

Following the same outline as discussed in the section on primary charging for a single grain, the motion of the particles now occurs in an effective potential $U_\alpha(r, z)$, using cylindrical coordinates in the presence of a magnetic field,

$$U_\alpha(r, z) = q_\alpha V(r, z) + \frac{1}{2} m_\alpha r^2 \left[\frac{J_\alpha}{m_\alpha r^2} - \frac{\Omega_\alpha}{2} \right]^2. \qquad (2.45)$$

Since the kinetic energy cannot be negative, a particle having a particular E_α and J_α is confined to a region of the (r, z)-plane for which $E_\alpha \geq U(r, z)$, i.e. inside the particle's magnetic bottle (Figure 2.5). These bottles have rotational symmetry about the z-axis.

In the collisionless limit particles cannot cross magnetic bottle boundaries. When the magnetic bottle of an electron or an ion intersects the dust grain, viewed as a probe, it is assumed that the particle is collected. This yields an upper bound for the actual charging current, if one ignores the effects of thermal motions at infinity [Parker and Murphy 1967]. When the thermal motion at infinity is taken into account, Laframboise and Sonmor [1993] derived an upper bound I_α^{max} for the current.

Another attempt was made by Sonmor and Laframboise [1991], who computed a large number of particle orbits by integrating Newton's equation of motion with a Coulomb potential ($\lambda_D^* \to \infty$), and derived the currents as functions of β_α and χ_α. Their conclusions were the following:

- When $\beta_\alpha \to 0$, magnetic effects can be neglected and the orbital motion limited theory is recovered.

- When χ_α increases, the collected current asymptotically approaches I_α^{max}, but very slowly. Hence this value will never be obtained under real dusty plasma conditions.

It has to be stressed that this kind of computation is not self-consistent, because then the equations of motion and Poisson's equation have to be solved together, which is a notoriously intractable problem. Hence it comes as no surprise that the number of papers dealing with self-consistent solutions is very limited. However, a numerical three-dimensional particle-in-cell analysis has been carried out for a cylindrical probe [Singh et al. 1994], who

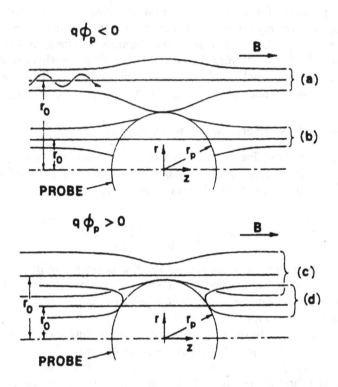

Figure 2.5. Magnetic bottles in the neighbourhood of a grain. *Reprinted from Rubinstein and Laframboise 1982 (Fig. 6 on p. 1178), with permission of the American Institute of Physics*

concluded that the actual current is adequately described by the upper bound limit, when the electric field gradient λ_D^* is sufficiently large. This happens even for realistic grain potentials as low as a few volts.

2.4.2 Combined effects of space charge and collisions

In a collisionless plasma, depending on the explicit expression of the electric potential, some of the magnetic bottles will be depopulated, namely those not connected to $|z| \to \infty$. For a collisional plasma, however, all bottles will contain particles, due to collisional scattering.

It can be shown [Laframboise and Sonmor 1993] that effects of collisions and space charge cannot be separated, not even in the limit of large mean free paths. Furthermore, a strictly collisionless theory cannot be really exact for finite Debye lengths. To see why, consider the depletion of particles at large distances from a spherical grain, caused by current collection. If

$B_0 = 0$, this depletion occurs equally in all directions for both ions and electrons, and therefore results in a spherically symmetric distribution of net space charge and therefore of potential.

On the other hand, when $B_0 \neq 0$, this depletion occurs predominantly along and adjacent to the grain's magnetic shadow. In other words, we expect in the collisionless limit that for plasma particles far away from the grain, both ion and electron density disturbances will become functions only of the cylindrical radius r. In contrast to the nonmagnetic case, these disturbances will have a different r-dependence for ions and electrons. Indeed, the smaller gyroradius of the electrons will engender electron depletion much more closely to the magnetic shadow itself, whereas ion depletion will be more widespread.

The resulting charge imbalances then produce potential disturbances which also will depend only on r, even far away from the grain. In the absence of collisions, no mechanism exists that would make charge density disturbances decay with increasing $|z|$, and the resulting potential disturbances must therefore also extend to infinity in both directions along the magnetic shadow. However, if the mean free path is finite, no matter how large, collisions will ultimately repopulate the depleted regions.

Hence the effects of finite Debye length and an ambient magnetic field are intimately interlocked, and the problem becomes very complicated. Until now there is no fully self-consistent analytical treatment, although there are simulation results, as described in this section. Almost all space plasmas are magnetized, but because of the intrinsic difficulties and lack of proper modelling, none of the papers reviewed further on have attempted to use more than the standard unmagnetized probe theory in the orbit limited form.

2.5 Dusty plasma experiments

After this rather lengthy and theoretical overview about dust charging processes, it is worth recalling very briefly some of the experimental results. These can be grouped in three different categories, to be discussed in the next subsections. First of all, on the more technological side come various aspects of surface plasma processing and etching in weakly ionized radio-frequency discharges. Then there are experimental verifications of the dust charging mechanisms and of typical low-frequency dust modes, which is obviously closer to the scope of the present monograph. Finally, exciting and important new research has gone into strongly coupled systems, where regimes occur, that are completely outside plasma physics as we knew it, but rejoin investigations into one component plasmas.

2.5.1 Plasma processing and etching

For typical radio-frequency discharges the driving frequency lies between the ion and electron plasma frequencies, which means that the electrons oscillate in a high frequency field, whereas the ions mainly feel a time averaged field. Such plasmas consist of distinct regions, a dark sheath near the electrodes, and a radiating plasma glow zone in the middle.

One of the major technical applications of radio-frequency discharges involves plasma-surface interactions, because they are very efficient sources for deposition of various layers, such as amorphous silicon in the fabrication of solar cells and transistors, or carbon layers used as anti-reflection and/or protective coatings. Furthermore, plasma chemistry in combination with high energy ions allows anisotropic etching of different substrates. This kind of etching can produce narrow and deep structures, which are not possible by wet chemical etching.

However, radio-frequency discharges also produce macroscopic dust particles, initially viewed as very harmful, since they contaminated the substrate, and therefore research connected with dust aimed at avoiding particle formation and/or contamination altogether. The negative charge on the particles is responsible for their trapping in the positive plasma glow, which in turn causes them to grow and coalescence, producing cauliflower-like grains. Despite many extensive precautions, such as clean rooms and sophisticated handling techniques in order to avoid processed wafer contamination by dust, their presence seemed inescapable. Indeed, particles can be formed from wafer or substrate material, and improving the quality of the vacuum cannot prevent nucleation and growth of dust particles, even in very clean rooms. The greatest fraction of yield-killing contaminants land on the wafer, not during handling, but under vacuum conditions during plasma processing.

Although this is an extremely interesting and important branch of dusty plasma physics, it is totally outside the focus of this monograph, however, and the interested reader is referred to a recent book dealing with these aspects in great detail [Bouchoule 1999].

2.5.2 Charging experiments and dust modes

In the experimental studies of dusty plasmas *per se*, two strands of research are intertwined and overlapping. One is concerned with the experimental verification of dust charging processes, and the other with observing wave modes due to charged dust. Although I can only briefly describe some of the salient properties of these experiments, their importance in understanding dusty plasmas cannot be underestimated. Indeed, these have greatly advanced our knowledge, and have extended it into directions which are only now being theoretically studied in earnest, where dense dusty plasmas lead us to new wave regimes [Rao 1999].

In typical experiments [Chu and I 1994; Thomas *et al.* 1994; Melzer *et al.* 1994; Pieper and Goree 1996], the particles are trapped in the sheath region above the lower electrode of parallel plate radio-frequency discharges, where the electric field can levitate the grains against gravity. In that region the dust grains interact through their Coulomb repulsion, and can form a more or less ordered structure, a dusty plasma lattice or crystal [Ikezi 1986]. Here the charge on the particles is a crucial parameter, both for particle trapping as well as for the formation of plasma crystals themselves.

Some experimental studies have focused on the charging processes [Barkan *et al.* 1994; Hazelton and Yadlowsky 1994; Walch *et al.* 1995]. The dust grains used are made of silicon, graphite (with a small secondary emission coefficient), copper (a metallic conductor) and glass (with a high secondary emission), and have a mean size distribution in the range of 1 to 125 μm. The dust density was low enough, so that the grains could be considered as not interacting and the number density of the dust was not specified. Only in Barkan *et al.* [1994] a value of $N_d = 10^{10}$ m^{-3} was mentioned. On the other hand, the effect on the charging of more closely packed grains has been investigated by Xu *et al.* [1993], using kaolin of two different sizes (1μm and 50μm), and aluminum oxide with nominal sizes of 0.3 μm and 0.01 μm.

In parallel experiments [Chu and I 1994; Thomas *et al.* 1994; Melzer *et al.* 1994] Coulomb lattices were duly formed, with lattice constants of the order of 10^{-4} m. Dust particles used were 7.0 μm diameter melamine/formaldehyde spheres [Thomas *et al.* 1994], for which mean particle velocity measurements yield kinetic temperatures estimated as $T_d = 310$ K, close to room temperature. Other possibilities include micrometer-sized SiO$_2$ particles [Chu and I 1994; Chu *et al.* 1994; I *et al.* 1996] or TIO$_2$ particles [Melzer *et al.* 1994], with a size of up to 30 μm, which were later replaced by highly monodisperse melamine/formaldehyde particles of a mean diameter $a = 9.4$ μm, with dust densities reaching $N_d = 10^9 \sim 10^{10}$ m^{-3}.

In all these experiments, electron energies were of the order $\kappa T_e = 0.2 \sim 5$ eV, at ion densities N_i between $10^{11} \sim 10^{15}$ m^{-3}, and with an appropriate charging theory this indicates negative dust charges in the range $9.8 \times 10^3 e \leq |Q_d| \leq 27.3 \times 10^3 e$. Later refinements were introduced to measure the charge on a single grain by laser-excited vertical resonances in plasma crystals [Homann *et al.* 1999], greatly extending earlier results [Melzer *et al.* 1994].

Wave dispersion provides another possibility for the determination of the dust charge. For this method, however, the whole wave dispersion has to be measured, and the dust charge is then obtained from a fit to the theoretical dispersion curves. This fitting is very sensitive to the particular wave modes chosen. Two different wave types have rapidly emerged.

One of these is the dust-acoustic wave, introduced by Rao *et al.* [1990]

and fully discussed in Chapter 5 in its proper context of low-frequency electrostatic modes in a dusty plasma as a whole. The low-frequency fluctuations observed in some experiments [Chu and I 1994; Chu *et al.* 1994] have been claimed by D'Angelo [1995] as the first detection of this dust-acoustic mode, also confirmed later by Barkan *et al.* [1995a], Thompson *et al.* [1997] and Merlino *et al.* [1998]. Further experiments on wave phenomena in weakly coupled systems were performed by Barkan *et al.* [1995b,1996], involving verification of other types of waves in dusty plasmas, like the electrostatic ion cyclotron and ion-acoustic waves. Again I refer to Chapter 5 for a description of these.

On the other hand, in strongly coupled dusty crystal plasmas, another mode is possible, the dust lattice wave first proposed by Melandsø [1996] and studied experimentally in a linear chain arrangement [Homann *et al.* 1997]. More about the dust lattice wave in Chapter 10, as this mode relies on the interaction between nearest neighbours in the chain, and thus obeys a quite different physical mechanism. The latter was already proposed by Ikezi [1986], who predicted that a system of charged dust particles embedded in a plasma would crystallize when the nearest-neighbour potential energy of the grains is large compared to their thermal energy, as can be accomplished by using large particles and cooling them by drag on a neutral gas background. This leads to very regular arrangements of micrometer size monodisperse spherical particles, discussed already in the context of some charging experiments.

Finally, before closing this brief review of experimental verifications of dusty plasma modes and relevant charging mechanisms, mention should be made of laboratory work on the electrostatic charging properties of simulated lunar dust regoliths [Horányi *et al.* 1995]. This is a prelude to experiments on real lunar dust, in order to understand the active electrostatic transport of dust on the Moon. These investigations bridge the gap between laboratory experiments and space observations, dealt with in the next chapter.

2.6   Single grain dynamics and planetary rings

At the end of this chapter, I want to touch upon the single particle description of charged dust grains, as their dynamics has been used to explain some of the features occurring in planetary rings, like spoke formation or finite thickness effects. The basic idea is that the behaviour of grains with similar charges can be inferred from their single particle dynamics, rather than through collective interactions and waves. Each grain is described as a test particle, perhaps in the averaged field of the others, and all grains are similar and carry the same charge. Electrostatic repulsion could then become important, and lead in planetary rings or cometary tails to levitation out of the equatorial plane and to a finite thickness for the rings, besides providing one of the possible (and competing) explanations about the formation of spokes in planetary rings.

For the motion of a single grain near a planet with an intrinsic magnetic field one starts from the equation of motion

$$\frac{d\mathbf{v}}{dt} = \frac{Q}{m}(\mathbf{E} + \mathbf{v} \times \mathbf{B}) - \frac{GM_p}{r^3}\mathbf{r}. \tag{2.46}$$

Here $\mathbf{v}$ refers to the velocity of a grain with position vector $\mathbf{r}$ originating in the center of the planet, and Q and m to the average grain charge and mass. The subscript d has been omitted on all grain variables. Furthermore, $\mathbf{E}$ and $\mathbf{B}$ are the electric and magnetic fields, M_p is the mass of the planet and G the gravitational constant. Only the Lorentz and gravitational forces have been included. One could easily add additional forces like collisions between grains and plasma, radiation pressure (which seems to be negligible in Jovian and Saturnian magnetospheres, but could be important for parts of the terrestrial magnetosphere), collisions with other dust grains, but these are usually not the dominant factors [Northrop 1992].

This equation of motion is usually transformed to a frame which is corotating with the planet, such that the velocity becomes $\mathbf{w} = \mathbf{v} - \mathbf{\Omega}_p \times \mathbf{r}$. The result is

$$\frac{d'\mathbf{w}}{dt} = \mathbf{w} \times \left(\frac{Q\mathbf{B}}{m} + 2\mathbf{\Omega}_p\right) + \nabla\left(\frac{1}{2}|\mathbf{\Omega}_p \times \mathbf{r}|^2 - \frac{GM_p}{r}\right). \tag{2.47}$$

The electric field is assumed to be generated by the rotation of the planetary magnetic dipole field, and hence drops out of the transformed equation.

Nevertheless, this equation cannot hold in the same form for planets where the dipole field is a poor approximation to the real field, as the quadrupole or higher order contributions will modify this picture, or when the magnetic dipole is not aligned with the rotation axis. For Saturn the above description is a fair approximation, as Saturn's magnetic field is adequately represented by a dipole aligned with the rotation axis but with a small offset to the north, whereas already at Jupiter's the magnetic field is tilted by 10° from the rotation axis. These properties will be further detailed in Chapter 3 dealing with the observations.

It is clear from the equations of motions for a single grain that there will be a competition between gravitational effects due to the attraction of the planet and electromagnetic effects. Very small grains will be trapped by the planetary magnetic field, much as plasma electrons and ions, and almost rigidly corotate with the planet. Very large grains, on the other hand, will follow gravitationally bound Keplerian orbits, without being distracted from them by electromagnetic forces. It is only for medium-sized grains that an interesting balance can occur between both effects.

As the spokes in the B ring of Saturn have been observed as being darker than the ring plane itself in back scattered light, but brighter in forward scattered light [Smith *et al.* 1981], the conclusion has been that these could be attributed to charged dust grains of micron size. Moreover, the apex of the spokes is at synchronous orbit. The synchronous orbit of a planet is defined as the orbit for which the Keplerian angular velocity defined through $\Omega_K^2 = GM_p/R^3$ (with R referring to the distance from the planet's center) equals the rotation angular velocity Ω_p of the planet.

This would tend to corroborate the idea that the dust grains essentially follow Keplerian orbits around the planet, whereas the plasma corotates with the planet, being trapped in the planetary magnetosphere. At the synchronous distance both the Keplerian and corotation frequencies are equal, and then the $\mathbf{E} \times \mathbf{B}$ drift on the dust grains forces these grains away, inward inside and outward outside the corotation distance [Northrop 1992].

Another dynamic effect which has been invoked is that repelling electrostatic forces between dust grains with the same charges would mean that planetary rings cannot collapse to very thin sheets, but would need to retain a finite thickness transverse to the equatorial plane. This could be important for Jupiter's and some of Uranus' rings, but maybe for others too, if the uncertainties in the relevant parameters are taken into account. The thickness of the rings is determined by a balance between the gravity component toward the central plane and the expanding electrostatic force of the dust grains. If the ambient plasma conditions are changed, natural oscillations occur in the thickness profiles, which become increasingly complicated for denser rings, and resonances may be found between these oscillations and the Doppler-shifted Keplerian frequencies. These were studied by Melandsø and Havnes [1991] in electrostatically supported dust rings around planets.

The description is essentially based upon a single test-particle approach to determine the charge on each grain, but uses an equilibrium vertical dust density distribution. It is found that such oscillations occur at a frequency $\Omega_K(L)\sqrt{3}$, with $\Omega_K(L)$ being the purely Keplerian frequency at distance L from the planetary center. For the tenuous clouds under consideration, the oscillation frequency is practically unaffected by charging delays, so that the dust charges can be supposed to remain constant. There is now scope for resonant phenomena to occur between the ring oscillations and plasma features in the magnetosphere which corotate with the planet. Therefore integers ℓ and n are needed such that

$$\mp n[\Omega_K(L) - \Omega_{CR}] = \ell\sqrt{3}\,\Omega_K(L). \qquad (2.48)$$

This leads to specific distances

$$L_{n,\ell} = L_{CR}\left(1 \pm \frac{\ell}{n}\sqrt{3}\right)^{\frac{2}{3}}. \qquad (2.49)$$

Here $\pm$ refer to whether we are outside or inside the corotation distance L_{CR}. Such resonances have been claimed to be responsible for certain gaps

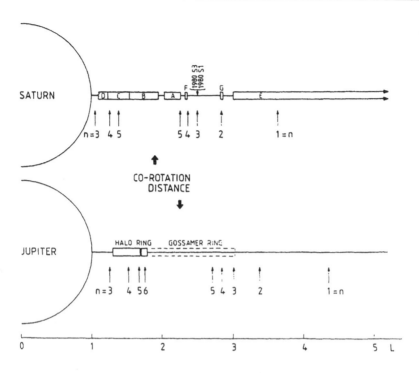

Figure 2.6. Radial distances of the major ($m = 1, n = 1, 2, \ldots$) inner and outer resonances of Jupiter and Saturn, shown together with major ring features. *Reprinted from Melandsø and Havnes 1991 (Fig. 3 on p. 5842), with permission of the American Geophysical Union*

or prominent features in the ring structures at Jupiter and Saturn, as shown in Figure 2.6. The whole treatment, of course, relies upon the dust being effectively charged rather than neutral. A more definite observational answer will have to wait until the Cassini-Huygens satellite is in orbit around Saturn.

CHAPTER 3

SPACE OBSERVATIONS

3.1 Generalities

As pointed out in the introductory chapter, dust, not just neutral but charged dust, is believed to be ubiquitous in nature, although opinions diverge about how important charged dust indeed is. However, before embarking in the following chapters on various aspects of wave phenomena in dusty space plasmas, where we are among the believers, it is extremely useful to collect some of the available data and separate what is really known from exhilarating flights of fancy. This is at the same time a sobering exercise, because many characteristics of dust grains in space plasma environments are still unknown, to say the least. Spurred by solar system observations that can only be explained by the essential presence of charged dust, since purely gravitational forces are clearly insufficient, theoreticians have leapt ahead of the available data and explored many exciting but maybe speculative avenues. These need to be calibrated to what is accepted observational knowledge, and it is for this purpose that the present chapter is included here, although I will also sin in the next chapters, when reviewing the different modes studied in the literature and thought to be important in dusty plasma research.

To start from home, dusty plasmas in the solar system are found in the Earth's magnetosphere, in cometary tails and comae, in planetary rings, as well as in the interplanetary medium or in circumsolar dust rings. There is little discussion about the abundance of fine dust in the solar system and its neighbourhood, and the dark bands of dust that block part of the Orion and other nebulae from view indicate that quite some dust must have been present in the nebula that coalesced to form the solar system. Although little of that primordial dust remains, new dust is constantly regenerated by colliding meteoroids, comets passing near the Sun and many other processes. Being immersed in plasmas, the dust is then charged.

Our knowledge about dusty plasmas mainly comes from remote sensing and from radiation reflected or emitted by the dust grains at wavelengths ranging from the ultraviolet to the microwave bands. The radiation method requires the inversion of the radiative transfer equation, which gives the collective behaviour in terms of the individual particle properties. The solution of this inverse problem is not unique, and depends on the plasma model, so that results have to be interpreted with great care. Much more can be obtained from *in situ* measurements, although the availability of such data is limited to specific solar system plasmas. Outside the solar system only remote sensing remains, as for most of astronomy. Luckily, experimental laboratory evidence is now emerging to complement this, and briefly discussed in the previous chapter.

In the light of the previous remarks, saying that space observations of dusty plasmas exist, is a statement to be used with extreme caution. More than the proverbial pinch of salt is needed, as the arguments are all indirect, even if compelling. It is also wise to keep in mind that no actual and direct space measurements of the charge of a dust grain have yet been made. The various detectors on board space missions essentially use impact parameters to deduce the grain characteristics. It is only very recently that direct methods to detect charged dust have been proposed, by using wire dipole antennae on a spacecraft as a kind of radio dust analyzer [Meuris *et al.* 1996]. Unfortunately, this interesting suggestion has yet to be tested on a real spacecraft.

In this chapter some observational evidence is reviewed on dusty plasmas in the solar system obtained mostly *in situ*, with the aim to bridge the gap that exists between theoretical dusty plasma concepts and their space applications. In addition, some other astrophysical plasmas will be discussed. Caution is advised when using data quoted in literature, without looking at the way in which they were obtained. A rather unsuspected example of this can be found in the famous and widely cited review article by Goertz [1989], giving some data without referring to sources. Nevertheless, numbers from this otherwise so eminent paper are very often quoted, if not used, in the literature, without any qualifications that they are estimates at best.

As seen in the previous chapter, the charging mechanisms are highly dependent on the local conditions. Therefore, the discussion cannot be limited to the dust properties (like grain size distribution, composition or charge) alone, but needs also to take the plasma parameters (like densities or temperatures) into account, as well as other characteristics (like relative dust-plasma velocities, magnetic field strengths or photon fluxes). Finally, a re-evaluation of the data sets can be used to quantify some of the mathematical models that were published during the last years. The interest in the physics of dusty plasmas has been increasing exponentially ever since, but in the end all models rely on the same too scarce observations for their validation.

In the next sections, I begin with some of the most obvious environments for dusty plasmas in astrophysics, going out from the Earth. These are noctilucent clouds, circumsolar dust rings, planetary rings, cometary comae and tails, and much further out, interstellar dust. I will make great use here of the valiant efforts by Peter Meuris to collect some of these data (and their references), and incorporate them in his PhD thesis [1997b].

3.2 Noctilucent clouds and magnetospheric dust

The closest examples of naturally occurring dusty plasmas have been documented in the polar mesosphere, at altitudes of 80 to 90 km. This is the coldest place in the Earth's environment, where the summer temperatures drop to 100 K, compared to a normal winter temperature above 200 K. Several phenomena occur, such as the noctilucent clouds and the polar mesospheric summer echoes, strong radar backscatter observed at frequencies from 50 MHz to 1.3 GHz. The noctilucent clouds are at high enough altitudes, so they can be illuminated the whole night by the Sun, which just disappears below the horizon during the summer months south of the arctic circle. *In situ* rocket measurements at these altitudes find layers of strong electron depletion and sometimes also layers of increased positive ion density. One of the possibilities is the occurrence of charged dust grains with a total charge density that is significant compared to the electron or ion component.

Two dust probes were launched in 1994 from the Andøya Rocket Range (Norway), with a dust and an electron probe on both payloads [Havnes *et al.* 1990,1996 and references therein]. The dust probes were designed to block out the electron and ion components at the mesopause, and to detect primary currents due to charged dust impacts and also secondary plasma production during these dust impacts. The results clearly show that during the polar mesospheric summer echoes and noctilucent clouds conditions large amounts of charged dust must be present, with average sizes of about 0.1 μm, at densities of several 10^9 m^{-3}. On different occasions both negatively and positively charged dust has been detected, even on the same rocket flight at different altitudes. In some parts of the layers the negative charges locked up in the dust is large enough to cause appreciable electron depletion, as shown in Figure 3.1.

Among the remaining uncertainties are the structure of the dust. Theories of ice grain formation cannot give the observed densities based on the available water vapor in the mesosphere. Positive charges rely on photoelectric charging as the dominant mechanism, and then pure ice has too high a work function, so that contamination with metals needs to be invoked [Havnes *et al.* 1996].

Other dust particles in the magnetosphere can be of cosmic origin, where cosmic is meant to denote extraterrestrial sources. The magnetosphere acts as a shield with an efficiency that depends on the size and velocity of the incoming dust particles. Simulations for dust grains of interplanetary, cometary and lunar origin indicate that magnetospheric effects reduce the flux of interplanetary and lunar dust with sizes below 0.1 μm, but the shielding is much less effective for cometary grains because of their much higher approach velocities [Juhász and Horányi 1999]. In addition, there is quite a lot of space debris that is either dragged onto elliptical orbits

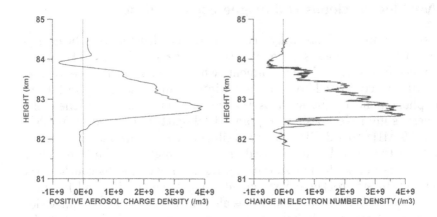

Figure 3.1. Positive aerosol and electron charge densities during the summer mesopause soundings. *Reprinted with permission from O. Havnes*

to enter the atmosphere or else ejected into interplanetary space. Hence, as will be discussed further on, it is not only at the larger planets that the magnetospheres shape the size and spatial distributions of small dust grains in planetary rings, but also in the Earth's magnetosphere and in circumsolar dust rings.

There are other, intriguing speculations of dusty plasmas occurring in unexpected places. One of these is the recent observation of enhanced positive cloud-to-ground lightning in thunderstorms over the United States in smoke-contaminated air, where smoke from forest fires in Mexico were advected north-east [Lyons *et al.* 1998]. The smoke from these massive fires spawned stronger, more sustained lightning than normal. Moreover, the positively charged bolts carry twice as much current as similar flashes in smoke-free storms. While it is too early to quantify this and similar contamination due to the transport of soot or volcanic ash over large distances, there is probably no doubt that charged particles play a major role here, literally typifying the fire element of the ancients Greeks.

3.3 Circumsolar dust rings and zodiacal light

Now we leave the Earth's immediate vicinity and first go towards the Sun. Measurements on the Ulysses and SOHO spacecraft indicate that there is dust sublimation close to the Sun, the dust itself streaming into the solar system from interstellar space, as detected by the Ulysses and Galileo missions [Grün *et al.* 1993]. Indeed, the motion of the solar system relative to the local interstellar medium causes a flux of neutral gas and dust into the solar system, at incoming velocities that greatly exceed the escape velocity.

The dust gets charged by different processes, and the smaller dust grains are deflected by the Lorentz force exerted by the interplanetary magnetic field, as the smaller particles have higher charge-to-mass ratios. Hence, selection effects based on size operate already near the heliopause, the transition region between the solar system and interstellar plasmas, and consequently alter the distribution of the dust at various distances from the Sun [Kimura and Mann 1998]. Light charged dust from outside the solar system is easily deflected, whereas heavier neutral grains penetrate deeper into the solar system and become charged there.

The picture leads to a dust cloud or disk close the Sun, intimately connected with the zodiacal light, a faint brightness of the sky that can be observed after dusk or before dawn in the direction of the zodiac, or the ecliptic plane, if you prefer. This zodiacal light is produced by sunlight scattered off the dust grains distributed along the ecliptic plane, hence the more recent name of circumsolar dust ring. The polarization, colour and intensity of this light yield information not only on the distribution but also about the properties of the dust grains themselves, indicating that besides the circumsolar dust complex there might also be an outer dust cloud in or beyond the Kuiper Belt region [Mann 1996].

Krivov et al. [1998a,b] have investigated these circumsolar dust rings at several solar radii, by choosing fractal aggregates consisting of either silicate or carbon as representative of dielectric and absorbing fluffy particles. A large array of forces acting on these charged dust grains were taken into account: solar gravity, direct solar radiation pressure, the Poynting-Robertson drag, sublimation and in particular, the Lorentz force. It is found that variations in the solar magnetic field during the solar cycle do not appreciably influence the dynamics and spatial distribution of carbon aggregates, which are thought to produce the peak features observed in the near-infrared F corona brightness. On the other hand, there might be fluctuations in the latitudinal distribution of silicate aggregates, but not sufficient to correlate with variations in brightness [Kimura et al. 1997]. The F corona corresponds to the innermost part of the zodiacal cloud, and observational and theoretical results still give controversial answers to this fascinating problem [Krivov et al. 1998b]. In addition, while all grains typically spiral inward from interstellar space, silicate particles sublimate at 2 to 3 solar radii, while carbon aggregates intensively sublimate at 4 solar radii. While the dust cloud is mainly concentrated in the ecliptic plane, observations also point towards a non negligible out-of-ecliptic distribution, as evidenced by Ulysses [Mann 1995].

All these considerations have several important consequences. Large dust clouds around other stars are of course very important in understanding planet formation, although that is not the case for the Sun any longer. Furthermore, the path of a future Solar Probe will bring it close to the Sun, roughly to about 4 solar radii where the dust concentration is largest, and hence might lead to severe impact damage, that one would like to assess reliably on beforehand.

3.4   Planetary rings

Next on our tour of dusty space plasmas are the planetary rings, locations where the occurrence of charged dust was first suspected in earnest since the astonishingly successful Voyager missions. Each of the giant planets (Jupiter, Saturn, Uranus and Neptune) is now known to be encircled by rings. Some of these, like the A, B and C rings of Saturn and the nine narrow Uranian rings, are rather optically thick and composed primarily of large bodies (1 cm to 10 m). However, every other ring system has been found to contain a large population of micron-sized dust, replenished on a continuous basis. While it has been realized for more than a century that planetary rings must be composed of a myriad of particles, even in the post-Voyager era we have not yet seen a single one. So we continue to rely on indirect and incomplete evidence, supplemented by theoretical considerations.

To evaluate the dust-plasma velocity, following Mendis *et al.* [1982], the gyration frequency of a charged dust grain in a rotating magnetosphere is taken to lie between the Keplerian frequency Ω_K and the corotation frequency. Which of the two bounds will be closer to reality depends on the charge-to-mass ratio of the charged grains. Highly charged and small enough grains are almost corotating, while massive grains are gyrating at the Keplerian frequency, as discussed already in the previous chapter.

The magnetic field in the neighbourhood of planetary rings is often modelled by a stationary dipole field. This assumes that the planetary magnetospheres are not distorted by ring currents, and that the magnetic dipole is more or less aligned with the rotation axis, like for the Earth, Jupiter and Saturn. In fact, the only good example of this is Saturn, because already in the case of Jupiter the magnetic field is tilted by about 10° from the rotation axis. Moreover, electric currents along field lines connecting Jupiter and Io and in a current sheet around Jupiter's equator distort the magnetic field sufficiently so as to make a simple offset dipole field a poor representation. Much worse, both Uranus and Neptune have magnetic fields which are highly tilted with respect to their rotation axes and offset from their centers by large fractions of the planetary radii. In particular, the 60° tilt of the Uranian dipole field gives rise to a rather messy magnetosphere with a marked rotational asymmetry, especially as the rotation axis of Uranus lies itself almost in the ecliptic plane [Anderson and Kurth 1989].

Radial distances, in what follows, will be expressed by the magnetic dipole shell parameter L, corresponding to the distance from the magnetic axis of the planet, measured in planetary radii. Planetary radii and masses can be found in Table 3.1. In what follows, great use will be made of

Planet	Radius (m)	Mass (kg)
Jupiter	71.5×10^6	1.90×10^{27}
Saturn	60.3×10^6	5.69×10^{26}
Uranus	25.6×10^6	8.69×10^{25}
Neptune	24.8×10^6	1.03×10^{26}

Table 3.1. Planetary radii and masses

the wealth of information contained in the series of review volumes dealing with Jupiter [Gehrels 1976], Saturn [Gehrels and Matthews 1984], Uranus [Bergstralh *et al.* 1991], and Neptune and Triton [Cruikshank 1996], unless specific other references are given. The discussions for the different planets will be structured along a common format, where after generalities a review is given first of the dust and then of the plasma properties, before wrapping up with other relevant characteristics.

And now it is time to move outward, starting from Jupiter and retracing the path of the Voyager missions that served us so well! Indeed, as far as the structure of the outer heliosphere is concerned, these veterans are still the only sources of *in situ* observations, whereas for Jupiter newer data from the Ulysses and Galileo spacecraft have allowed us to refine our picture.

3.4.1 Jupiter

Jupiter's ring system was discovered in a single image from the Voyager 1 flyby in 1979, and subsequently imaged in greater detail by Voyager 2. The Ulysses and Galileo dust detectors have been handicapped for various reasons, like a malfunctioning antenna in case of the latter. The first Galileo images of Jupiter's ring system were acquired in November 1996. For Ulysses the closest approach is 6.3 Jovian radii from the center of the planet, outside the rings, and also Galileo stayed outside the rings. Because one fears for the integrity of the spacecraft systems, passages through the denser rings are avoided at all cost, thus preventing the direct data acquisition that the dusty plasma community would have hoped for and is in dire need of. In the future, other methods will have to be devised to probe planetary rings from farther afield. One of these is the possible detection of Mach cones [Havnes *et al.* 1995] formed by the wake of larger boulders in the rings, when lighter dust grains are deflected. It is hoped that the Cassini mission will yield here valuable information, by actually discovering such wakes and measuring their opening angles from a safe distance.

The ring system of Jupiter appears to contain three rather distinct components. The main ring is relatively thin, and near its boundary is a halo, a vertically extended cloud of material. Showalter *et al.* [1985] were the first to note the presence of an even fainter ring extending outward from the main ring, in a single Voyager image. The optical depth of this "gossamer" ring appears to have a peak near the location of the synchronous orbit, which suggests a significant interaction with the local plasma. Particles in Jupiter's rings probably do not stay there too long, due to atmospheric and magnetic drag. Therefore, if the rings are permanent features, they must be continuously replenished. The small satellites Metis and Adrastea, which orbit within the ring, are the obvious sources for ring material, unless the gyrophase drift due to temperature gradients in the plasma concentrates small grains close to synchronous orbit [Northrop *et al.* 1989]. More recent data suggest that faint rings may accompany all small inner satellites of the Jovian planets [Burns *et al.* 1999], based on what the Galileo observations have indicated for the rings at Jupiter [Ockert-Bell *et al.* 1999].

The only *in situ* measurements were obtained by the meteorite penetration detectors on board Pioneers 10 and 11. Pioneer 10 registered 11 impacts of dust particles with masses $m \geq 2 \times 10^{-12}$ kg and Pioneer 11 registered 2 dust particles with masses $m \geq 2 \times 10^{-11}$ kg [Humes 1980].

Dust properties

The main ring is scattering substantially more light into high phase angles than into small ones, indicating a considerable dust concentration. Interpretation of the photometry of the Jovian system is hindered because the imaging represents the intensity of a line of sight through both ring and halo. For the forward scattered component, a description by means of a power law integrated normal to the ring plane has been given. As power law distributions will be encountered in different contexts, I briefly add here that this is a way to describe the grain size distribution by means of 4 parameters through

$$n_d(a)da = Ca^{-\beta}da, \qquad (3.1)$$

in a restricted radius domain $[a_{\min}, a_{\max}]$. Here $n_d(a)da$ denotes the differential number density of grains per unit volume with radius between a and $a + da$, β is the power law index and C a normalization constant. For the Jovian rings, $C = 1.7 \pm 0.1 \times 10^{-11}$ and $\beta = 2.5 \pm 0.5$, for grains with radii between 0.3 μm up to 100 μm. Backscattering information reveals a strikingly consistent similarity with Amalthea and provides a good indication that there are rough, macroscopic bodies present, with surface properties quite similar to those of Amalthea [Showalter *et al.* 1987].

The halo, a broad faint torus, is even more difficult to analyze photometrically, because the measured intensity is composed of contributions throughout the halo region along the line of sight. Different arguments [Showalter *et al.* 1987] indicate that the halo consists of a dust distribution with a slightly steeper power law, or one in which some of the larger grains have been lost.

The gossamer ring has only been detected in backscattering for a limited range of phase angles. Although the variance of the ring intensity suggests diffraction by grains of radius $a \simeq 1.5$ μm, it is not possible to exclude the presence of either larger or smaller particles [Showalter *et al.* 1985].

Earth based measurements of the Jovian ring reflectance spectrum have been performed [see Greenberg and Brahic 1984]. The absence of absorption features rules out water, methane and ammonia ice as major ring constituents, and a silicate or carbonaceous composition has therefore been inferred. In addition, there are now indications that the interplanetary and interstellar grains are streaming through the Jovian magnetosphere, where their orbits are perturbed by the Lorentz force, so that submicron grains can be captured into stable orbits, forming a diffuse exogenic dust band outside the main Jovian ring [Colwell *et al.* 1998].

Plasma parameters and other characteristics

Gehrels [1976] give an order of magnitude for the plasma parameters in the plasmasphere as $N_e = N_p = 10^8$ m^{-3}, with $T = 100$ eV. These values were quoted ever since, as by Goertz [1989]. Unfortunately, there are no further plasma data available for the Jovian rings, and none are closer than $L = 5$.

Jupiter's plasmasphere extends to $L \sim 10$ and its dipole axis is tilted by 9.6°. The dipole moment is given by $M = 1.5 \times 10^{27}$ A m^2 [Parks 1991]. The plasma is assumed to rotate rigidly with a speed $v_\alpha(r) = \Omega_J \times r$, with $\Omega_J = 1.76 \times 10^{-4}$ rad s^{-1} [Ness 1994].

3.4.2   Saturn

As I had already the occasion to point out on several occasions in the preface and introductory chapter, some peculiar features in the ring system of Saturn caused a major boost in the development of the physics of dusty plasmas. Indeed, during the Voyager spacecraft flybys, radial structures (quickly referred to as spokes) in the B ring, and braided structures in the F ring (braids) were discovered. These features could not, contrary to general belief up to then, be explained in purely gravitational terms, and hence a substantial development of the knowledge of the physics of dusty plasmas became necessary.

But dust is not only present in these spectacular features. The innermost D ring seems to occupy most of the region between the C ring and the planet's cloud tops. The distinct brightening of the ring at the highest phase angles clearly indicates diffraction by micron-sized dust. Traveling outward, localized high dust densities are to be expected in the spoke regions (B ring). The next ring to show a preponderance of dust is a narrow ringlet near the middle of the Encke gap in the A ring. The F ring itself was the first narrow and longitudinally variable ring observed. Finally, Saturn

Figure 3.2. A picture of the Saturn ring system at a large phase angle. *Copyright JPL and NASA*

has two very faint outer rings, designated G and E. Interest in these two rings has increased recently because of the potential hazard they may pose to the Cassini orbiter. The Saturn ring system is shown in Figure 3.2.

Dust properties

Our knowledge of particle properties has been greatly enhanced by space-craft observations (Pioneer 11, Voyagers 1 and 2). These allowed views of the rings in a broad range of phase angles extending to about 160°, fairly close to forward scattering. The main rings are typically backscattering, which means that overall they contain only a small area fraction of dust, at most. The photometric behaviour of the optically thin rings is quite different from that of the main rings, as these diffuse rings are brighter at high phase angles. This clearly indicates the presence of micron-sized dust particles.

All of the high-quality spectral reflectance data, needed to resolve the particle composition of the rings, were obtained for the A and B rings [Gehrels and Matthews 1984]. These spectra have been used to establish that water ice is a major component of the ring material [Clark and McCord 1980; Clark 1980]. The A and B ring particles are probably primarily water ice, possibly containing clathrate hydrates, although minor amounts of colored material (possibly sulfur or iron-bearing silicate compounds) are also present. It is likely that the overall ring composition is not too different from that of the various icy satellites [Gehrels and Matthews 1984].

- Little is known about the dust properties of the **D ring**, because of its low optical depth, very likely less than 10^{-3} everywhere [Wilson 1991]. It could consist merely of dust grains, based on the few Voyager pictures where it can be seen [Showalter et al. 1991].

- One of the most intriguing features observed in the Saturnian ring system by both Voyager 1 and 2 were the nearly-radial, wedge-shaped features in Saturn's **B ring**. The wedges become wider towards the planet. Because of their radial extension these structures were called **spokes** [Collins et al. 1980], with an inner boundary at $\sim 1.72\ R_S$ and an outer boundary at approximately the outer edge of the B ring. A typical spoke pattern is seen in Figure 3.3. High resolution images show that the leading and trailing edges of the spokes have distinctly different angular velocities, the leading edge basically having a Keplerian velocity, whereas the trailing near-radial edge moves approximately at the corotational velocity. Against the background of the B ring they appear dark in backscattered light, and bright in forward scattered light, indicating that they are composed of micron and submicron-sized grains [Grün et al. 1983]. The observed deviation of the angular velocity of the spoke features from the Keplerian value can yield the charge-to-mass ratio of the spoke particles. A lower limit is given on the spoke particle size of about 0.01 μm [Thomsen et al. 1982]. This is important when I will discuss mass and size distributions and the impact of these in Chapter 9.

The spokes were explained as follows. Consider a macroscopic ring particle in the neighbourhood of which a dense plasma cloud develops. This cloud might be caused by meteor impact on the ring particles, or by other mechanisms like lightning discharges. The dense plasma causes small dust grains that reside on the macroscopic ring particle to get electrostatically levitated. Due to the gravitational forces, these dust grains will drift relative to the plasma. Indeed, the plasma particles with their very low charge-to-mass ratio corotate with the planet and its magnetic field, while the dust grains have angular ve-

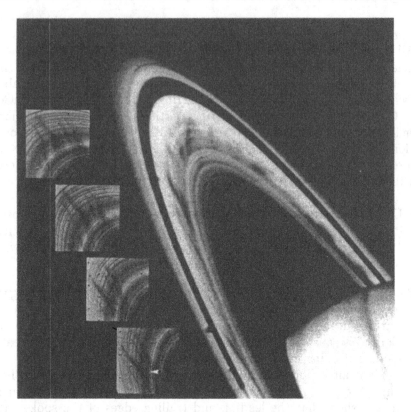

Figure 3.3. These Voyager 2 images show a distinct pattern of "spokes" in the B Ring (in reversed gray scale in forward scattered light). *Copyright JPL and NASA*

locities between the Keplerian and corotation velocities, as explained earlier. This generates charge separation and hence an electric field $\mathbf{E}$ and corresponding $\mathbf{E} \times \mathbf{B}$ drifts in the plasma. Outside synchronous orbit this results in a drift away from the planet, while inside the synchronous orbit the $\mathbf{E} \times \mathbf{B}$ drift is directed towards the planet [Northrop 1992]. As long as the drifting plasma remains dense enough, dust can be levitated, which marks the radial trail of the plasma. An observer will see this dust trail very much like the dust cloud left behind by a car racing along a dusty road [Goertz and Morfill 1983]. Other explanations have involved density waves [Bliokh and Yaroshenko 1985; Bliokh *et al.* 1995] and will be further discussed in Chapter 5 dealing with electrostatic dust modes.

- The **Encke gap** brightens significantly at high phase angles, again indicating a preponderance of micron-sized dust. This ring has never received the detailed investigation that it warrants, although an ample body of Voyager data exists.

- The **F ring** has been studied by the Voyager imaging instruments, by occultation measurements and by photo polarimetry. Showalter *et al.* [1992] analyzed the photographic pictures using a semi-empirical theory for scattering by randomly oriented, nonspherical particles. The contribution of "Saturn-shine" was included in the incident radiation field. The power law index of the dust ($a_{min} = 0.001$ μm and $a_{max} = 20$ μm) and the fractional contribution f of the dust to the total optical depth were treated as free parameters, leading to $\beta = 4.6 \pm 0.5$ and $f \geq 98\%$.

  The braids, kinks and clumps seen in only a few pictures of the F ring are the only features in the ring system which have up to this time defied a generally agreed upon explanation. Avinash and Sen [1994], the first to our knowledge to tackle the braiding problem in qualitative detail, explain this phenomenon based on the stability that occurs by the balance of the pinch pressure due to the dust ring current and the electrostatic pressure. Purely gravitational theories rely upon shepherding satellites and the resonances their motions induce in the rings. However, possible kinks and clumps formed in this way are washed out too quickly for braiding to occur. A combined gravito-electrodynamical approach does not work either, as it requires high dust charges, that can only occur for isolated single grains, not for those in an ensemble, as was discussed in Chapter 2. In addition, to complicate the picture, Showalter [1998] indicates that centimeter-sized meteoroid impacts in the F ring are too short lived to relate to the longer lived clumps.

- The very faint and narrow **G ring** has been studied photometrically [Showalter and Cuzzi 1993], when Voyager 2 crossed the ring plane, very close to the G ring, at 2.86 R_S. Local dust impacts could be retrieved [Gurnett *et al.* 1983; Tsintikidis 1994; Meyer-Vernet *et al.* 1998].

  There are only two Voyager photographic images available that show the G ring visible to the naked eye. Showalter and Cuzzi [1993] proposed a power law index of 6.0 ± 0.2 with $a_{min} = 0.03$ μm and $a_{max} \simeq 0.5$ μm. Gurnett *et al.* [1983], however, found $\beta = 7$ for $a_{min} = 0.3$ μm and $a_{max} = 3$ μm, a result that was revised by Tsintikidis [1994], who obtained a dust number density of the order of 10^{-2} m^{-3}, with a mass threshold ranging from 10^{-14} to 5.4×10^{-12} kg. Rough continuity was assumed between the different dust distributions, leading to an inferred value of $\beta < 3.5$ for grains between half a micron and a few microns. The dust distribution integrated radially along the G ring and normal to the ring plane is determined by $C = 5.8 \pm 0.3$ and $\beta = 6$ [Showalter and Cuzzi 1993].

- The **E ring** has been studied photometrically [Showalter *et al.* 1991], by local impact ionization [Meyer-Vernet *et al.* 1996] and through plasma wave detection by Tsintikidis *et al.* [1995]. A very narrow distribution of slightly nonspherical particles with sizes 1.0 ± 0.3 μm provides the best fit of the photometry [Showalter *et al.* 1991], in perfect agreement with the local analysis of Meyer-Vernet *et al.* [1996]. From photometric data we can only obtain integrated densities along the line of sight. The number density of the dust at $L = 6.1$ is given by Meyer-Vernet *et al.* [1996] as 4.3×10^{-3} m^{-3}, whereas Tsintikidis *et al.* [1995] find slightly larger grains, but with number densities of the same order.

Plasma parameters and other characteristics

The ion parameters measured by the Pioneer instruments differ substantially from those reported by Voyager. Whether this indicates a real variation in Saturn's magnetosphere or reflects a difficulty in analyzing the Pioneer data is uncertain. Outside $L = 12$ plasma spectra change rapidly [Richardson 1995], but inside this limit, where the rings are, the plasma environment can be considered as steady state. Azimuthal symmetry was assumed, a reasonable assumption because time scales for ionization of neutrals and losses of ions are of the order of weeks and even months, much longer than the 10-hour rotation period of Saturn. Plasma densities are extrapolated along magnetic field lines, using the equation for force balance parallel to the magnetic field.

The plasma consists of a heavy ion group (O^+, OH^+, H_2O^+, H_3O^+ and N^+), all of which have masses near 16 proton masses and could not be resolved in separate components by the Voyager instruments, with a density of $1 \sim 100 \times 10^6$ m^{-3}. A light ion population was also found, consisting merely of protons with a density of $0.1 \sim 10 \times 10^6$ m^{-3}. The electron distribution functions are clearly non-Maxwellian, having a cold (thermal) component with Maxwellian shape and a hot (suprathermal) non-Maxwellian component. In the neighbourhood of the rings, the hot electrons are by far outnumbered by the thermal electrons [Sittler *et al.* 1983], with the total electron density in the ring plane in the range of $1 \sim 100 \times 10^6$m^{-3}.

For the special case of the spokes in the B ring the situation is totally different. Following the spoke formation model of Goertz and Morfill [1983], the plasma density in the spokes is expected to be much higher than the surrounding plasma density. Such densities could be caused by meteor impacts on the larger B ring particles and can be several orders of magnitude bigger than the general B ring plasma density. Even densities as high as $N_e = 10^{21}$ m^{-3} are not impossible [Goertz and Morfill 1983].

Finally, Saturn's plasmasphere extends to $L \sim 10$ and its dipole axis is within $1°$ of being parallel to the rotation axis. The dipole moment is given by $M = 4.7 \times 10^{25}$ A m^2 and the field is very well described by that model [Ness 1994]. As I did while discussing Jupiter, the plasma is assumed to rigidly rotate with a speed $v_\alpha(r) = \Omega_S \times r$, with $\Omega_S = 1.64 \times 10^{-4}$ rad s^{-1} [Ness 1994].

3.4.3 Uranus

The nine classical rings of Uranus have high optical depths, and hence very little dust. The narrow λ ring, discovered in the backscattered Voyager images, is somewhat of an exception. In one of the Voyager images taken at high phase angles, the rings were far brighter than their environment, indicating that dust is the major constituent. An additional broad ring is visible in another Voyager image at a 90° phase angle, inside all other rings. Unfortunately, very little can be determined about the particle properties of this ring from a single view, although a predominance of dust is strongly suspected. Between the δ and λ ring a region is encountered which is mostly forward scattering, called the dust belt [Bergstralh et al. 1991]. More recent reprocessing of the Voyager data seems to indicate that the Uranian λ ring has marked longitudinal brightness variations, symptomatic of arcs and clumps [Showalter 1995], not unlike the braiding and clumps observed in the F ring of Saturn.

Dust properties

Local impact ionization analysis of the Voyager 2 data [Gurnett et al. 1987; Meyer-Vernet et al. 1986] was done for the ring plane crossing at $L = 4.40$, farther away from the planet than the rings are. Grains of a few μm (1 to 8 μm) struck the spacecraft, having a maximum number density of 10^{-4} to 10^{-3} m^{-3}. A photometric analysis was made by Ockert et al. [1987]. The brightness distribution is dominated by backscattering, which means that mostly macroscopic grains are present. The fractional area in dust-sized particles in and around the nine main rings was analyzed and found to be less than 2×10^{-3} m^{-3}. Furthermore, the observations were consistent with an average dust particle size of 1.0 ± 0.3 μm, with a power law index $\beta = 2.5 \pm 0.5$. The particles are quite dark and scatter light like rough dark bodies such as the Moon or Callisto. The gray color of the ring particles could be explained by chondrites with a coating of carbon [Bergstralh et al. 1991]. Here chondrites refer to a refractory organic phase, containing mostly low-mass elements like C, H, O, N, hence the name.

Plasma parameters and other characteristics

The closest encounter with the planet took place in January 1986, when Voyager 2 passed the planet at a magnetic shell with $L = 4.59$, so that the rings were not really visited. The plasma population at $L = 4.59$ and locations farther away from the planet is modelled by a thermal Maxwellian and a hot non-Maxwellian distribution for the electrons, and by a hot, an intermediate and a warm ion population.

Any extrapolation of these data has to be carried out very carefully, because of the different orientations of the rotation and magnetic axes [Bergstralh *et al.* 1991]. The dipole axis of the Uranian plasmasphere is tilted by almost $59°$. The dipole moment is given by $M = 3.8 \times 10^{24}$ A m^2. Furthermore, the dipole offset (planet center to dipole center) equals $L = 0.3$ [Parks 1991]. Assuming rigid body rotation in the magnetosphere indicates that the plasma rotates with a speed $v_\alpha(r) = \Omega_U r$, with $\Omega_U = 1.01 \times 10^{-4}$ rad s^{-1} [Ness 1994].

3.4.4   Neptune

The Voyager cameras revealed a number of new dust rings around Neptune. The largest set of data available on the rings of Neptune consists of about 800 images collected by Voyager 2, in addition to a handful of detections by ground based observations during the 1980s. Again, the Voyager 2 spacecraft crossed the ring plane outside the rings, for evident reasons as always.

Photometric modelling indicates that all of these rings contain a mixture of dust and larger bodies with comparable optical properties. Especially the Adams and Le Verrier rings contain a significant fraction of dust, comparable to that observed in Saturn's F ring. The Lassell ring appears to have a different mix of particle sizes from the Le Verrier or Adams rings. Azimuthal arcs on scales of the order of $1°$ to $10°$ were found in the Adams ring, and within these arcs several long linear features, believed to be discrete clumps or denser-than-average regions of the arc. It was advanced that they represent an accumulation of dust-sized grains, although a purely gravitational explanation is possible [Cruikshank 1996]. Hence much work remains to be done.

The low light levels at Neptune, and the extremely low albedo of the rings, lead to uncommonly long exposures, causing the photos to be badly smeared and so photometric analyses need a careful interpretation. A further complication arises from the fact that the expected symmetry plane for particles on gravitationally dominated orbits, called the Laplace plane, is not generally coincident with Neptune's equatorial plane, because of perturbations from a massive moon, Triton, on an inclined orbit.

Dust properties

The results from different Voyager experiments [Pedersen *et al.* 1991; Gurnett *et al.* [1991] differ for the precise peak locations, vertical thicknesses

Figure 3.4. This Voyager image shows the faint Neptunian rings, where the light of the planet had to be blocked out to render the rings visible. *Copyright JPL and NASA*

and dust volume densities. The first paper mentioned here indicates a power law for the dust grain distribution with $\beta = 4$, $a_{\min} = 1.6$ μm and $a_{\max} = 10$ μm, during the outbound crossing at $R_N = 4.26$, where R_N is the radius of Neptune. The other experiment detected grains in the range of $5 \sim 10$ μm, with an uncertainty of a factor of 2 to 3. The densities are given as 10^{-2} m^{-3} for the outbound crossing at $R_N = 4.26$ and as 10^{-3} m^{-3} for the inbound crossing at $R_N = 3.45$.

Data on the ring optical depth have been used by Ferrari and Brahic [1994] to derive additional constraints on the refractive indices of the ring particles, leading to the conclusion that the refractive indices best match a "dirty ice" composition, rather than a highly absorbing carbon-rich material. However, refractive indices of this magnitude could also point to a variety of silicates. Even the occurrence of silicates cannot be ruled out, although the fit to the data is not very good.

Plasma parameters and other characteristics

The analysis of the Voyager plasma data can be found in Cruikshank [1996]. The spacecraft crossed the ring plane in the magnetic shell with $L = 3.45$, and hence outside the ring plane. The overall magnetic field can be characterized by an offset tilted dipole model with moment 2×10^{24} A m^2, with a dipole offset of $L = 0.55$ and a dipole axis tilted by $47°$ [Cruikshank 1996]. The plasma rigidly rotates with a speed $v_\alpha(r) = \Omega_N \times r$, with $\Omega_N = 1.08 \times 10^{-4}$ rad s^{-1} [Ness 1994].

3.5   Cometary plasmas

Other intriguing objects containing solar system dust are cometary environments. The appearance of dust grains is very obvious in one of the
optically largest structures that can be admired in the sky: cometary dust
tails. The tail, pushed away from the Sun by solar pressure, is now quite
well understood and can reveal lots of important scientific information on
the dust properties.

To have an idea of the global morphology of the cometary environment,
we briefly discuss the interaction with the solar wind. Heavy cometary neutral species (atoms and molecules) are ionized by solar ultraviolet radiation
or charge exchange with the solar wind ions, and are then assimilated into
the magnetized solar wind. The continuous mass loading of the inflowing
solar wind by the newly created ions causes the solar wind to decelerate
and warm up, but this deceleration is possible only as long as the mean
molecular weight of the plasma remains below a critical value, at which a
weak bow shock forms ahead of the comet. Downstream from the shock, the
mass-loaded subsonic solar wind continues to interact with the cometary
atmosphere, penetrating into a region of ever increasing neutral density.

Strong deceleration occurs at a boundary called the cometopause, where
significant momentum is transferred by collisions from the outflowing cometary neutrals to the solar wind ions. Inside this transition region, collisions
dominate and the solar wind decelerates rapidly and cools due to charge
exchange processes with the less energetic cometary neutrals, while the
magnetic field compresses to form a magnetic barrier. A tangential dis-
continuity interface (ionopause) forms at the inner edge of the magnetic
barrier and separates two plasmas: the purely cometary plasma and the
mass-loaded solar wind plasma. Inside the ionopause there is a magnetic
field-free cavity. Further properties can be found in the many papers gath-
ered in the proceedings of an important conference devoted to comets in
the post-Halley era [Newburn et al. 1991].

Prior to the 1986 appearance of comet 1P/Halley, all attempts to determine the physical properties of cometary dust were limited to remote
observations and the analysis of various particles captured by the Earth's
magnetosphere. Here again difficulties related to the inverse problem show
up. For example, deconvolution of the observed brightness to the local dust
number density involves knowledge of the size distribution, which might be
changing spatially and/or temporally. Moreover, spatial and temporal effects are difficult to disentangle. Within a period of less than three weeks
in March 1986, Halley's comet was encountered by six spacecraft belonging
to four space agencies and carrying a total complement of 50 experiments
[Mendis 1988]. Three of the spacecraft passed within 10,000 km of the
nucleus, providing us with the first opportunity to investigate the full size
range of dust particles. From all those spacecraft, Giotto was the only one

going through the entire cometosheath including the inner cometosheath and the ionopause, and as we saw, part of the equipment did not survive this close encounter unharmed.

Dust properties

Dust particles ejected by the cometary nucleus will be accelerated by the expanding gas coma, until they reach a final expansion velocity that increases with decreasing particle size. This velocity also depends on the heliocentric distance r_h, increasing at decreasing r_h. Once a dust grain leaves the region of dust-gas interaction, the only forces acting on it in the heliocentric frame are the solar radiation pressure and the solar gravity, since attraction by the small cometary nucleus is negligible. This leads to the so-called fountain model in cometocentric bipolar coordinates, where the nucleus is at the origin and one of the axes is along the heliocentric radius vector. The particle trajectories are parabolas, the envelope of which is a paraboloid of revolution with focus at the nucleus [Newburn et al. 1991].

As a first approximation, the number density of the dust grains is proportional to

$$N_d(r_h, a, r) \propto \frac{P(r_h, a)}{r^2} \qquad (r < A(r_h, a)) \qquad (3.2)$$

at a distance r from the cometary nucleus, if $P(r_h, a)$ is the dust production rate and $A(r_h, a)$ the apex distance for a grain with radius a. It was found that the dust release rates in the coma decreases with the increase of heliocentric distance as a power law

$$P(r_h, a) \propto r_h^{-\beta}, \qquad (3.3)$$

with $\beta = 3.0 \pm 0.7$. This is in agreement with other models, all yielding values in the neighbourhood of $\beta = 4$ ([Singh et al. 1997] and references therein). The fountain model for the dust dynamics does not include any orbital motion by the comet, but gives good predictions of dust grain flux magnitudes [McDonnell et al. 1992].

Although at first glance the fountain model seems to match the available data [see Newburn et al. 1986], reality is more complicated because other effects also play a role. The fountain model is that where the dust grains leave the nucleus in all directions, but are blown away by the incoming solar wind. Measurements lead to the following remarks.

- Both the VEGA and Giotto spacecraft detected packets of dust. Small dust particles arrived in sequences of events, preceded and followed by relatively long time gaps during which no events were measured.

- Small dust grains abound, and the smallest particles were detected much farther away than expected from the fountain model. The dust grains were well outside their bounding paraboloids, if we assume that they were slightly absorbing, as detected by the VEGA dust detectors [Mendis 1988].

- In the inner regions of the coma, the dust particle number density varies substantially stronger than r^{-2}, and the apex boundaries were not as sharp as modelling anticipated.

- There is a tendency for the mass density of the dust grains to decrease with increasing size. For these very small particles, electromagnetic forces will play a dominant important role, and these effects will be discussed in later chapters. The mass density of the grains themselves seems to be in the range of 1000 to 2500 kg m^{-3}. Indeed, grains with atomic (C+O)/(Mg+Si+Fe) ratios in the range of 0.01 to 10 (silicate dominated) have densities around 2500 kg m^{-3}, while those with larger ratios (CHON-dominated) have a mean density of 1000 kg m^{-3}. Hence CHON dominated grains are fluffy, while the silicate dominated ones are more compact [see Newburn $et\ al.$ 1991]. Most dust particles recorded by VEGA consist of a fluffy silicate core, covered by refractory, icy fluffy organics.

The dust size distribution is best fitted by a power law, the index of which seems to vary with mass of the dust grains and with time. Hence indices range from $\beta = 3.3$ for VEGA 2 to $\beta = 4.1$ for Giotto observations, although modelling would suggest $\beta = 3.6$ from 10^{-13} all the way up to 10^{-19} kg particles in the coma. For larger dust particles ($m \geq 10^{-9}$ kg), the spectrum is much flatter, with $\beta = 2.7$ [see Newburn $et\ al.$ 1991]. At small distances from the nucleus, the relative number of small, submicron particles is high. The relative contribution of such particles decreases with increasing distance from the nucleus. Furthermore, the slope of the mass spectrum decreases for small masses. At large distances from the nucleus, one observes a clearly pronounced separation of two dust groups, into smaller ($m < 10^{-17}$ kg) and larger ($m < 10^{-13}$ kg) particles.

Other dust size distributions have been proposed, like [Singh $et\ al.$ 1992]

$$n(a)da = g_0 \left[1 - \frac{a_{\min}}{a}\right]^M \left[\frac{a_{\min}}{a}\right]^N, \qquad (a \geq a_0),$$

$$n(a)da = 0, \qquad\qquad\qquad\qquad (a < a_0). \qquad (3.4)$$

The radius $a_{\min}$ of the smallest grain is taken to be 0.1 μm, and g_0 is a normalization constant, with $N(= 4.2)$ defining the slope for the largest grains (N plays the role of power law index for grains with $a \gg a_{\min}$), and $\log M = 1.13 + 0.62 \log r_{h\{AU\}}$.

It is commonly believed that comets were formed within the solar system from the same initial dust and gas nebula as were the larger bodies, such as the planets. However, the small size of the comet implies the absence of large-scale mixing processes, and one may thus expect that the isotopic, chemical and molecular characteristics of the presolar material are best preserved in comets. The chemical properties of individual dust grains were revealed by the mass spectrometers on board Giotto and both VEGA's, and it was concluded that Halley's comet is composed of two components: a refractory organic phase consisting of chondrites, and a Mg-rich silicate phase. The chondrite component is probably coating the silicate cores. Nevertheless, because the observations cannot readily be reproduced, the data should be interpreted very cautiously and models with too far reaching consequences are to be avoided.

Dust densities are given for distances up to 2×10^8 m, for the different mass channels ranging from 10^{-13} to 10^{-20} kg per dust particle. The dust population consists mainly of submicron grains, with number densities of the order of 10^{-3} m^{-3} at a distance of 10^8 m. The first dust particles were already detected by VEGA 1 at 260×10^6 m and by VEGA 2 at 320×10^6 m [Vaisberg et al. 1986].

The ICE flyby at comet 21P/Giacobini-Zinner was discussed by Boehnhardt and Fechtig [1987]. For silicate grains the equilibrium potentials vary between +6 and +10 V within the bow shock and outside the cometopause. Within the cometopause, however, these potentials are steadily decreasing to slightly negative numbers (-0.1 V), due to a high density of low-energy plasma electrons. Carbon grains charge to a potential between +5 V and -19 V outside the cometopause, but inside quickly achieve negative potentials (-0.1 V).

Boehnhardt and Fechtig [1987] also discuss the Giotto flyby at Halley. Outside the cometopause, grain charging is dominated by photoelectrons. Silicate grains charge in the range $6 \sim 10$ V, while carbon grains have an equilibrium potential of $4 \sim 5$ V. The spacecraft potentials are typically +7 V in the solar wind, reduce to about 2.5 V in the outer coma, and drop to 1 V at closest approach [Grard et al. 1989].

Plasma parameters

Plasma parameters from the VEGA, Giotto and SUISEI missions are given by Grard et al. [1989], Johnstone [1991] and Mukai et al. [1986], respectively. However, only Giotto was going through the entire cometosheath and the ionopause.

- The kinetic energy of the electrons, measured at distances from 80×10^6 to 800×10^6 m, has mean values of 0.53 eV and 0.49 eV, for VEGA 1 and 2 respectively [Grard et al. 1989]. Such average electron kinetic energies are not typical at all for the solar wind, where values 30 times larger are generally observed. After correction of spurious effects, caused by photo-emission and nitrogen releases, the electron density for VEGA 1 is on average 60×10^6 m^{-3} at distances of 300×10^6 to 10^9 m, roughly between the bow shock and the cometopause. In the same distance range, the VEGA 2 measurements give a density of 40×10^6 m^{-3}. Both spacecraft measure an electron density of 10^9 m^{-3} closer to the nucleus (at 16×10^6 m).

- Other authors [reported in Johnstone 1991] consider four different plasma populations. Besides the electron population, a hot and a cold water group population and a proton population are covered for distances between 10^8 and 10^9 m from the nucleus, which is inside the bow shock. The proton density is rather constant and assumed to be $20 - 30 \times 10^6$ m^{-3}, with a temperature of about 35 eV. Also the high energy water group ions have a rather constant number density of $0.3 \sim 0.7 \times 10^6$ m^{-3}, and a temperature of about 4.3 keV. The low energy water group population, on the other hand, changes considerably with cometocentric distance. The number density at 10^8 m is of the order of 5×10^9 m^{-3}, and decreasing for larger distances to 10^6 m^{-3}. Energies are 1.7 keV and 80 eV, respectively.

- An abrupt change in the plasma parameters was registered at 450×10^6 m away from the nucleus, in the plasma density, which probably represents the bow shock. Just after the bow shock the density was 35×10^6 m^{-3}. At a distance of 10^9 m, the density had dropped to 15×10^6 m^{-3} [Mukai et al. 1986].

Other characteristics

The Giotto data for the magnetic field during the ionopause crossing of comet Halley can be found in Newburn et al. [1991]. Outside the cometary ionopause, which separates the outflowing cometary ions from the inflowing, contaminated solar wind plasma, the plasma is magnetized at $B \simeq 20$ nT, while inside the ionopause the magnetic field vanishes completely. This interplanetary magnetic field varies near Earth's orbit from a few nT at quiet times to more than 10 to 20 nT in solar disturbances [Parks 1991]. The relative dust-plasma velocity is of the order of several, up to 100 m s^{-1} [Richter et al. 1991].

3.6 Interplanetary dust

On our trip through the solar system dust was already encountered when discussing the circumsolar dust rings. Interplanetary dust has been intensively studied by a number of detection methods. Several theoretical models

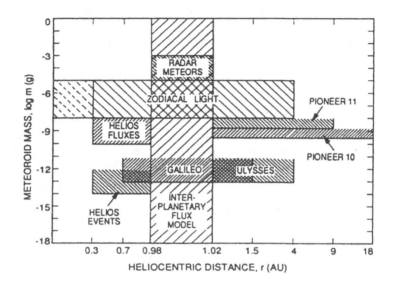

Figure 3.5. The coverage in mass and heliospheric distance for the data sets used. *Reprinted from Divine 1993 (Fig. 1 on p. 17032, with permission of the American Geophysical Union*

were worked out, the most comprehensive one being given by Divine [1993], the results of which are shown in Figure 3.5. The basic assumptions are that solar gravitation is the only force entering the picture, and that the system is symmetric both in ecliptic latitude and longitude. Five distinct Keplerian dust populations successfully match all data, and they range from 10^{-21} kg to 10^{-3} kg in mass per particle, and from less than 0.1 AU to 20 AU in heliocentric distance. No time dependencies are considered, although for the submicron dust flux a 22-year cycle, closely matching the true solar cycle, has been suggested [Grün *et al.* 1997]. The latter authors improve upon the Divine model by taking the complete and newer Galileo and Ulysses data sets into account. Although Divine used purely Keplerian dynamics to derive impact rates, Grün *et al.* [1997] added radiation pressure to study the dynamics of micron-sized dust. There is also now an additional interstellar dust population penetrating the solar system on hyperbolic trajectories.

Dust properties

The following relevant data sets have been used to interpret the interplanetary dust parameters.

- A first knowledge of the properties of interplanetary dust grains comes from meteor and meteorite observations, as well as lunar cratering records, and also from impact detectors aboard several spacecraft orbiting the Earth. A review of the available data [Grün *et al.* 1985] indicates threshold mass ranges from 10^{-21} to 0.1 kg, and the flux is presented as the average over a spinning flat plate.

- Data from the impact detectors aboard Pioneers 10 and 11 [Humes 1980] are represented as spin-averaged penetration fluxes, averaged over unequal intervals in heliospheric distance between 1 and 18 AU within 3.1° of the ecliptic plane for Pioneer 10, and 1 and 9 AU within 15° of the ecliptic plane for Pioneer 11. There were 95 penetrations recorded by Pioneer 10, and 87 penetrations by Pioneer 11.

- Part of the available data on dust grain properties of the Helios space mission can be found in Grün *et al.* [1980]. The range between 0.31 and 0.98 AU is studied in the ecliptic plane, and the grain flux is given, averaged over intervals of 0.1 AU.

- Galileo data are represented as spin-averaged penetration fluxes, averaged over unequal intervals in heliospheric distance between 0.88 and 2.3 AU in the ecliptic plane.

- Ulysses dust data are also represented as spin-averaged penetration fluxes, averaged over unequal intervals in heliospheric distance between 1.0 and 5.4 AU, over a wide range of ecliptic latitudes between −80° and 80°.

- Furthermore, there are radar observations and observation of zodiacal light in the range $0.3 \sim 4.0$ AU.

The five populations proposed by Divine [1993] are the core, eccentric, asteroidal, inclined and halo populations. Closer to the Sun, the core and inclined populations dominate, further away the halo and asteroidal. The dust number density for grains with sizes between 0.46 μm to 10 μm decreases roughly as $1/r_h^2$ for $r_h < 2$ AU. The dust number density is of the order of 10^{-8} m^{-3} in the neighbourhood of the Earth. For $r_h > 2$ AU, the halo population comes into play, and the power law decrease is less steep.

Plasma parameters and other characteristics

The plasma in the interplanetary medium is the solar wind, with highly variable parameters. Reasonable averages for the electron, proton and helium ion densities, are given for low, average and high solar activity, respectively, by $N_e = (0.4; 6.5; 100) \times 10^6$ m^{-3}, $N_p = (0.4; 6.2; 75) \times 10^6$ m^{-3} and $N_\alpha = (0; 0.3; 25) \times 10^6$ m^{-3}, in the ecliptic plane at 1 AU [Mukai 1981].

These densities decrease by an inverse square law with heliocentric distance. The electron temperature can be taken as $(5 \times 10^3; 2 \times 10^6; 10^6)$ K, varying with heliocentric distance according to a power law. The power index at high latitudes was measured by the Ulysses spacecraft ($q = 0.81 \sim 1.03$) [Goldstein et al. 1996], whereas $q = 1.22$ was found by Mukai [1981] in the ecliptic plane. The solar wind velocity is of the order of $(200 \times 10^3; 400 \times 10^3; 900 \times 10^3)$ m s^{-1}.

The interplanetary magnetic field varies near Earth's orbit from a few nT at quiet times to more than 10 to 20 nT during solar disturbances. The magnetic field can be considered to vary as $B^2 \sim (1+r_h^2)/r_h^4$, thus expressing the Parker spiral field model, and readily verified by the Voyager spacecraft [Burlaga et al. 1984]. The magnetic field for $r_h \gg 1$ AU will change as r_h^{-1}.

3.7 Interstellar dust clouds

Once outside the heliopause and associated bow shocks, we leave the solar system behind us and penetrate in the interstellar plasma, where charged dust is as present as everywhere else in the universe.

As seen when discussing the penetration of interstellar dust grains into the inner regions of the heliosphere close to the Sun, our interest in charged dust grains is manifold. Not only do such dust grains repopulate various cosmic features like the zodiacal cloud, but there might be much more profound influences at work. Several authors have suggested that the physical conditions inside dense molecular clouds might favour the formation of amino acids and complex organic polymers. There now exists both astronomical and laboratory evidence supporting this idea, and a recent review of the role of interstellar grains as amino acid factories and the origin of life on Earth can be found in Sorrell [1997].

One of the fundamental problems in discussing interstellar dust is the near absence of information about its charged or neutral nature. Some estimates for the dust charges in interstellar clouds seem to point to average fractional charges, and then the interpretation is that some grains are charged while others are not.

In order to get a feeling for the parameters involved, I give some simple average estimates for interstellar dust [Whitten 1994; Evans 1996]. Assume that the interstellar dust (charged and/or neutral) has a total density of 10^{-1} m^{-3}, and an average mass for water-ice, micron-sized grains of $m_d = 4 \times 10^{-15}$ kg. The dust charges are totally unknown, as there are no reliable estimates here. The interstellar magnetic field is of the order of $B_0 = 3 \times 10^{-10}$ T, and the dust temperature is $T_d \simeq T_n = 30$ K, which is to be compared to an electron temperature of $T_e = 10^4$ K. Little more can be said for the moment, at least not in relation to the waves in dusty space plasmas that we will be discussing in the coming chapters in more theoretical detail.

To sum up, I gather some of the relevant data about cosmic environments in Table 3.2. Further details can be found in review papers [Goertz 1989; Mendis and Rosenberg 1994; Horányi 1996; Hartquist *et al.* 1997].

	N_i (m^{-3})	T (K)	N_d (m^{-3})	d/λ_D	a (μm)
Noctilucent clouds	10^9	150	10^7	0.2	0.1
Zodiacal dust	10^6	10^5	10^{-6}	5	10
Saturn's E ring	10^7	$10^5 - 10^6$	$0.01 - 0.1$	1	1
Spokes (B ring)	$10^5 - 10^8$	10^4	10^6	0.01	1
Braids (F ring)	$10^7 - 10^8$	$10^5 - 10^6$	10^7	10^{-3}	1
Comet Halley	$10^8 - 10^9$	10^4	$0.01 - 0.1$	10	$0.1 - 10$
Interstellar clouds	10^3	10	0.1	0.3	$0.01 - 10$

Table 3.2. Plasma and dust parameters in typical cosmic environments

CHAPTER 4

MULTISPECIES FORMALISM AND WAVES

4.1 General framework

After having set the scene by discussing charge mechanisms and space observations, I go on and come to the main topic of this monograph, a review of wave phenomena in dusty space plasmas studied during the last decade. At the same time, I will try to filter out of all the predictions made those which might really be relevant to particular astrophysical situations.

Many of the relevant papers have been written in the context of a fluid description, which usually has the advantage of being physically more transparent, being in terms of macroscopic variables like number densities, fluid velocities and pressures. Of course, a complex physical system like a plasma, and even more so, a dusty plasma, consists in reality of a huge number of different particles, and these determine the microscopic information, at its most basic level. How does one then go from the microscopic to the macroscopic description, in other words, how does one average?

Let me first sketch, for standard plasmas, the way this is dealt with, before going into mathematical details, and before discussing specific complications arising in dusty plasmas. If you look into a general plasma physics textbook, you find that the full, general starting point is either the Klimontovich or the Liouville equation for some density in the phase space of all particles. Together with the microscopic Maxwell equations, this constitutes an exact description of a plasma, provided one knows the initial positions and velocities of all particles. However, there has never been any hope of dealing in a tractable way with that many individual particles, and quantum mechanics would even forbid us the precision required for the initial conditions. But there is more, one is not really interested in that kind of overwhelming, unstructured information, but in determining how the macroscopic quantities behave.

Hence, the averaging needed goes in steps. First, the full distribution function is replaced by reduced distribution functions, each one obeying a member of the BBGKY hierarchy of equations, named after Bogoliubov, Born, Green, Kirkwood and Yvon. You can find further references in plasma physics textbooks. The BBGKY equations form a chain of coupled equations, formally equivalent to the starting point, the Klimontovich or Liouville equation, if they are treated and solved all together. Yet again, this is in practice unworkable. The next step is the truncation of the BBGKY hierarchy to the lowest order, a kinetic equation for the single particle distribution function, that incorporates some information about the average interactions between two particles and thus involves plausible but *ad hoc* hypotheses. Finally, one can integrate over velocity space to obtain moments of the single particle distribution, and that leads to the usual fluid equations. Although all moment equations together are completely equivalent to the kinetic equation they relate to, mathematical tractability demands a new truncation to lower order equations, like the continuity, momentum and pressure equations. Thus new, additional closure hypotheses of a different kind are needed.

Because dusty plasmas are even more complex than ordinary plasmas like a hydrogen plasma, it is worth recalling some of the steps of the reduction process from the Liouville to a kinetic equation, without giving all the intermediate details. As said already, the latter can be found in plasma physics textbooks, and I will follow that of Krishan [1999] in the next section. Once we see these basic steps before us, I can give indications about the main differences and difficulties, when trying to describe dusty plasmas along similar lines.

The remainder of the chapter then consists in deriving the multispecies fluid equations I shall use to describe linear and nonlinear wave phenomena. General dispersion laws will be given here, and the discussion of various modes deferred to the next chapters, in a logical progression from simpler to more complicated situations, as I see it.

4.2 From Liouville to kinetic equations

The time evolution of a system of N particles is given by Hamilton's equations of motion

$$\frac{\mathrm{d}\mathbf{x}_i}{\mathrm{d}t} = \frac{\partial H}{\partial \mathbf{p}_i}, \qquad \frac{\mathrm{d}\mathbf{p}_i}{\mathrm{d}t} = -\frac{\partial H}{\partial \mathbf{x}_i}, \tag{4.1}$$

where H is the Hamiltonian, function of the canonically conjugate coordinates $\mathbf{x}_i$ and momenta $\mathbf{p}_i$, for $i = 1, \ldots, N$. The mechanical state of such a system can be represented by a single point in a phase space of dimension $6N$, by specifying all components of $\mathbf{x}_i$ and of $\mathbf{p}_i$. The motion of such a point is governed by Hamilton's equations (4.1). Many different microscopic configurations might correspond to the same macroscopic state of the system. A collection of such systems is called a Gibbs ensemble, and an ensemble is represented in phase space by a number of points. Rather than describe these points separately, one introduces a phase space probability $f_N(\mathbf{x}_1, \ldots, \mathbf{x}_N, \mathbf{p}_1, \ldots, \mathbf{p}_N, t) d\mathbf{x}_1 \ldots d\mathbf{x}_N d\mathbf{p}_1 \ldots d\mathbf{p}_N$ giving the probability of finding a system in an elementary volume $d\mathbf{x}_1 \ldots d\mathbf{x}_N \, d\mathbf{p}_1 \ldots d\mathbf{p}_N$ around $(\mathbf{x}_1, \ldots, \mathbf{x}_N, \mathbf{p}_1, \ldots, \mathbf{p}_N)$ at any instant t. The time evolution of such probabilities is given by

$$\frac{df_N}{dt} = \frac{\partial f_N}{\partial t} + \sum_{i=1}^{N} \frac{\partial f_N}{\partial x_i} \cdot \frac{dx_i}{dt} + \sum_{i=1}^{N} \frac{\partial f_N}{\partial p_i} \cdot \frac{dp_i}{dt}. \qquad (4.2)$$

If the points are followed in their motion, together with the phase space volume element they occupy, one obtains the Liouville equation, expressing the conservation of phase space probability. It can be written as

$$\frac{df_N}{dt} = \frac{\partial f_N}{\partial t} + \sum_{i=1}^{N} v_i \cdot \frac{\partial f_N}{\partial x_i} + \sum_{i=1}^{N} a_i \cdot \frac{\partial f_N}{\partial v_i} = 0, \qquad (4.3)$$

where microscopic velocities v_i are used, rather than the conjugate momenta p_i. The acceleration a_i of each particle is due to external forces as well as to the interactions with the other particles. Thus (4.3) incorporates the microscopic laws of motion

$$v_i = \frac{dx_i}{dt}, \qquad a_i = \frac{dv_i}{dt}. \qquad (4.4)$$

Now I introduce reduced probabilities f_k for k particles, by integrating f_N over the positions and velocities of the other $N - k$ particles. In this way, one is concerned with the information about k particles, regardless of where the other $N - k$ particles might be or what velocities they have. Equations for the reduced probabilities can be obtained from (4.3) by integration over the appropriate coordinates and velocities, and the resulting set of coupled equations is known as the BBGKY hierarchy [Nicholson 1983]. Tacitly, all particles are assumed interchangeable, which is really not the case for astrophysical charged dust. At the lowest level, the evolution of the one-particle distribution function f_1 is governed by

$$\frac{\partial f_1}{\partial t} + v_1 \cdot \frac{\partial f_1}{\partial x_1} + \int a_1 \cdot \frac{\partial f_N}{\partial v_1} d^3 x_2 \ldots d^3 x_N d^3 v_2 \ldots d^3 v_N = 0, \qquad (4.5)$$

assuming that the different distribution functions vanish at infinity. Since the acceleration contains external sources as well as interactions with other particles, I can decompose a_1 as

$$a_1 = a_{ext} + \sum_{i=2}^{N} a(x_1, v_1; x_i, v_i) \equiv a_{ext} + \sum_{i=2}^{N} a_{1i}. \qquad (4.6)$$

That allows me to rewrite (4.5) as

$$\frac{\partial f_1}{\partial t} + v_1 \cdot \frac{\partial f_1}{\partial x_1} + a_{ext} \cdot \frac{\partial f_1}{\partial v_1} = -\sum_{i=2}^{N} \int a_{1i} \cdot \frac{\partial f_2(x_1, v_1; x_i, v_i; t)}{\partial v_1} d^3 x_i d^3 v_i. \qquad (4.7)$$

Everything now crucially depends on the assumptions of how the plasma particles interact and how they are correlated in the two-particle distribution function f_2. If there are no direct interactions (4.7) reduces to the collisionless Boltzmann equation,

$$\frac{\partial f}{\partial t} + \mathbf{v} \cdot \frac{\partial f}{\partial \mathbf{x}} + \mathbf{a}_{\text{ext}} \cdot \frac{\partial f}{\partial \mathbf{v}} = 0, \tag{4.8}$$

written for simplicity without particle subscripts.

What if there are interactions? Recall that $f_2(\mathbf{x}_1, \mathbf{v}_1; \mathbf{x}_i, \mathbf{v}_i; t)$ is the joint probability of finding one particle at $(\mathbf{x}_1, \mathbf{v}_1)$ and the other at $(\mathbf{x}_i, \mathbf{v}_i)$, at a given time t. A formal decomposition into

$$f_2(\mathbf{x}_1, \mathbf{v}_1; \mathbf{x}_i, \mathbf{v}_i; t) = f_1(\mathbf{x}_1, \mathbf{v}_1, t) f_1(\mathbf{x}_i, \mathbf{v}_i, t) + g_{12}(\mathbf{x}_1, \mathbf{v}_1; \mathbf{x}_i, \mathbf{v}_i; t) \tag{4.9}$$

is always possible, where g_{12} now represents the pair correlations. Thus (4.7) becomes

$$\frac{\partial f}{\partial t} + \mathbf{v} \cdot \frac{\partial f}{\partial \mathbf{x}} + (\mathbf{a}_{\text{ext}} + \mathbf{a}_{\text{self}}) \cdot \frac{\partial f}{\partial \mathbf{v}} = \frac{\delta f}{\delta t}, \tag{4.10}$$

where subscripts have again been omitted,

$$\mathbf{a}_{\text{self}} = \sum_{i=2}^{N} \int \mathbf{a}_{1i} f_1(\mathbf{x}_i, \mathbf{v}_i, t) d^3 \mathbf{x}_i d^3 \mathbf{v}_i \tag{4.11}$$

represents the mean acceleration due to all the other particles and is therefore called the self-consistent contribution, and $\delta f / \delta t$ is the change in f due to pair interactions g_{12}. If the latter can be neglected, the Vlasov equation

$$\frac{\partial f}{\partial t} + \mathbf{v} \cdot \frac{\partial f}{\partial \mathbf{x}} + (\mathbf{a}_{\text{ext}} + \mathbf{a}_{\text{self}}) \cdot \frac{\partial f}{\partial \mathbf{v}} = 0 \tag{4.12}$$

is obtained. It is seen that the difference between the collisionless Boltzmann equation (4.8) and the Vlasov equation (4.12) resides in the precise interpretation of the acceleration terms. There are many more examples of kinetic equations, according to the hypotheses one introduces about the interactions between the particles of like or of different species, but all obey the standard form

$$\frac{\partial f}{\partial t} + \mathbf{v} \cdot \frac{\partial f}{\partial \mathbf{x}} + \mathbf{a} \cdot \frac{\partial f}{\partial \mathbf{v}} = \frac{\delta f}{\delta t}. \tag{4.13}$$

4.3   Specific complications for dusty plasmas

Before I discuss the transition from kinetic theory to fluid equations, it is worth remembering that a dusty plasma is not just a plasma with two or more ion species. One of the basic differences between dust grains and other plasma particles is that the charge of the grains can be very large, is not fixed and very much depends on the surrounding plasma characteristics. In addition, the dust grains themselves are large compared to electrons and typical ions, and certainly in astrophysical situations come in a whole range of sizes. For such a system of non-identical particles, each of which at any time may change "personality" through capture of electrons and or ions, the most one can hope for is that overall system charge is conserved. But the uniqueness and the "multiple personalities" of the dust grains forbid a rigorous reduction of the Liouville equation via the BBGKY hierarchy to a system of kinetic equations, such as was briefly recalled in the previous section. No wonder that a proper kinetic treatment of dusty plasmas is in a state of infancy, many efforts and interim results notwithstanding.

The implications are far reaching, because so far the dusty plasma community has tended to disagree as how to go about it, and also about the extent to which deviations from existing, simple minded theories are really needed, desirable or even unavoidable. From many papers it emerges that the standard multispecies picture has allowed us to understand different aspects of dusty plasmas in broad lines. Thus, this rather simple minded approach will be followed in the next chapters. Underlying assumptions are that charged dust is treated as another ionic species, ideally as point particles with fixed charges. Later, *ad hoc* modifications are introduced to model charge fluctuations, effects of finite size and/or size distributions.

Nevertheless, a minority of the experts rightfully judge that more is needed to fully understand dusty plasma physics in all its complexity, even though for many applications full sophistication is neither needed nor warranted. Difficulties start already at incorporating the charging and interaction effects of dust grains from the very beginning. As we saw in Chapter 2, the charging process itself is not easily amenable to detailed analytical treatment, except in simple limiting cases. Indeed, the charging process in general seems so difficult to describe in a closed analytical form, that we might never achieve satisfaction here. Let me point out some of the directions in which efforts have gone.

At the level of new kinetic theories, Coppa *et al.* [1996] have given a treatment of charged particles of variable shape, blobs as they were called, extending the classic procedure for point and for fixed-shape particles. Surprisingly enough, the resulting equation for linear forces closely resembles the Vlasov equation, with suitable reinterpretations.

Next, Tsytovich and De Angelis have taken upon themselves a formidable task, that of giving a consistent formulation of general dusty plasma kinetic equations, and of the corresponding collision integrals, taking into account that the dust charges fluctuate. Unfortunately, only the first paper of a series is available yet [Tsytovich and De Angelis 1999]. Due to the large scale differences between the dust on the one hand and the electrons and ions on the other, the latter are described by ordinary kinetic equations. Dust charge fluctuations not only alter known type of collisions, such as dust-plasma particle collisions, but also introduce new type of integrals related to the charging process which determine the width of the distributions. In addition, there are new interactions such as dust-dust attractions, due to shadowing effects of the plasma fluxes on the dust particles, that can only be described within a kinetic approach, since the shadowing occurs in a definite angular interval of the plasma particle velocity space. However, this shadowing effects is still rather controversial, and by no means generally accepted by the dusty plasma community.

This paper is really the start of a whole program, and we have to wait until more progress is achieved. Attraction forces between charged dust grains and their associated Debye spheres, rather than purely Coulomb repulsion, are still under investigation [Vladimirov and Tsytovich 1998].

Other problems are associated with the coupling between dust grains, which can lead to plasma regimes quite different from the usual ones. Wang and Bhattacharjee [1997,1998] have studied pair correlations in strongly coupled dusty plasmas. The kinetic integral equations are solved in two opposite limits. For long-wavelength modes, the pair correlation exhibits the classic Debye-Hückel or Yukawa behaviour, known from shielding effects in ordinary plasmas, where the average interaction potential between two like shielded charges decays exponentially with their separation. Quite to the contrary, the short-wavelength regime is appropriate when the characteristic spatial scales are of the order of the inter dust distance. Then the static structure function for strongly coupled dust grains exhibits almost liquid or crystal-like behaviour, provided the average unshielded Coulomb energy largely exceeds the kinetic or thermal energies. Strongly coupled plasmas are of considerable interest, because of potential applications to white dwarf matter, interiors of heavy planets, and various laboratory experiments [Rosenberg and Kalman 1997; Kaw and Sen 1998].

After this brief overview of basic and as yet unresolved difficulties in the modelling of charged dust, I prefer to continue with a pedestrian approach, where charged dust is treated merely as one or more additional ionic species. There is still some time before a consensus emerges as how to improve on this, and what is sketched further on is the usual multispecies framework.

4.4 Macroscopic fluid equations

4.4.1 Macroscopic averages

A fluid description uses the physical space and time coordinates, x and t respectively, as independent coordinates. Compared to the kinetic theories surveyed in the preceding sections, velocity becomes one of the macroscopic and dependent quantities, as are density and pressure, given or to be determined at each physical point and for each instant in time as averages over the whole fluid. In such a fluid picture, different velocities between neighbouring particles are just averaged out. From the (single particle) distribution function $f(\mathbf{x}, \mathbf{v}, t)$ in phase space, the usual mass or number densities in physical space are recovered, upon integration over all velocities. It is somewhat easier to work now with mass densities, which I define here by

$$\rho = \int f d^3 v, \tag{4.14}$$

which immediately leads to the average or fluid velocity $\mathbf{u}$ as

$$\mathbf{u} = \langle \mathbf{v} \rangle = \frac{1}{\rho} \int f \mathbf{v} d^3 v. \tag{4.15}$$

In more general terms for microscopic quantities Q, such as velocity or energy, I define the corresponding average or macroscopic variable as

$$\langle Q \rangle = \frac{1}{\rho} \int f Q d^3 v, \tag{4.16}$$

which is consistent with the definitions of ρ and $\mathbf{u}$.

In order to get an evolution equation for $\langle Q \rangle$, I multiply (4.13) with Q and integrate over all $\mathbf{v}$

$$\int Q \left(\frac{\partial f}{\partial t} + \mathbf{v} \cdot \frac{\partial f}{\partial \mathbf{x}} + \mathbf{a} \cdot \frac{\partial f}{\partial \mathbf{v}} \right) d^3 v = \Delta Q, \tag{4.17}$$

where I have defined the interaction term as

$$\Delta Q = \int Q \frac{\delta f}{\delta t} d^3 v. \tag{4.18}$$

The first two integrals in (4.17) immediately give

$$\int Q \frac{\partial f}{\partial t} d^3 v = \frac{\partial}{\partial t} (\rho \langle Q \rangle) - \rho \left\langle \frac{\partial Q}{\partial t} \right\rangle,$$
$$\int Q \mathbf{v} \cdot \frac{\partial f}{\partial \mathbf{x}} d^3 v = \frac{\partial}{\partial \mathbf{x}} \cdot (\rho \langle \mathbf{v} Q \rangle) - \rho \left\langle \mathbf{v} \cdot \frac{\partial Q}{\partial \mathbf{x}} \right\rangle. \tag{4.19}$$

For the last term on the left-hand side of (4.17) the standard assumption is that the distribution function vanish at the boundaries in phase space, meaning simply that there are no particles at infinite or very large distances nor with infinite or very large velocities, and so

$$\int Q \mathbf{a} \cdot \frac{\partial f}{\partial \mathbf{v}} d^3 v = -\rho \left\langle \frac{\partial}{\partial \mathbf{v}} \cdot (\mathbf{a} Q) \right\rangle. \tag{4.20}$$

Collecting then all the results one finds

$$\frac{\partial}{\partial t}(\rho\langle Q\rangle) + \frac{\partial}{\partial x}\cdot(\rho\langle \mathbf{v}Q\rangle) = \rho\left\langle\frac{\partial Q}{\partial t} + \mathbf{v}\cdot\frac{\partial Q}{\partial x} + \frac{\partial}{\partial \mathbf{v}}\cdot(\mathbf{a}Q)\right\rangle$$
$$+ \Delta Q. \tag{4.21}$$

Because in dusty plasmas mass and momentum are not necessarily conserved between species when dust charges fluctuate, it is prudent to keep track of the source, interaction or collision terms, as they may be called.

4.4.2 Continuity equations

To derive the continuity equations I put $Q = 1$, and find from (4.21) that

$$\frac{\partial \rho}{\partial t} + \frac{\partial}{\partial x}\cdot(\rho\,\mathbf{u}) = S. \tag{4.22}$$

These are continuity equations with sink or source terms given, in a very formal way, by

$$S = \int \frac{\delta f}{\delta t} d^3\mathbf{v}. \tag{4.23}$$

If the interactions between particles do not affect their number or mass densities, conservation of mass means implies that $S = 0$.

4.4.3 Equations of motion

For the equations of motion, I first introduce the peculiar velocity,

$$\mathbf{v}' = \mathbf{v} - \langle\mathbf{v}\rangle = \mathbf{v} - \mathbf{u}, \tag{4.24}$$

with zero average, put $Q = \mathbf{v}'$ and define the pressure tensor as

$$\mathsf{P} = \rho\langle\mathbf{v}' \otimes \mathbf{v}'\rangle = \int f(\mathbf{v} - \mathbf{u}) \otimes (\mathbf{v} - \mathbf{u})d^3\mathbf{v}. \tag{4.25}$$

The pressure tensor is symmetric by definition, and $\otimes$ refers to the tensor product of two vectors. Furthermore, the interaction between different particles might not conserve momentum, and hence I put

$$\mathbf{M} = \frac{1}{\rho}\int(\mathbf{v} - \mathbf{u})\frac{\delta f}{\delta t}d^3\mathbf{v}, \tag{4.26}$$

more explicit forms of which will be given at a later stage. Since $\mathbf{v}'$ depends on x and t through $\mathbf{u}$, I need to compute in (4.21) that

$$\left\langle\frac{\partial \mathbf{v}'}{\partial t} + \mathbf{v}\cdot\frac{\partial \mathbf{v}'}{\partial x}\right\rangle = -\left(\frac{\partial \mathbf{u}}{\partial t} + \mathbf{u}\cdot\frac{\partial \mathbf{u}}{\partial x}\right). \tag{4.27}$$

In addition, for velocity-independent forces it is seen that

$$\left\langle \frac{\partial}{\partial \mathbf{v}} \cdot \mathbf{a} \otimes \mathbf{v}' \right\rangle = \mathbf{a}. \tag{4.28}$$

A similar result is true for the velocity-dependent magnetic part of the Lorentz force, corresponding to

$$\mathbf{a} = \frac{q}{m}(\mathbf{v} \times \mathbf{B}), \tag{4.29}$$

as here

$$\frac{\partial}{\partial \mathbf{v}} \cdot \mathbf{a} = 0. \tag{4.30}$$

Substituting all this in (4.21) gives the desired equation of motion

$$\frac{\partial \mathbf{u}}{\partial t} + \mathbf{u} \cdot \frac{\partial}{\partial \mathbf{x}} \mathbf{u} + \frac{1}{\rho} \frac{\partial}{\partial \mathbf{x}} \cdot \mathbf{P} = \langle \mathbf{a} \rangle + \mathbf{M}. \tag{4.31}$$

By $\langle \mathbf{a} \rangle$ is meant that in the microscopic expression for $\mathbf{a}$ every $\mathbf{v}$ is replaced by $\mathbf{u}$, but I really think of the magnetic part of the Lorentz force, of course.

4.4.4 Pressure equations

For the equation determining the pressure, put $Q = \mathbf{v}' \otimes \mathbf{v}'$ and compute first that

$$\rho \langle \mathbf{v}Q \rangle = \rho \langle \mathbf{v} \otimes \mathbf{v}' \otimes \mathbf{v}' \rangle = \mathbf{u} \otimes \mathbf{P} + \mathbf{Q}, \tag{4.32}$$

with the heat flow tensor $\mathbf{Q}$ defined as

$$\mathbf{Q} = \rho \langle \mathbf{v}' \otimes \mathbf{v}' \otimes \mathbf{v}' \rangle = \int f(\mathbf{v} - \mathbf{u}) \otimes (\mathbf{v} - \mathbf{u}) \otimes (\mathbf{v} - \mathbf{u}) d^3 \mathbf{v}. \tag{4.33}$$

Furthermore,

$$\left\langle \frac{\partial}{\partial t}(\mathbf{v}' \otimes \mathbf{v}') \right\rangle = -\left\langle \frac{\partial \mathbf{u}}{\partial t} \otimes \mathbf{v}' + \mathbf{v}' \otimes \frac{\partial \mathbf{u}}{\partial t} \right\rangle = 0, \tag{4.34}$$

and

$$\rho \left\langle \mathbf{v} \cdot \frac{\partial}{\partial \mathbf{x}}(\mathbf{v}' \otimes \mathbf{v}') \right\rangle = -\rho \left\langle \left(\mathbf{v} \cdot \frac{\partial}{\partial \mathbf{x}} \mathbf{u}\right) \mathbf{v}' + \mathbf{v}' \left(\mathbf{v} \cdot \frac{\partial}{\partial \mathbf{x}} \mathbf{u}\right) \right\rangle$$
$$= -\mathbf{P} \cdot \frac{\partial}{\partial \mathbf{x}} \mathbf{u} - \left(\mathbf{P} \cdot \frac{\partial}{\partial \mathbf{x}} \mathbf{u}\right)^T. \tag{4.35}$$

For the third term on the right-hand side of (4.21) there follows that

$$\left\langle \mathbf{a} \cdot \frac{\partial}{\partial \mathbf{v}}(\mathbf{v}' \otimes \mathbf{v}') \right\rangle = \langle \mathbf{a} \otimes \mathbf{v}' + \mathbf{v}' \otimes \mathbf{a} \rangle. \tag{4.36}$$

If **a** is independent of velocity, then this contribution vanishes, whereas for the magnetic part of the Lorentz force I get

$$\rho \left\langle \mathbf{a} \cdot \frac{\partial Q}{\partial \mathbf{v}} \right\rangle = -\frac{q}{m} \left[\mathbf{B} \times \mathsf{P} + (\mathbf{B} \times \mathsf{P})^T \right]. \qquad (4.37)$$

The interaction between different particles or species might influence the pressure changes, with a source term

$$\mathbf{\Pi} = \int (\mathbf{v} - \mathbf{u}) \otimes (\mathbf{v} - \mathbf{u}) \frac{\delta f}{\delta t} d^3 \mathbf{v}. \qquad (4.38)$$

Substitution of all the expressions thus obtained in (4.21) gives

$$\frac{\partial \mathsf{P}}{\partial t} + \frac{\partial}{\partial \mathbf{x}} \cdot (\mathbf{u} \otimes \mathsf{P} + \mathsf{Q}) + \mathsf{P} \cdot \frac{\partial}{\partial \mathbf{x}} \mathbf{u} + \left(\mathsf{P} \cdot \frac{\partial}{\partial \mathbf{x}} \mathbf{u} \right)^T$$
$$+ \frac{q}{m} \left[\mathbf{B} \times \mathsf{P} + (\mathbf{B} \times \mathsf{P})^T \right] = \mathbf{\Pi}. \qquad (4.39)$$

In principle, one needs to continue like this and determine the equation for the heat flow, which in turn will involve a moment of higher order to which there corresponds no simple physical quantity, and so on and on. The above derivations rely on averages of the underlying kinetic equation, and it is obvious that as soon as the distribution function is known, all those averages, like the mean velocity or the pressure, can be calculated explicitly. But this would require the solution of the corresponding kinetic equation and that is quite another mathematical problem, in which major complications can arise from the interaction terms.

The infinite chain of moment equations formally contains the same information as the original kinetic equation from which it was derived. However, as pointed out in the beginning of this chapter, this is not workable, and hence the chain of moment equations is closed by assuming something about the pressure or any of the higher order moments. I will give examples of closures further on, but address here the usual neglect of the divergence of the heat flow tensor in the pressure equation. Saying that

$$\left| \frac{\partial}{\partial \mathbf{x}} \cdot \mathsf{Q} \right| \ll \left| \frac{\partial \mathsf{P}}{\partial t} \right| \qquad (4.40)$$

gives an order of estimate for typical components Q and P of these tensors as

$$\left| \frac{Q}{P} \right| \ll \frac{\omega}{k}. \qquad (4.41)$$

Since Q involves a higher order moment than P in terms of the microscopic velocities, Q/P is a measure for the thermal velocities c_s, and one notes that the closure of the chain of moment equations essentially implies that $c_s \ll \omega/k$, in other words that the phase velocities considered be large compared to the thermal velocities. Otherwise, in principle, one has to consider a full kinetic treatment, at least for that particular species.

Nevertheless, I will in the next chapters sometimes use the fluid equations beyond their apparent range of validity, when considering massless electrons and/or ions, described by Boltzmann distributions, as explained in Chapter 5 for ion-acoustic or dust-acoustic modes.

4.4.5 Set of basic fluid equations

Reverting to the more traditional way of writing the basic equations and introducing the species index gives for the continuity equations

$$\frac{\partial}{\partial t} n_\alpha + \nabla \cdot (n_\alpha \mathbf{u}_\alpha) = S_\alpha, \qquad (4.42)$$

where the index α refers to the species under consideration, with number density $n_\alpha = \rho_\alpha/m_\alpha$. The source or sink terms S_α, redefined here per mass unit, are not further specified for the time being, but are assumed to vanish in equilibrium for all species, and for the dust always. The reason is that the dust number density is not affected by the dust loosing or picking up some charges. Coalescing or breaking up of the dust might very well be important, but has not been incorporated in the treatments discussed further on.

Similarly, the equations of motion are

$$\left(\frac{\partial}{\partial t} + \mathbf{u}_\alpha \cdot \nabla\right) \mathbf{u}_\alpha + \frac{1}{n_\alpha m_\alpha} \nabla \cdot \mathsf{P}_\alpha = \frac{q_\alpha}{m_\alpha}(\mathbf{E} + \mathbf{u}_\alpha \times \mathbf{B}) - \nabla \psi + \mathbf{M}_\alpha. \quad (4.43)$$

Besides the Lorentz force I have also introduced possible self-gravitational effects through the gradient of the corresponding potential ψ.

To close the set of equations, I will need to assume something about how the pressures behave. Usually, this will involve for simplicity scalar and barotropic pressures, where the pressure of each species depends only on its density:

$$\mathsf{P}_\alpha = p_\alpha(n_\alpha) \mathbf{1}. \qquad (4.44)$$

With scalar pressures there arises a conceptual difficulty. In principle the equations of motion (4.43) can be obtained from the collisionless Vlasov equation (4.12). Scalar pressures arise from some isotropization for which collisions between particles of the same species are needed, yet at the same time collisions between particles of different species are totally neglected. In other words, why do particles of species α collide between themselves, but studiously avoid all particles of other species? The use of scalar pressures thus implies some form of collisional effects, and great care has to be exercised before the collisionless descriptions are taken at face value.

If there is a need to explicitly consider anisotropic pressures, which is the more natural case in really collisionless plasmas, I use (4.39) in the form

$$\frac{\partial \mathsf{P}_\alpha}{\partial t} + \nabla \cdot (\mathbf{u}_\alpha \otimes \mathsf{P}_\alpha + \mathsf{Q}_\alpha) + \mathsf{P}_\alpha \cdot \nabla \mathbf{u}_\alpha + (\mathsf{P}_\alpha \cdot \nabla \mathbf{u}_\alpha)^T$$
$$+ \frac{q_\alpha}{m_\alpha} \left[\mathbf{B} \times \mathsf{P}_\alpha + (\mathbf{B} \times \mathsf{P}_\alpha)^T \right] = \mathbf{\Pi}_\alpha. \tag{4.45}$$

At this level, the set of basic equations is often closed by neglecting the divergence of the heat flow tensor, which amounts to considering the processes adiabatic in a generalized sense.

4.4.6 Conservation of charge

Due to the possible variability of the dust charges, there is now also an equation expressing the conservation of charge in the plasma as a whole,

$$\frac{\partial}{\partial t} \sum_\alpha n_\alpha q_\alpha + \nabla \cdot \sum_\alpha n_\alpha q_\alpha \mathbf{u}_\alpha = 0. \tag{4.46}$$

This can be rewritten with the help of the continuity equations (4.42) as

$$\sum_{\alpha \neq dust} q_\alpha S_\alpha + \sum_{dust} n_{dust} \left(\frac{\partial}{\partial t} + \mathbf{u}_{dust} \cdot \nabla \right) q_{dust} = 0. \tag{4.47}$$

An explicit distinction has been made between the electron and plasma ions, for which the charges are obviously fixed but their number densities are not, and the charged dust grains, where the opposite holds.

4.5 Maxwell's equations

The electromagnetic fields occurring in the equations of motion (4.43) and in the anisotropic pressure equations (4.45) obey Maxwell's equations

$$\nabla \times \mathbf{E} + \frac{\partial}{\partial t} \mathbf{B} = 0,$$

$$c^2 \nabla \times \mathbf{B} = \frac{\partial}{\partial t} \mathbf{E} + \frac{1}{\varepsilon_0} \sum_\alpha n_\alpha q_\alpha \mathbf{u}_\alpha,$$

$$\nabla \cdot \mathbf{E} = \frac{1}{\varepsilon_0} \sum_\alpha n_\alpha q_\alpha,$$

$$\nabla \cdot \mathbf{B} = 0, \tag{4.48}$$

and the current and charge densities have already been expressed for a multispecies plasma.

4.6 Waves in multispecies plasmas

4.6.1 Introductory remarks

In the following chapters, I will primarily be looking at those wave phenomena which have the most chances of clearly showing the influence of the charged dust. That is, either modes at the lower end of the frequency spectrum, where the dust can give some distinctive contribution, by modifying existing or creating novel waves, or else particular effects which have to do with the variability of the grain charges or with their mass and size distributions. I will try to distinguish each time between these approaches.

However, the classification of the wealth of waves that can exist in an ordinary plasma is already by no means an easy nor straightforward task, especially if more than the customary electron-ion plasmas are considered. As there are many textbooks where one can find a description of these different linear wave modes, and on the other hand most of you might not be too familiar with the expressions for multispecies plasmas, I prefer to give in this chapter a rather general derivation of the dispersion laws involved. To be fully general, this will include self-gravitational effects, although I will only discuss these in Chapter 8.

Before all this, however, it is worth thinking about other complicating factors. One of these is collisions between the plasma species, because then the different fluid equations are all linked. A similar coupling occurs in dusty plasmas even in a collisionless regime, due to the grain charging mechanism that leads to density fluctuations and momentum exchanges. Hence, it is more prudent to postpone the discussion of fluctuating dust charges to Chapter 7, and I will concentrate in the next two chapters on the modifications induced by charged dust at fixed charges. As you will see, there are already enough novel effects to merit such a at first sight simple and naive treatment.

For simplicity of the subsequent derivations, I will only consider wave propagation along the z-axis, so that $\nabla = \mathbf{e}_z \partial/\partial z$. On the other hand, the static magnetic field $\mathbf{B}_0$ will be in the x, z-plane, with $\mathbf{B}_0 = B_0(\sin \vartheta \mathbf{e}_x + \cos \vartheta \mathbf{e}_z)$, where ϑ is the angle between the directions of wave propagation and the external magnetic field. This will allow us to discuss the full transition from parallel to purely perpendicular wave propagation when needed.

For this space dependence the continuity equations are, from (4.42),

$$\frac{\partial n_\alpha}{\partial t} + \frac{\partial}{\partial z}(n_\alpha u_{\alpha z}) = 0, \tag{4.49}$$

and the equations of motion from (4.43),

$$\frac{\partial \mathbf{u}_\alpha}{\partial t} + u_{\alpha z}\frac{\partial \mathbf{u}_\alpha}{\partial z} = \frac{q_\alpha}{m_\alpha}(\mathbf{E} + \mathbf{u}_\alpha \times \mathbf{B}) - \frac{1}{n_\alpha m_\alpha}\frac{\partial p_\alpha}{\partial z}\mathbf{e}_z - \frac{\partial \psi}{\partial z}\mathbf{e}_z. \tag{4.50}$$

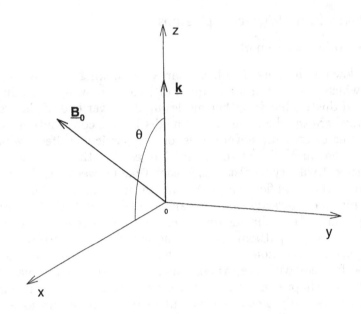

Figure 4.1. Directions of wave propagation and external magnetic field

The precise composition of these dusty plasmas will be specified in later chapters, after I have derived the linear dispersion law in general. The pressures are supposed to obey a barotropic law,

$$p_\alpha = p_\alpha(n_\alpha), \tag{4.51}$$

as discussed already. This form is sufficiently general to include many interesting cases, like isothermal or adiabatic pressure variations, and even deviations from the ideal gas laws like we find in the Van der Waals prescriptions. Moreover, it is perfectly possible to chose a different pressure behaviour for electrons, ions and dust grains.

The description is closed by Maxwell's equations, written here as

$$\mathbf{e}_z \times \frac{\partial \mathbf{E}}{\partial z} + \frac{\partial \mathbf{B}}{\partial t} = \mathbf{0}, \tag{4.52}$$

$$c^2 \mathbf{e}_z \times \frac{\partial \mathbf{B}}{\partial z} = \frac{\partial \mathbf{E}}{\partial t} + \frac{1}{\varepsilon_0} \sum_\alpha n_\alpha q_\alpha \mathbf{u}_\alpha, \tag{4.53}$$

$$\varepsilon_0 \frac{\partial E_z}{\partial z} = \sum_\alpha n_\alpha q_\alpha, \tag{4.54}$$

$$\frac{\partial B_z}{\partial z} = 0, \tag{4.55}$$

together with the gravitational Poisson's equation

$$\frac{\partial^2 \psi}{\partial z^2} = 4\pi G \sum_\alpha n_\alpha m_\alpha. \tag{4.56}$$

Whether one is looking at linear or at nonlinear waves, the underlying assumption is that there exists a stationary state, which is perturbed by the waves, of smaller or larger amplitudes. In view of the many complicating factors, a brief discussion of this stationary state is in order.

4.6.2 Equilibrium discussions

For most of the applications I shall review, a homogeneous equilibrium state is assumed to exist. Of course, really nonhomogeneous stationary states are closer to physical reality, but far more difficult to treat, and I will relegate them to Chapters 8 and 10, lest we get entangled in too many difficulties all at once.

What is implied by a homogeneous equilibrium? In ordinary plasmas, while not often explicitly stated, the model is that of an infinite system. To put the idea on a semi-quantitative foundation, I can think of an infinite plasma as large sphere, say, whose size R is very much larger than the maximum wavelength λ_{max} under study ($\lambda_{max} \ll R$). Equivalently, $k_{min} R \gg 1$, if k_{min} is the corresponding wavenumber. The implication is that physical boundary effects and/or conditions can, to a good approximation, be ignored for the wavelengths under consideration. As the wavelength λ becomes larger and approaches R, boundary effects have an increasing effect on the allowable waveform and can no longer be neglected. Moreover, a homogeneous equilibrium model assumes that λ_{max} is also very much smaller than the length scale L over which the system changes appreciably.

Another consequence is that the wave problem can be treated as an initial value problem, where an initial perturbation to the plasma parameters (such as charge density) in the vicinity of the origin is permitted to evolve in time. The localization of the initial (perturbed) conditions is crucial for the solution of the problem via Fourier-Laplace or double Fourier (suitably modified to enforce causality) transform methods. Simpler methods of solution that prescribe parameter dependencies of the form $\exp i(kz - \omega t)$, as introduced below, really amount to the same procedure.

In a dusty plasma, several additional complications occur. A first difficulty arises from the modelling of charged dust as heavy ions, because the existence of a constant equilibrium charge is tacitly implied here. Given the many discussions about the interactions between and the charging of dust grains, and the lack of unanimity about whether complex plasmas like dusty plasmas in nature are open or closed systems, the notion of an equilibrium charge is by no means trivial. But let me be optimistic and assume that this indeed exists.

The following hurdle to be taken crops up when self-gravitational forces are included in the momentum equations, coupled to the gravitational Poisson equation (4.56). Since gravitation, contrary to electromagnetic forces, cannot be shielded but is always attractive, gravitational collapse of extended regions with distributed masses is inevitable, unless counteracted upon by thermal agitation or repulsion due to other causes. In principle, there is no way to make the gravitational potential disappear in the zeroth order, as one normally can do for the electrostatic potential or other electric field effects. What that really means is that a truly homogeneous equilibrium is impossible, and even more so an infinite one.

Nevertheless, some important physical insight might be gained by invoking the so-called "Jeans swindle". By this subterfuge is meant that the zeroth order of the self-gravitational potential is completely ignored, when looking at local perturbations. Before drawing precise conclusions about (Jeans) instability or not of the modes thus described, it would seem natural to check that the perturbations are indeed local in the said sense. And here one runs into a methodological quandary. For more complicated plasma models one has not been able to find the stationary state, so that I now take the Jeans swindle to mean the following. Local perturbations are studied, based on the tacit assumption that the medium is uniform over a distance of several wavelengths, because the basic state cannot intrinsically be uniform on a larger scale. That modes are indeed local, however, cannot be checked afterwards, because the basic state cannot be determined, except for some simple models which can be worked through. Usually, there is no way out of this dilemma, and I will return to the Janus aspects of the Jeans instabilities in Chapter 8.

Nevertheless, progress is only possible by starting from a stationary, homogeneous state, as the linearization procedure discussed in the next subsection will assume. Nonlinear waves are not amenable to investigation by general methods as linear problems are, and hence I will have to deal with nonlinear modes at various places in the next chapters.

4.6.3   Linearization procedure

Following standard procedure, I linearize the relevant equations around a homogeneous equilibrium, by putting

$$
\begin{aligned}
n_\alpha(z,t) &\rightarrow N_\alpha + \delta n_\alpha(z,t), \\
p_\alpha(z,t) &\rightarrow P_\alpha + \delta p_\alpha(z,t), \\
\mathbf{u}_\alpha(z,t) &\rightarrow \mathbf{U}_\alpha + \delta \mathbf{u}_\alpha(z,t), \\
\mathbf{B}(z,t) &\rightarrow \mathbf{B}_0 + \delta \mathbf{B}(z,t), \\
\mathbf{E}(z,t) &\rightarrow \mathbf{E}_0 + \delta \mathbf{E}(z,t), \\
\psi(z,t) &\rightarrow \delta \psi(z,t),
\end{aligned}
\tag{4.57}
$$

and neglecting squares and products of the small fluctuations involving δ. Global charge and current neutrality in equilibrium would mean that

$$\sum_\alpha N_\alpha q_\alpha = 0,$$

$$\sum_\alpha N_\alpha q_\alpha \mathbf{U}_\alpha = \mathbf{0}. \tag{4.58}$$

From equilibrium considerations, I see that the momentum equations (4.50) indicate that for all species

$$\mathbf{E}_0 + \mathbf{U}_\alpha \times \mathbf{B}_0 = \mathbf{0}, \tag{4.59}$$

and the perpendicular velocity components are equal, so that

$$\mathbf{U}_\alpha = \frac{\mathbf{E}_0 \times \mathbf{B}_0}{B_0^2} + U_{\alpha\|}\frac{\mathbf{B}_0}{B_0}. \tag{4.60}$$

The perpendicular part of the equilibrium velocities and hence also the zeroth order electric field can be eliminated, by going to a so-called de Hoffman-Teller frame, moving across the static magnetic field with a global $\mathbf{E}_0 \times \mathbf{B}_0$ velocity. That leaves only the velocity components parallel to the static magnetic field to be considered in further applications.

The basic set thus consists of linear equations, and the next step is to Fourier analyze all equations, or equivalently, assume that all wave quantities vary in a plane-wave fashion as

$$\delta f(z,t) = \tilde{f}\exp(ikz - i\omega t), \tag{4.61}$$

and subsequently neglect for simplicity of notation the tildes in referring to the amplitudes or the Fourier transformed quantities.

If I want to include equilibrium streaming between the different plasma constituents, so as to be able to discuss beam-plasma or streaming instabilities, the expressions get horribly complicated. It is perfectly possible to derive these, but hardly any meaningful discussion can be given. Hence there seems little point in being too general and the treatment is split in two.

First, I include the possibility of zero-order streaming between the plasma species but restricted to wave propagation parallel to the external magnetic field, at $\vartheta = 0$. The other possibility is to look at all possible angles of wave propagation, but leave the equilibrium drifts out of the treatment.

4.7 Linear parallel modes and equilibrium streaming

First of all, the linearized continuity equation yields

$$n_\alpha = N_\alpha \frac{k u_{\alpha z}}{\omega - kU_\alpha}. \tag{4.62}$$

Combining this with (4.50)–(4.52), suitably linearized and Fourier transformed, allows me to express the fluid velocities as linear functions of the components of the wave electric field and the self-gravitational potential:

$$u_{\alpha x} = \frac{q_\alpha(\omega - kU_\alpha)}{m_\alpha \omega} \cdot \frac{i(\omega - kU_\alpha)E_x - \Omega_\alpha E_y}{(\omega - kU_\alpha)^2 - \Omega_\alpha^2},$$

$$u_{\alpha y} = \frac{q_\alpha(\omega - kU_\alpha)}{m_\alpha \omega} \cdot \frac{\Omega_\alpha E_x + i(\omega - kU_\alpha)E_y}{(\omega - kU_\alpha)^2 - \Omega_\alpha^2},$$

$$u_{\alpha z} = i\frac{q_\alpha(\omega - kU_\alpha)}{m_\alpha[(\omega - kU_\alpha)^2 - k^2 c_{s\alpha}^2]}E_z + \frac{k(\omega - kU_\alpha)}{(\omega - kU_\alpha)^2 - k^2 c_{s\alpha}^2}\psi. \tag{4.63}$$

Sound velocities $c_{s\alpha}$ have been introduced in a very general way through $m_\alpha c_{s\alpha}^2 = [dp_\alpha/dn_\alpha]_{N_\alpha}$. The latter notation means that the derivative is evaluated for the equilibrium densities, consistent with the customary definition of the sound velocity in gases. Later, these need to be specified in terms of temperatures and possibly other parameters. Furthermore, the gyrofrequencies $\Omega_\alpha = q_\alpha B_0/m_\alpha$ include the sign of the charges.

Since there is a clear decoupling between the components transverse and longitudinal to the external magnetic field, I first deal with the former, for which I find in combination with the transverse projections of the linearized form of (4.53) that

$$\begin{pmatrix} D_{xx} & iD_{xy} \\ -iD_{yx} & D_{yy} \end{pmatrix} \cdot \begin{pmatrix} E_x \\ E_y \end{pmatrix} = 0, \tag{4.64}$$

where

$$D_{xx} = D_{yy} = \omega^2 - c^2 k^2 - \sum_\alpha \frac{\omega_{p\alpha}^2(\omega - kU_\alpha)^2}{(\omega - kU_\alpha)^2 - \Omega_\alpha^2},$$

$$D_{xy} = D_{yx} = \sum_\alpha \frac{\omega_{p\alpha}^2(\omega - kU_\alpha)\Omega_\alpha}{(\omega - kU_\alpha)^2 - \Omega_\alpha^2}. \tag{4.65}$$

In previous chapters we already encountered the plasma frequencies, given through $\omega_{p\alpha}^2 = N_\alpha q_\alpha^2/\varepsilon_0 m_\alpha$. The dispersion law hence amounts to

$$D_{xx} \pm D_{xy} = 0, \tag{4.66}$$

or more explicitly,

$$\omega^2 = c^2 k^2 + \sum_\alpha \frac{\omega_{p\alpha}^2 (\omega - kU_\alpha)}{\omega - kU_\alpha \pm \Omega_\alpha}. \tag{4.67}$$

The modes obeying this dispersion law are transverse, electromagnetic right- and left-hand circularly polarized (RHCP, LHCP), respectively, following the usual sign convention. It is obvious that pressure and self-gravitational effects do not come into play here, as the associated gradients work parallel to the external magnetic field, giving only rise to longitudinal motions of the plasma constituents.

The multispecies treatment does not qualitatively change this discussion compared to the standard electron-ion case without streaming, and I can refer to plasma astrophysics textbooks [Baumjohann and Treumann 1996, 1997; Krishan 1999] for further details concerning the mode polarization and related aspects.

The other modes will combine the parallel component of Ampère's law (4.53) with the gravitational Poisson's equation (4.56), to yield

$$\begin{pmatrix} D_{zz} & iD_{z\psi} \\ iD_{\psi z} & D_{\psi\psi} \end{pmatrix} \cdot \begin{pmatrix} E_z \\ (k/\sqrt{4\pi\varepsilon_0 G})\psi \end{pmatrix} = 0, \tag{4.68}$$

with

$$D_{zz} = 1 - \sum_\alpha \frac{\omega_{p\alpha}^2}{(\omega - kU_\alpha)^2 - k^2 c_{s\alpha}^2},$$

$$D_{z\psi} = D_{\psi z} = \sum_\alpha \frac{\omega_{p\alpha}\omega_{J\alpha}}{(\omega - kU_\alpha)^2 - k^2 c_{s\alpha}^2},$$

$$D_{\psi\psi} = 1 + \sum_\alpha \frac{\omega_{J\alpha}^2}{(\omega - kU_\alpha)^2 - k^2 c_{s\alpha}^2}. \tag{4.69}$$

The Jeans frequencies have been introduced through $\omega_{J\alpha}^2 = 4\pi G N_\alpha m_\alpha$. As a general remark, $\omega_{p\alpha}\omega_{J\alpha}$ is a convenient way to rewrite

$$N_\alpha q_\alpha = \left(\frac{\varepsilon_0}{4\pi G}\right)^{1/2} \omega_{p\alpha}\omega_{J\alpha}, \tag{4.70}$$

and thus whenever you encounter a single $\omega_{p\alpha}$ it has to be interpreted as including the sign of the charge. The dispersion law for the longitudinal modes is thus

$$D_{zz}D_{\psi\psi} + D_{z\psi}^2 = 0, \tag{4.71}$$

or more in detail,

$$\left(1 - \sum_\alpha \frac{\omega_{p\alpha}^2}{(\omega - kU_\alpha)^2 - k^2 c_{s\alpha}^2}\right)\left(1 + \sum_\alpha \frac{\omega_{J\alpha}^2}{(\omega - kU_\alpha)^2 - k^2 c_{s\alpha}^2}\right)$$

$$+ \left(\sum_\alpha \frac{\omega_{p\alpha}\omega_{J\alpha}}{(\omega - kU_\alpha)^2 - k^2 c_{s\alpha}^2}\right)^2 = 0. \tag{4.72}$$

These longitudinal, electrostatic Langmuir-Jeans modes will be discussed in detail, both in the next chapter dealing with electrostatic modes but in the absence of self-gravitational forces, and in Chapter 8 with its specific emphasis on the latter.

4.8   Arbitrary angles of wave propagation without streaming

4.8.1   General expressions

Working now at general angles of wave propagation ϑ but leaving out all equilibrium streaming effects between the different plasma species, I combine the linearized and Fourier analyzed forms of (4.49)–(4.52), which gives rather involved expressions for the velocity components,

$$u_{\alpha x} = \frac{q_\alpha}{m_\alpha \mathcal{L}_\alpha}\left\{i\omega(\omega^2 - k^2 c_{s\alpha}^2 - \Omega_\alpha^2 \sin^2\vartheta)E_x - \Omega_\alpha(\omega^2 - k^2 c_{s\alpha}^2)E_y \cos\vartheta\right.$$

$$\left. - i\omega\Omega_\alpha^2 E_z \sin\vartheta\cos\vartheta\right\} - \frac{\omega k \Omega_\alpha^2}{\mathcal{L}_\alpha}\psi\sin\vartheta\cos\vartheta,$$

$$u_{\alpha y} = \frac{q_\alpha}{m_\alpha \mathcal{L}_\alpha}\left\{\Omega_\alpha(\omega^2 - k^2 c_{s\alpha}^2)E_x \cos\vartheta + i\omega(\omega^2 - k^2 c_{s\alpha}^2)E_y\right.$$

$$\left. - \omega^2\Omega_\alpha E_z \sin\vartheta\right\} + i\frac{k\omega^2\Omega_\alpha}{\mathcal{L}_\alpha}\psi\sin\vartheta,$$

$$u_{\alpha z} = \frac{q_\alpha}{m_\alpha \mathcal{L}_\alpha}\left\{-i\omega\Omega_\alpha^2 E_x \sin\vartheta\cos\vartheta + \omega^2\Omega_\alpha E_y \sin\vartheta\right.$$

$$\left. + i\omega(\omega^2 - \Omega_\alpha^2 \cos^2\vartheta)E_z\right\} + \frac{\omega k(\omega^2 - \Omega_\alpha^2 \cos^2\vartheta)}{\mathcal{L}_\alpha}\psi, \tag{4.73}$$

where

$$\mathcal{L}_\alpha = \omega^2(\omega^2 - \Omega_\alpha^2) - k^2 c_{s\alpha}^2(\omega^2 - \Omega_\alpha^2 \cos^2\vartheta). \tag{4.74}$$

Substituting (4.73) in the linearized forms of (4.53) and (4.56) then yields

$$\begin{pmatrix} D_{xx} & D_{xy} & D_{xz} & D_{x\psi} \\ D_{yx} & D_{yy} & D_{yz} & D_{y\psi} \\ D_{zx} & D_{zy} & D_{zz} & D_{z\psi} \\ D_{\psi x} & D_{\psi y} & D_{\psi z} & D_{\psi\psi} \end{pmatrix} \cdot \begin{pmatrix} E_x \\ -iE_y \\ \omega E_z \\ i(\omega k/\sqrt{4\pi\varepsilon_0 G})\psi \end{pmatrix} = 0, \tag{4.75}$$

from which the general dispersion law follows as

$$\det[D_{ij}] = 0. \tag{4.76}$$

The elements of the symmetric dispersion tensor are given by

$$D_{xx} = \omega^2 - c^2 k^2 - \omega^2 \sum_\alpha \frac{\omega_{p\alpha}^2 (\omega^2 - k^2 c_{s\alpha}^2 - \Omega_\alpha^2 \sin^2 \vartheta)}{\omega^2(\omega^2 - \Omega_\alpha^2) - k^2 c_{s\alpha}^2 (\omega^2 - \Omega_\alpha^2 \cos^2 \vartheta)},$$

$$D_{xy} = \omega \sum_\alpha \frac{\omega_{p\alpha}^2 \Omega_\alpha (\omega^2 - k^2 c_{s\alpha}^2)}{\omega^2(\omega^2 - \Omega_\alpha^2) - k^2 c_{s\alpha}^2 (\omega^2 - \Omega_\alpha^2 \cos^2 \vartheta)} \cos\vartheta,$$

$$D_{xz} = \omega \sum_\alpha \frac{\omega_{p\alpha}^2 \Omega_\alpha^2}{\omega^2(\omega^2 - \Omega_\alpha^2) - k^2 c_{s\alpha}^2 (\omega^2 - \Omega_\alpha^2 \cos^2 \vartheta)} \sin\vartheta \cos\vartheta,$$

$$D_{x\psi} = -\omega \sum_\alpha \frac{\omega_{p\alpha} \omega_{J\alpha} \Omega_\alpha^2}{\omega^2(\omega^2 - \Omega_\alpha^2) - k^2 c_{s\alpha}^2 (\omega^2 - \Omega_\alpha^2 \cos^2 \vartheta)} \sin\vartheta \cos\vartheta,$$

$$D_{yy} = \omega^2 - c^2 k^2 - \omega^2 \sum_\alpha \frac{\omega_{p\alpha}^2 (\omega^2 - k^2 c_{s\alpha}^2)}{\omega^2(\omega^2 - \Omega_\alpha^2) - k^2 c_{s\alpha}^2 (\omega^2 - \Omega_\alpha^2 \cos^2 \vartheta)},$$

$$D_{yz} = -\omega^2 \sum_\alpha \frac{\omega_{p\alpha}^2 \Omega_\alpha}{\omega^2(\omega^2 - \Omega_\alpha^2) - k^2 c_{s\alpha}^2 (\omega^2 - \Omega_\alpha^2 \cos^2 \vartheta)} \sin\vartheta,$$

$$D_{y\psi} = \omega^2 \sum_\alpha \frac{\omega_{p\alpha} \omega_{J\alpha} \Omega_\alpha}{\omega^2(\omega^2 - \Omega_\alpha^2) - k^2 c_{s\alpha}^2 (\omega^2 - \Omega_\alpha^2 \cos^2 \vartheta)} \sin\vartheta,$$

$$D_{zz} = 1 - \sum_\alpha \frac{\omega_{p\alpha}^2 (\omega^2 - \Omega_\alpha^2 \cos^2 \vartheta)}{\omega^2(\omega^2 - \Omega_\alpha^2) - k^2 c_{s\alpha}^2 (\omega^2 - \Omega_\alpha^2 \cos^2 \vartheta)},$$

$$D_{z\psi} = \sum_\alpha \frac{\omega_{p\alpha} \omega_{J\alpha} (\omega^2 - \Omega_\alpha^2 \cos^2 \vartheta)}{\omega^2(\omega^2 - \Omega_\alpha^2) - k^2 c_{s\alpha}^2 (\omega^2 - \Omega_\alpha^2 \cos^2 \vartheta)},$$

$$D_{\psi\psi} = -1 - \sum_\alpha \frac{\omega_{J\alpha}^2 (\omega^2 - \Omega_\alpha^2 \cos^2 \vartheta)}{\omega^2(\omega^2 - \Omega_\alpha^2) - k^2 c_{s\alpha}^2 (\omega^2 - \Omega_\alpha^2 \cos^2 \vartheta)}. \tag{4.77}$$

These results will be used and discussed in greater detail in Chapters 5 and 8, where oblique modes and self-gravitational effects are reviewed.

4.8.2 Oblique electrostatic modes

There is a special case, however, that the waves can still be considered electrostatic, but at oblique propagation. In that case I put $\mathbf{E} = -\nabla\phi$ and get from (4.73) with $E_x = E_y = 0$ and $E_z = -ik\phi$ that

$$u_{\alpha z} = \frac{\omega k}{K_\alpha}\left(\frac{q_\alpha}{m_\alpha}\phi + \psi\right), \tag{4.78}$$

where

$$\mathcal{K}_\alpha = \omega^2 - k^2 c_{s\alpha}^2 - \frac{\omega^2 \Omega_\alpha^2 \sin^2 \vartheta}{\omega^2 - \Omega_\alpha^2 \cos^2 \vartheta}. \tag{4.79}$$

Hence I find for the perturbed densities that

$$n_\alpha = \frac{k^2 N_\alpha}{\mathcal{K}_\alpha} \left(\frac{q_\alpha}{m_\alpha} \phi + \psi \right), \tag{4.80}$$

which I can use in both Poisson's equations, the electrostatic and the grav-itational one. This yields then a dispersion law

$$\left(1 - \sum_\alpha \frac{\omega_{p\alpha}^2}{\mathcal{K}_\alpha}\right) \left(1 + \sum_\alpha \frac{\omega_{J\alpha}^2}{\mathcal{K}_\alpha}\right) + \left(\sum_\alpha \frac{\omega_{p\alpha} \omega_{J\alpha}}{\mathcal{K}_\alpha}\right)^2 = 0, \tag{4.81}$$

quite analogous to (4.72). A more thorough discussion will be given in the next chapter, when dealing with oblique electrostatic modes, and in Chapter 8 when self-gravitational effects are included.

4.8.3 Perpendicular propagation

Finally, at strictly perpendicular propagation ($\vartheta = 90°$) the dispersion co-efficients in (4.77) indicate that the dispersion law factorizes into the one for the ordinary mode, coming from $D_{xx} = 0$,

$$\omega^2 = c^2 k^2 + \sum_\alpha \omega_{p\alpha}^2, \tag{4.82}$$

while the remainder is then referring to the extraordinary mode.

It is worth remembering that the ordinary mode only decouples from the extraordinary mode in the strict absence of equilibrium streaming among the plasma components. In that case the ordinary mode is characterized by a wave electric field parallel to the external magnetic field, although the value of that magnetic field is not of any importance. Neither is the ordinary mode affected by pressure or self-gravitational forces, as long as the pressures are scalar and isotropic, as then all gradients act in a direction orthogonal to the wave quantities. All the latter effects, however, will very much influence the extraordinary mode, as we shall see in Chapter 6 in general and in Chapter 8 when self-gravitation is considered.

CHAPTER 5

ELECTROSTATIC MODES

5.1 Introduction

Many of the papers about waves in dusty plasmas are dealing with electrostatic modes, so it is natural to start our overview of dusty plasma with these, and leave the electromagnetic waves for the next chapter. Electrostatic waves are easily generated, for example, when a perturbation creates a charge imbalance in a fluid element that was originally charge neutral. Such an imbalance will accelerate electrons (and also ions and other charged particles) in the neighbourhood of the charged fluid element, resulting in charges oscillating back and forth, essentially involving the electric field and hence called electrostatic. The oscillating magnetic field is either absent or can be neglected. In that case, Faraday's law (4.52) shows that the electric field $\mathbf{E}$ is parallel to the direction of the wave vector $\mathbf{k}$.

As seen in the previous chapter for modes that propagate parallel to an externally applied, static magnetic field, there is a decoupling between the longitudinal, electrostatic and the transverse, electromagnetic modes. A similar decoupling occurs in unmagnetized plasmas. Special cases of electrostatic modes include Langmuir, dust ion-acoustic and dust-acoustic waves, as discussed below. However, when the propagation is oblique with respect to the external magnetic field, Ampère's law (4.53) indicates that there are restrictions on the allowable currents. I discuss these in the section dealing with oblique modes at the end of this chapter.

Moreover, the distinction between parallel longitudinal and transverse modes is based on a linear description, where it is obviously true. For nonlinear modes the picture is more subtle. Parallel longitudinal modes will remain so, whereas even originally transverse modes will usually generate a longitudinal component at the nonlinear level through the $\mathbf{u} \times \mathbf{B}$ term in the Lorentz force. This is another reason to postpone the discussion of electromagnetic modes to the next chapter and restrict ourselves in the present chapter to electrostatic modes.

The structure of this chapter is as follows. First comes a brief reminder of some of the parallel electrostatic waves in ordinary plasmas, in order of decreasing frequencies. Since typical dust frequencies are so small, the descent in frequency naturally goes on when modifications to existing plasma modes are discussed, or new dust modes introduced. Before going on to oblique modes, some elements are given of nonlinear wave theory, because there is a marked difference in behaviour between parallel and oblique modes at this level. The final part of the chapter is then devoted to oblique modes, both to the linear as to the nonlinear aspects.

5.2 Linear parallel modes in ordinary plasmas

5.2.1 Langmuir waves

So let us embark on our journey towards lower and lower frequencies, reflecting the plasma response of heavier and heavier particles to the wave disturbances. Although difficult, since so much has been written that fits in here, I want to keep this chapter to a reasonable length, and start as simple as possible, treating the dust as an additional heavy ion species with constant charges. Once dusty plasma waves are understood at this level, we can look at the complications due to charge fluctuations, which also came later in the development of dusty plasma wave theory. Similarly, possible self-gravitation between the heavy dust grains is relegated to a later chapter.

The general dispersion law for parallel electrostatic modes was derived in the previous chapter. Without self-gravitation the purely electrostatic part of (4.72) gives

$$\sum_\alpha \frac{\omega_{p\alpha}^2}{(\omega - kU_\alpha)^2 - k^2 c_{s\alpha}^2} = 1. \tag{5.1}$$

I now give a brief discussion of this dispersion law in ordinary plasmas, before turning to dusty plasmas. This is certainly not meant to be exhaustive, and you can find all about it in general textbooks on plasma waves. However, what I recall is meant to put dusty plasma waves in a proper perspective, trying to instill a logical order.

The simplest and oldest application of (5.1) is to a plasma where the mobile electrons support the waves in a fixed, neutralizing background of cold ions, without any equilibrium streaming between the two species ($U_{e,i} = 0$). This is tantamount to looking at high-frequency waves, where high frequencies are defined in such a way that the ions cannot really respond to the perturbations, but are quasi-immobile. Mathematically, one takes the $m_i \to \infty$ limit, hence $\omega_{pi} \to 0$, which yields

$$\omega^2 = \omega_{pe}^2 + k^2 c_{se}^2. \tag{5.2}$$

Such Langmuir waves or plasma oscillations obviously can only occur at frequencies above the electron plasma frequency ω_{pe}.

When I go to lower frequencies, the ion dynamics also have to be taken into account, and in the absence of relative streaming between the electrons and the ions I now get

$$\frac{\omega_{pe}^2}{\omega^2 - k^2 c_{se}^2} + \frac{\omega_{pi}^2}{\omega^2 - k^2 c_{si}^2} = 1, \tag{5.3}$$

which gives

$$\omega^2 \simeq \omega_{pe}^2 + k^2 \left(c_{se}^2 + \frac{m_e c_{si}^2}{m_i} \right), \tag{5.4}$$

for the high-frequency branch of the full dispersion law. This is quite similar to (5.2), except for small thermal changes, and I have of course neglected ω_{pi} compared to ω_{pe}. There is also a low-frequency branch, known as ion-acoustic modes, and addressed in a subsequent subsection.

5.2.2 Buneman instabilities

Another interesting case is when the electrons stream with respect to the ions. Neglecting all thermal effects to keep the discussion simple, I find that

$$\frac{\omega_{pe}^2}{(\omega - kU_e)^2} + \frac{\omega_{pi}^2}{\omega^2} = 1. \tag{5.5}$$

The left-hand side of this dispersion law formally goes to infinity when $\omega \rightarrow 0$ or when $\omega \rightarrow kU_e$. This is schematically indicated in Figure 5.1. In between these two extreme values, it reaches a local minimum in ω_m

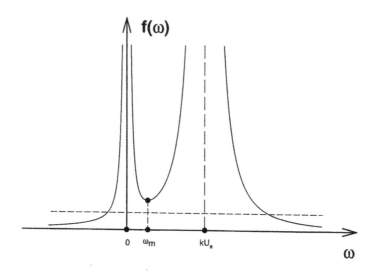

Figure 5.1. Dispersion diagram for Buneman modes

obeying

$$\frac{\omega_{pe}^2}{(\omega_m - kU_e)^3} + \frac{\omega_{pi}^2}{\omega_m^3} = 0, \tag{5.6}$$

in other words when

$$\omega_m = kU_e \sqrt[3]{\frac{m_e}{m_i}} \left(1 + \sqrt[3]{\frac{m_e}{m_i}}\right)^{-1}. \tag{5.7}$$

Unstable roots can occur provided this local minimum is higher than 1,

$$\frac{\omega_{pe}^2}{(\omega_m - kU_e)^2} + \frac{\omega_{pi}^2}{\omega_m^2} > 1, \tag{5.8}$$

and this condition defines the threshold as

$$k^2 < \frac{\omega_{pe}^2}{U_e^2} \left(1 + \sqrt[3]{\frac{m_e}{m_i}}\right)^3. \tag{5.9}$$

This is a two-stream or Buneman-type instability [Buneman 1958; Stix 1962,1992]. By expanding ω in (5.5) in powers of $\sqrt[3]{m_e/m_i}$ I find that the unstable root is given by

$$\omega = \frac{1 + i\sqrt{3}}{2} \sqrt[3]{\frac{m_e}{2m_i}} \omega_{pe}, \tag{5.10}$$

at values for k of the order ω_{pe}/U_e. Hence the instability occurs because the Doppler shifted frequency that the streaming electrons see is small enough to be in the ion plasma frequency range, and both species can resonate. The growth rate of the instability is given by

$$\gamma = \frac{\sqrt{3}}{2} \sqrt[3]{\frac{m_e}{2m_i}} \omega_{pe}, \tag{5.11}$$

of the order of the real part. Both are small compared to the electron plasma frequency.

5.2.3 Ion-acoustic modes

At lower frequencies, the electrons can follow the perturbations induced by the waves in an almost instantaneous balance between the electrostatic and pressure effects, so that

$$\frac{1}{n_e} \frac{\partial p_e}{\partial z} \simeq e \frac{\partial \phi}{\partial z}. \tag{5.12}$$

For isothermal electrons obeying the ideal gas law, I can put

$$\frac{\partial p_e}{\partial z} = \kappa T_e \frac{\partial n_e}{\partial z} = m_e c_{se}^2 \frac{\partial n_e}{\partial z}, \tag{5.13}$$

and hence obtain a Boltzmann type electron density distribution

$$n_e = N_e \exp \frac{e\phi}{\kappa T_e}. \tag{5.14}$$

At the same time I will suppose the ions to be almost cold ($T_i \ll T_e$). Both assumptions together delineate a domain for the phase velocities of the waves (ω/k) as $c_{si} \ll \omega/k \ll c_{se}$, and the dispersion law (5.1) becomes in this approximation

$$\frac{\omega_{pi}^2}{\omega^2} - \frac{\omega_{pe}^2}{k^2 c_{se}^2} = 1. \tag{5.15}$$

A note of caution should be sounded here. As seen in the previous chapter, the infinite chain of moment equations is formally equivalent to the kinetic description they are obtained from, if all equations are dealt with. This is, of course, totally unwieldy and even more complicated than working with kinetic equations themselves, hence the truncation of this set of moment equations to the first two or three members in a fluid picture. The neglect of higher-order moments like the heat flow assumes, whether explicitly stated or not, that the phase velocity of the waves is large compared to the thermal velocities. However, from kinetic theory arguments one can immediately derive the Boltzmann distribution (5.14) in the opposite limit, that of phase velocities large compared to the thermal velocities.

There is thus an apparent contradiction, that (5.1) has been derived in the fluid limit of small thermal velocities, whereas (5.15) has made use of the opposite limit for the electrons, as if a kinetic theory treatment had been given for these particles. Formally, one can reconcile both descriptions by taking the limit of massless electrons, where $m_e \to 0$ while keeping $m_e c_{se}^2$ constant, where I am obviously thinking of $m_e c_{se}^2 = \kappa T_e$, but can easily be more general. It is in this spirit that I will on many occasions treat the electrons and/or the ions as massless or quasi-inertialess, starting from the fluid equations, without explicitly going back to the correct kinetic approach for these particles.

Introducing now the electron Debye length $\lambda_{De} = c_{se}/\omega_{pe}$ and the ion-acoustic velocity $c_S = \lambda_{De} \omega_{pi}$ allows us to rearrange (5.15) as

$$\left(\frac{\omega}{k}\right)^2 = \frac{c_S^2}{1 + k^2 \lambda_{De}^2} \simeq c_S^2. \tag{5.16}$$

The last expression is a valid approximation in the small wavenumber or long wavelength limit, defined as $k\lambda_{De} \ll 1$, where wavelengths are larger than the electron Debye length. Incidentally, this is another of the conditions for the valid use of a fluid description of plasma waves. If the electrons obey an ideal gas law, then $c_{se}^2 = \kappa T_e/m_e$, $\lambda_{De}^2 = \kappa T_e \varepsilon_0/N_0 e^2$ (with $N_0 = N_e = N_i$) and

$$\left(\frac{\omega}{k}\right)^2 \simeq c_S^2 = \frac{\kappa T_e}{m_i}. \tag{5.17}$$

The ion-acoustic velocity combines the electron temperature with the ion mass, and for most plasmas $c_{si} \ll c_S \ll c_{se}$ is valid. The physical interpretation of these low frequency ion-acoustic modes is that they are driven by electron pressure, whereas the ions provide the necessary inertia to set up wave motions, and as pressure and/or density perturbations these are longitudinal (electrostatic) in nature.

The above ideas can be extended to more than one lighter and hotter species in a multispecies plasma [Verheest and Hellberg 1997]. An effective Debye length λ_D is needed, given through

$$\frac{1}{\lambda_D^2} = \sum_{\text{lighter species}} \frac{1}{\lambda_{Ds}^2}, \tag{5.18}$$

and from the plasma frequencies of the colder and heavier species, I can define a global plasma frequency through

$$\omega_p^2 = \sum_{\text{heavier species}} \omega_{ps}^2. \tag{5.19}$$

A typical acoustic-like velocity for the lowest frequency modes in this system is introduced through

$$c_S = \lambda_D \omega_p. \tag{5.20}$$

These ideas will be useful when I look further on at the development of dust-acoustic modes, where the heavy, charged dust grains provide the inertia and the lighter species, in that case not only electrons but also the usual plasma ions, the thermal or pressure effects.

Note that it is not really necessary to invoke strict Boltzmann distributions for the lighter species, as is done in many papers, which has sometimes confused the issues about the ranges of validity of solutions thus obtained, especially when magnetic fields are present. All I really need are the thermal velocities, however defined. In addition, inertial effects for the lighter and hotter species can always be eliminated at intermediate stages in the computations by letting $m_e \to 0$ but keeping $m_e c_{se}^2$ finite, with similar limits for other species when needed.

5.3 Dusty plasma modes

5.3.1 Pioneering work on dust modified modes

In the previous section I have recalled that in ordinary plasmas at the lowest frequencies the phase velocity is of the order of the ion-acoustic velocity c_S. The physical picture is that both the electron pressure and the ion inertia keep the waves going. The ideas developed in the previous section have to be modified when charged dust grains are present, with their typically very small characteristic frequencies. And now we really embark on our journey towards dusty plasma modes!

Going back in history, to see how ideas have matured, I find that one of the first to treat collective modes in what was then called a microparticle plasma cloud were James and Vermeulen [1968]. They elaborated on earlier theoretical work for linear multispecies waves that had no specific applications in mind, however [see e.g. Verheest 1967]. Since the particle sizes vary over quite a range, James and Vermeulen used a cold plasma multispecies model, where each fluid represents all particles of a given size and with the same equilibrium streaming. Cold plasma waves in this framework obey

$$\sum_\alpha \frac{\omega_{p\alpha}^2}{(\omega - kU_\alpha)^2} = 1, \tag{5.21}$$

which leads to the possibility of wave damping in a multibeam model akin to macroscopic Landau damping, in the spirit of the Dawson modes [Dawson 1960; Stix 1962]. In those days, no space applications were intended.

One of the next milestones is the extensive review by Sodha and Guha [1971] of colloidal plasmas, as they were called then. This contains a lot of information on the ionization of colloidal particles, but not on wave phenomena. Somehow, the subject of microparticle or colloidal plasmas escaped from the mainstream of plasma physics research.

Much later, motivated by spacecraft discoveries, attention returned to truly dusty plasmas, at first for linear electrostatic modes where the dust species are assumed to have constant charges. The dust is treated either as an immobile neutralizing background (as the protons are in the simple descriptions of electron plasma waves) or as one or more additional species of negative ions, but with very much lower characteristic frequencies.

In the first category, I find work by De Angelis *et al.* [1988] on ion-acoustic waves in an unmagnetized plasma. The massive, immobile but charged dust grains are surrounded by a statistical distribution of plasma particles. This results in spherical electrostatic ion waves propagating around each grain in the dusty plasma. The choice of ion modes for this first look at the plasma response was motivated by low-frequency electrostatic noise enhancement associated with Halley's comet, as observed by the Vega and Giotto space probes in the region of increased dust, with a peak below the ion plasma frequency. Although the data could also be interpreted in terms of dust impact on the electrical probes, the experiments were sufficiently different on both Vega satellites to make it clear that at least part of the noise enhancement is really due to electrostatic modes, trapped in the dust region with increased amplitudes.

The motivation for the work clearly comes from astrophysical observations, but the authors did not try to get quantitative agreement with these, so that the work in itself is only qualitatively and in a rather intuitive way connected with dusty plasmas in space. Similar remarks can rightfully be addressed to many of the other papers in this field, where the interest is in very low frequency phenomena, on the scales introduced by the extremely low charge-to-mass ratios of charged dust.

Goertz briefly touched upon electrostatic modes in the presence of drifting dust grains in planetary environments, in his general review of dusty plasmas in the solar system [1989]. This was leading to variations on the familiar two-stream instability, which I briefly touched upon when recalling the Buneman modes. In particular, Goertz refers to previous work by Bliokh and Yaroshenko [1985] on electrostatic waves in Saturn's rings. The latter authors had considered a multistream model to account for the very many narrow rings and gaps in the B ring. Waves of dust charge density are at the same time waves of material density, leading to a possible connection with the observed spokes. The model is very close to the pioneering but apparently forgotten paper by James and Vermeulen [1968], as Bliokh and Yaroshenko [1985] did not mention this reference. Although the modes considered can probably not be strong enough to fully account for the spokes as observed, the idea of dust-driven plasma waves provides an intriguing mechanism for ion heating at the expense of the gravitational energy of the orbiting dust grains.

The treatment of the modes considered by Bliokh and Yaroshenko [1985] has been vastly expanded in their subsequent book [Bliokh et al. 1995]. However, because of the emphasis on self-gravitational effects in this book, I will postpone a further discussion to Chapter 8, where many of the fascinating aspects of self-gravitation are put in some perspective.

The real impetus of research into dust-acoustic modes was given by Rao et al. [1990], who included the dynamics of a tenuous dust fluid but used Boltzmann distributions for the electrons and the ions. They coined the name for the novel dust-acoustic modes at low enough frequencies and with phase velocities far below the ion-acoustic velocity. Due to the wide gap in frequencies, these waves are quite different and well separated from the familiar ion-acoustic modes, which can also exist in dusty plasmas, with modifications due to the resulting imbalance between the number of free electrons and ions. The latter waves are then referred to as dust-ion-acoustic modes, discussed in the next paragraph.

Of course, in this logical progression in going to lower and lower frequencies, the dust-acoustic modes are only newer applications of a much more general framework, but were not predicted or anticipated earlier. The fact that the name "dust-acoustic" has easily caught on, testifies to the innovative coining of it. The subsequent explosion in dusty plasma research also shows that the ideas of Rao *et al.* [1990] came at the right time.

5.3.2 Dust-ion-acoustic waves

There are two immediate ways in which the presence of charged dust grains make themselves felt, depending on the frequency regime. One possibility is that I still look at phase velocities below the electron thermal velocity, but larger than the ion and now also the dust thermal velocities, so that $c_{sd}, c_{si} \ll \omega/k \ll c_{se}$. There are now two sluggish species, the ions and the dust grains, moved by the electron pressure. Such waves have been coined dust-ion-acoustic modes by Shukla and Silin [1992], and their dispersion law follows from (5.1) as

$$\frac{\omega_{pi}^2 + \omega_{pd}^2}{\omega^2} - \frac{\omega_{pe}^2}{k^2 c_{se}^2} = 1. \tag{5.22}$$

Since the electron, ion and dust plasma frequencies are related through

$$\omega_{pd}^2 = \frac{Z_d m_i}{m_d} \omega_{pi}^2 - \frac{Z_d m_e}{m_d} \omega_{pe}^2, \tag{5.23}$$

and for all known dusty plasmas $Z_d m_e/m_d \ll Z_d m_i/m_d \ll 1$, I can safely neglect ω_{pd}^2 compared to ω_{pi}^2 in the dispersion law. Another way of looking at these modes is to view the dust grains as an immobile background, and modifications to the standard ion-acoustic modes occur due to electron depletion and high dust charges. With the usual assumption of Boltzmann electrons I then find that the phase velocity c_{dia} of the dust-ion-acoustic modes is approximated by

$$c_{dia}^2 \simeq \frac{N_i}{N_e} c_S^2. \tag{5.24}$$

At serious levels of electron depletion due to capture by the charged dust, it could be that N_e is small compared to N_i, so as to make c_{dia} proportionally larger than the usual ion-acoustic velocity c_S. This has also been observed experimentally for dusty plasmas [Barkan *et al.* 1996], in accordance with similar results for negative-ion plasmas [D'Angelo *et al.* 1966].

However, in order to be consistent with all approximations, I need that

$$c_{dia} \ll c_{se} \qquad \Longrightarrow \qquad \frac{m_e}{m_i} \ll \frac{N_e}{N_i}, \tag{5.25}$$

so that the electron depletion cannot be too severe. For a further discussion of this mode and the inherent modifications to the Landau damping of it, I refer to Shukla and Silin [1992].

In a way related to this approach, where the dust dynamics does not yet come into play, Salimullah and Sen [1992] examined the dielectric properties of a plasma studded with charged dust. The resulting inhomogeneous electric field significantly influences the dispersion properties of the plasma, even though the plasma Debye length is much smaller than the average grain separation. At low frequencies important modifications of the ion-acoustic branch occur, as well as a new ultra low-frequency mode arising from ion oscillations in the static dust distribution. The latter is not to be confounded with the dust-acoustic mode, as all dust dynamics has been left out of the description, and cannot be recovered in a fluid picture.

5.3.3 Dust-acoustic waves

Rather than studying only the changes wrought to ion-acoustic modes by some form of electron depletion, however interesting in themselves, it has proved very stimulating to open up a whole new phase velocity regime. Now we go well below the ion thermal speed and look at phase velocities such that $c_{sd} \ll \omega/k \ll c_{si}, c_{se}$. This is the true regime of the dust-acoustic modes [Rao et al. 1990], with dispersion law following from (5.1) as

$$\frac{\omega_{pd}^2}{\omega^2} - \frac{\omega_{pe}^2}{k^2 c_{se}^2} - \frac{\omega_{pi}^2}{k^2 c_{si}^2} = 1. \tag{5.26}$$

Not only the electrons but also the ions might be described here by Boltzmann distributions, although this is strictly speaking not really required. The phase velocity c_{da} of the dust-acoustic modes is approximated by

$$c_{da} = \lambda_D \omega_{pd}, \tag{5.27}$$

and involves the global Debye length, which could be heavily influenced by ion screening effects.

Results similar to those of Rao et al. [1990] were obtained by D'Angelo [1990], who looked at the electrostatic ion and dust cyclotron and ion-acoustic modes, and who later obtained a beautiful experimental verification [Barkan et al. 1995a], shown in Figure 5.2. The connection to earlier results on wave modes in plasmas containing negative ions or in plasmas with two positive ion species is given. Later, Shukla [1992] reviewed linear low-frequency electrostatic and electromagnetic modes. The possible relevance to astrophysical and cometary plasmas is pointed out, without going into specific details.

By averaging over a random distribution of dust particles, Forlani et al. [1992] have shown how plasma fluctuations are modified and point to

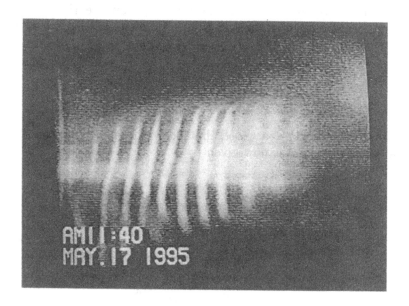

Figure 5.2. Experimental verification of the dust-acoustic wave. *Reprinted from Barkan et al. [1995a] (Fig. 2 on p. 3564), with permission of the American Institute of Physics*

the possibility of wave damping due to beating of the wave with the dust density fluctuations. This has nothing to do with damping due to variable dust charges, which was not included but will be discussed in the Chapter 7. A similar reasoning will hold for the electromagnetic dispersion law.

5.3.4 Two-stream and Buneman instabilities

Before discussing the various contributions to the two-stream and Buneman instabilities, I give a general outline which is very close to what I recalled for the classic Buneman streaming instability between ions and electrons. As long as the electrons stream with respect to the ions and the dust, the modifications to the Buneman instability are slight, and mostly come from electron depletion when negative charges are sitting on the grains.

On the other hand, the streaming could be between the ions and the dust grains, at such low frequencies that the electrons can be treated as Boltzmann distributed. Following part of the expose of Ishihara [1998], the dispersion law is then

$$\frac{\omega_{pi}^2}{(\omega - kU_i)^2} + \frac{\omega_{pd}^2}{\omega^2} = 1 + \frac{1}{k^2\lambda_{De}^2}, \qquad (5.28)$$

analogous to (5.5) but with important changes on the right hand side. There are corresponding changes to Figure 5.1. The left hand side reaches a local minimum in

$$\omega_m = kU_i \sqrt[3]{\frac{\omega_{pd}^2}{\omega_{pi}^2}} \left(1 + \sqrt[3]{\frac{\omega_{pd}^2}{\omega_{pi}^2}}\right)^{-1}. \tag{5.29}$$

Unstable roots can occur provided this local minimum is higher than $1 + 1/k^2\lambda_{De}^2$. The threshold is now given by

$$(1 + k^2\lambda_{De}^2)U_i^2 < \lambda_{De}^2\omega_{pi}^2 \left(1 + \sqrt[3]{\frac{\omega_{pd}^2}{\omega_{pi}^2}}\right)^3, \tag{5.30}$$

and the unstable root is then

$$\omega = \frac{1 + i\sqrt{3}}{2} \sqrt[3]{\frac{\omega_{pi}}{2\omega_{pd}}} \omega_{da}. \tag{5.31}$$

The dust-acoustic frequency is defined here as $\omega_{da} = k\lambda_{De}\omega_{pd}/\sqrt{1 + k^2\lambda_{De}^2}$, and the instability occurs close to the resonance $U_i \simeq \lambda_{De}\omega_{pi}$.

Variations on this theme have been studied in dusty plasmas by different authors. Havnes [1980] already examined the conditions for the onset of the instability in a plasma model in which two charged dust distributions stream relative to each other. Small grains (below micrometer in size) are brought to rest with respect to the surrounding gas in very short distances, while larger grains are practically unaffected. This braking mechanism is applied to interstellar gas clouds, for which one observes a large depletion of several elements at low cloud velocities, a depletion which decreases with cloud velocity. Hence this two-stream instability may be important for grain destruction in high velocity clouds, as it effectively brakes the small grains but leaves the larger ones intact, thereby creating two populations of grains which stream relative to each other with the cloud velocity. Collisions between the grains in the two populations may thus become sufficiently frequent and energetic to ensure the destruction of a large fraction of the total grain content. This could be important when shocks travel through interstellar gas, setting up grain separation and subsequent destruction. Grains thus destroyed are returned to the cloud in gas form, explaining why some elements are observed to be underabundant in the gas of low-velocity clouds but have near normal abundances in clouds of high velocity.

In a later paper [Havnes 1988], streaming is considered between the solar wind and cometary dust grains. At the high dust densities which may be found in cometary comae, the dust grains can be important charge carriers, leading to electron depletion. The resulting instability drastically enhances the coupling between the solar wind and the dust, which favours small dust grains to be swept along with the solar wind plasma. A high neutral gas density prevents the onset of this instability. Turbulent drag could form dust striae in cometary tails, leading to narrower ones in a high-velocity solar wind, whereas one would expect wider tails in lower velocity solar winds, where gravity, radiation pressure and expansion dominate. It is estimated that dust-solar wind interaction would be higher for comets at larger heliocentric distances, outside the inner part of the solar system. Further discussion of the interaction of cometary material with the solar wind is deferred to the next chapter, as electromagnetic modes are more important for the development of low-frequency turbulence.

Bharuthram *et al.* [1992] have analyzed how the dust grains influence two-stream instabilities between electrons and ions, or can themselves form a drifting beam, as in planetary rings. In all cases the presence of dust grains enhances the growth of the instabilities, as well as the velocity ranges over which the instability can occur. Stationary dust, as in the magneto-sphere of Neptune, also modifies the propagation properties, although then the planetary magnetic field might have to be taken into account, and low-frequency electromagnetic phenomena come into play.

A standard Vlasov approach to ion- and dust-acoustic instabilities was given by Rosenberg [1993]. The instabilities are caused by streaming be-tween the ions and/or electrons with respect to the dust grains, as can be the case in the rings of Saturn beyond the corotation distance, where the charged grains are slower than the magnetospheric plasma. The growth rates are estimated to be larger than the collisional damping rates, effec-tively producing an instability. This work was later extended by Bharuthram and Pather [1996], who numerically investigated the kinetic description of the dust-acoustic instability for a whole range of parameters.

Rosenberg [1993] also mentioned a weighting of the dust at different sizes and charges, to find that the charge goes as the radius a of the dust grains, whereas the mass goes as a^3. Hence if $N_d \sim a^{-p}$, then the plasma frequency squared goes as a^{-p-1}, with a consequent weighting toward grains of smaller radius. This is one of the few papers in which grains of different sizes are discussed and I will come back to it in Chapter 9.

In related papers, Rosenberg and Krall discuss drift instabilities, when there is a local electron density gradient opposite in sign to a dust density gradient [Rosenberg and Krall 1994], or two-stream instabilities [Rosenberg and Krall 1995]. Different equilibria are reviewed, according to whether the dust gyroradius is smaller or larger than the density scale length. Electrons and ions are magnetically confined, whereas the dust grains would be elec-trostatically coupled and a zeroth order electric field appears. Fluctuations in the dust charges have been omitted. As an application, it would seem that the edges of the ringlets in the F ring of Saturn are stable against the excitation of this high-frequency mode, whereas the low-frequency two-stream instability could be of importance in the E ring.

Winske *et al.* [1995] give a extensive numerical treatment of the dust-ion-acoustic instability, caused by a drift of plasma ions with respect to the charged dust. The instability saturates by trapping some of the ions, and is slightly weaker when the dust grains have a range of sizes, charges and masses. It is argued that this process could contribute to ion heating and diffusion observed in the inner magnetosphere of Saturn.

Another possible application of drift instabilities is in understanding the enhanced backscatter from the Space Shuttle exhaust. Bernhardt *et al.* [1995] conjectured that dust-acoustic modes are excited in the plasma created by the condensation of the water vapour of the expanding shuttle exhaust plume into ice grains, which are then subsequently charged. Bharuthram and Rosenberg [1998] have shown that for the reported parameters a drift instability is indeed possible. Because the ion drift speed in the dust frame is larger than the ion thermal speed, the instability is of the Buneman type.

5.4   General nonlinear wave theory

5.4.1   Arbitrary amplitudes and Sagdeev potential

For waves of larger amplitudes, we can no longer apply the linearization schemes of the previous sections, but have to deal with the equations in their full nonlinear glory, and difficulties! This, of course, cannot really be done in all generality, but needs to be looked at for specific problems. Because this will occur at various places in this monograph, I prefer to give here a broad picture of the main methods used, so as to set the scene and clarify misunderstandings occurring in the literature, before turning to the specifics for dusty plasmas. In doing so, the systematic expose of Verheest and Hellberg [1999] will be followed. Intermediate details have been spelled out, to allow later modifications to be pointed out, when I look at additional complications like fluctuating dust charges, or other wave types.

One of the successful approaches in the study of electrostatic solitons and double layers has been the Sagdeev or pseudo potential method. By solitons and double layers I mean two types of nonlinear solitary wave structures which propagate without change in form through the medium, either as stable humps or dips in some parameter, or else as shock-like transitions from one value to another. If there are several of these waves, they have to obey certain interaction criteria, which make them come out of these interactions essentially unchanged [Ablowitz and Clarkson 1991]. Inspired by these particle-like properties, the name "soliton" was coined, although it is sometimes loosely but incorrectly used just to denote a solitary wave. Because of their general properties and intrinsic robustness, solitons and double layers have cropped in many different fields of physics, after they were first observed as waves on shallow water. This monumental experimental discovery stood for a long time on its own, until much later appropriate theoretical model equations were developed. Examples of these will be given, once I am through the generalities.

In the Sagdeev potential approach, one works in a frame which follows the nonlinear structure. Hence all quantities depend on the space variable z and the time variable t through a moving frame combination $\zeta = z - Vt$, where V is the velocity of the structure. The basic equations are by now quite familiar and include the continuity equations,

$$\frac{\partial n_\alpha}{\partial t} + \frac{\partial}{\partial z}(n_\alpha u_\alpha) = 0, \tag{5.32}$$

the equations of motion,

$$\frac{\partial u_\alpha}{\partial t} + u_\alpha \frac{\partial u_\alpha}{\partial z} = -\frac{1}{n_\alpha m_\alpha}\frac{\partial p_\alpha}{\partial z} - \frac{q_\alpha}{m_\alpha}\frac{\partial \phi}{\partial z}, \tag{5.33}$$

the barotropic pressure variations,

$$p_\alpha = p_\alpha(n_\alpha), \tag{5.34}$$

and the electrostatic Poisson's equation,

$$\varepsilon_0 \frac{\partial^2 \phi}{\partial z^2} + \sum_\alpha n_\alpha q_\alpha = 0. \tag{5.35}$$

All these become just ordinary differential equations in the co-moving frame, to be integrated with respect to ζ.

Since I am investigating hump-like rather than periodic structures, I assume that all integrated quantities attain their equilibrium values for large enough ζ, in other words, far enough from the central hump/dip or transition the medium is almost undisturbed and homogeneous. This will be used in the form

$$\lim_{\zeta \to \infty} n_\alpha \to N_\alpha,$$

$$\lim_{\zeta \to \infty} p_\alpha \to P_\alpha,$$

$$\lim_{\zeta \to \infty} u_\alpha \to U_\alpha,$$

$$\lim_{\zeta \to \infty} \phi \to 0, \tag{5.36}$$

and all derivatives with respect to ζ vanish in the same limit. The continuity equation (5.32) can immediately be integrated to yield

$$u_\alpha = V - (V - U_\alpha)\frac{N_\alpha}{n_\alpha}. \tag{5.37}$$

In the equations of motion (5.33) the velocities are eliminated, leading to

$$\frac{(V - U_\alpha)^2 N_\alpha^2}{n_\alpha^3} \frac{dn_\alpha}{d\zeta} - \frac{1}{n_\alpha m_\alpha} \frac{dp_\alpha}{d\zeta} - \frac{q_\alpha}{m_\alpha} \frac{d\phi}{d\zeta} = 0. \tag{5.38}$$

A formal integration would give

$$\frac{(V - U_\alpha)^2}{2} \left(1 - \frac{N_\alpha^2}{n_\alpha^2}\right) - \frac{1}{m_\alpha} \int_{N_\alpha}^{n_\alpha} \frac{1}{n_\alpha} \frac{dp_\alpha}{dn_\alpha} dn_\alpha - \frac{q_\alpha}{m_\alpha} \phi = 0, \tag{5.39}$$

which in principle determines $n_\alpha(\phi)$, and through (5.34) also $p_\alpha(\phi) = p_\alpha(n_\alpha(\phi))$. Explicit expressions for these, however, can only be obtained in the most simple cases of isothermal ($p_\alpha \sim n_\alpha$, at constant T_α) or adiabatic ($p_\alpha \sim n_\alpha^3$) pressure variations, or for cold constituents, when there are no p_α terms. Examples of such power law dependencies ($p_\alpha \sim n_\alpha^{\gamma_\alpha}$) lead to the classic relation that $\ln n_\alpha \sim \phi$ when $\gamma_\alpha = 1$ and inertia is neglected. Explicit evaluation is also possible when $\gamma_\alpha = 3$ [Verheest et al. 1996]. For cold constituents I find from (5.39) that

$$\frac{(V - U_\alpha)^2}{2} \left(1 - \frac{N_\alpha^2}{n_\alpha^2}\right) = \frac{q_\alpha}{m_\alpha} \phi, \tag{5.40}$$

so that

$$n_\alpha = N_\alpha \left(1 - \frac{2q_\alpha \phi}{m_\alpha (V - U_\alpha)^2}\right)^{-1/2}, \tag{5.41}$$

with typical square root expressions. Another recently studied possibility is that of a nonideal Van der Waals equation of state [Rao 1998],

$$(p_d + An_d^2)(1 - Bn_d) = n_d \kappa T_d. \tag{5.42}$$

The corrections involve A and B which can be computed in terms of the critical parameters by requiring that

$$\frac{\partial p_d}{\partial n_d} = 0, \qquad \frac{\partial^2 p_d}{\partial n_d^2} = 0, \tag{5.43}$$

so that

$$A = \frac{9\kappa T_{dc}}{8n_{dc}}, \qquad B = \frac{1}{3n_{dc}}, \tag{5.44}$$

and the subscript c refers to values at the critical point. The Van der Waals behaviour accounts for the finite sizes of the dust grains in two ways. The volume reduction coefficient, translating the fact that the grains are no longer point particles, increases the linear phase speed, because a volume reduction increases the gas pressure and hence also the restoring force driving the dust-acoustic waves. On the other hand, the enhanced molecular cohesion between the grains works in the opposite way. For the nonlinear modes to be described further on, these effects can easily be incorporated in the general model I are discussing here.

Returning now to the general thread of the expose, it is seen that regardless of what can be done with (5.39), I can always multiply (5.38) with $n_\alpha m_\alpha$ and sum over all α. Eliminate $\sum_\alpha n_\alpha q_\alpha$ with the help of Poisson's equation (5.35) and integrate the overall equation with respect to ζ, to obtain

$$\frac{\varepsilon_0}{2}\left(\frac{d\phi}{d\zeta}\right)^2 + \sum_\alpha (P_\alpha - p_\alpha) + \sum_\alpha (V - U_\alpha)^2 N_\alpha m_\alpha \left(1 - \frac{N_\alpha}{n_\alpha}\right) = 0. \quad (5.45)$$

This is the energy integral for the one-dimensional motion of a classical particle in a given potential, where ϕ plays the role of the space coordinate, ζ represents time and ε_0 mass. In this picture the potential,

$$\Psi(\phi) = \sum_\alpha (P_\alpha - p_\alpha) + \sum_\alpha (V - U_\alpha)^2 N_\alpha m_\alpha \left(1 - \frac{N_\alpha}{n_\alpha}\right), \quad (5.46)$$

is known as the Sagdeev or pseudo potential. Having obtained this, I would like to point out that we did not encounter an obstacle in principle, as soon as barotropic pressures were introduced, but in more complicated cases there can obviously be great or even insurmountable practical difficulties in actually evaluating the explicit expressions for $n_\alpha(\phi)$ and $p_\alpha(\phi)$.

The discussion of possible solutions is now reminiscent of what happens in classical mechanics when a conservative force field is given. Between two successive single roots of $\Psi(\phi)$, such that in between $\Psi(\phi) < 0$, ϕ is periodic in ζ and leads in our picture to periodic nonlinear waves. These, however, do not obey our assumptions of vanishing influences at infinity.

Next, when a single root on one side of the interval considered, and a double root at the other end, a solitary wave is generated. Due to initial conditions incorporated in $\Psi(\phi)$, the double root is in $\phi = 0$ and it takes an infinitely long "time" to get away from it, after which ϕ reaches then a maximum or minimum in ϕ_m, and thereafter returns to 0, again taking infinitely long to get back. Hence, the conditions for solitons to exist are that

$$\Psi(0) = 0, \qquad \frac{d\Psi}{d\phi}\bigg|_{\phi=0} = 0,$$

$$\Psi(\phi_m) = 0, \qquad \frac{d\Psi}{d\phi}\bigg|_{\phi=\phi_m} \neq 0. \quad (5.47)$$

Moreover, ϕ_m is a single root away from 0, such that there are no other roots for $\Psi(\phi)$ in between. In addition to the conditions on the Sagdeev

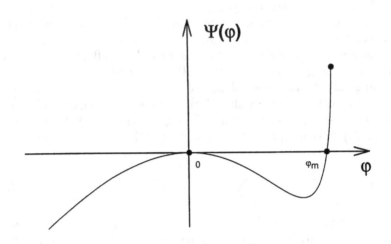

Figure 5.3. Shape of the Sagdeev potential to allow for the occurrence of solitons

potential and its first derivative, there is also the requirement that the double root in zero corresponds to a local maximum in ϕ, which imposes for the second derivative that

$$\left.\frac{\mathrm{d}^2\Psi}{\mathrm{d}\phi^2}\right|_{\phi=0} < 0. \tag{5.48}$$

The shape of the Sagdeev potential is shown in Figure 5.3. When plotting ϕ as a function of ζ, one obtains the typical hump ($\phi_m > 0$, as illustrated in Figure 5.4, or dip-like ($\phi_m < 0$) soliton behaviour, at least as far as the electrostatic potential is concerned. However, the proper terminology for these solitons is not unambiguous, as one sometimes refers to increases or decreases in one of the relevant densities, and whether one views the solitons as compressive or rarefactive then very much depends on which species is being considered. In electron-ion plasmas one usually takes the electron density as the reference, so that compressive solitons are hump-like for the electron density, but the reverse of course is true for the ions. Verheest and Hellberg [1997] gave a discussion for cold multispecies dust grains in the presence of a number of Boltzmann (inertialess) species.

For double layers one needs two successive double roots, so that in the parlance of classical mechanics ϕ can transit from one value to another without coming back, with a rather sharp and shock or kink-like transition between the two. In addition, the double roots correspond to local maxima.

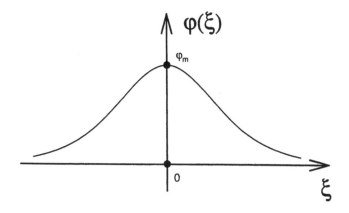

Figure 5.4. Typical shape of a compressive soliton

The conditions on the Sagdeev potential now are

$$\Psi(0) = 0, \qquad \frac{d\Psi}{d\phi}\bigg|_{\phi=0} = 0,$$

$$\Psi(\phi_m) = 0, \qquad \frac{d\Psi}{d\phi}\bigg|_{\phi=\phi_m} = 0, \qquad (5.49)$$

and a typical shape of the Sagdeev potential is shown in Figure 5.5. When actually discussing the possibility of dust-acoustic double layers later on, it has to be borne in mind that the double layers conditions impose rather stringent restrictions on the form of the allowable Sagdeev potentials, and in particular these might be impossible to fulfill in the approximate cases

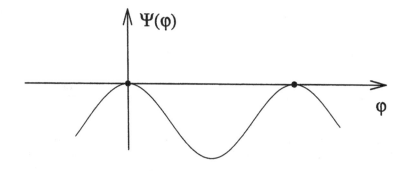

Figure 5.5. Shape of the Sagdeev potential for the occurrence of double layers

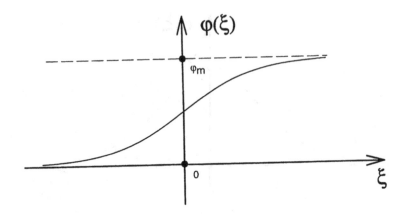

Figure 5.6. Typical shock or kink solution

which I will discuss in the framework of singular perturbation theory. When plotting ϕ as a function of ζ, one obtains shock-like structures, as illustrated in Figure 5.6.

As an illustration, ion-acoustic solitons follow the same assumptions underlying the linear description, namely that the electrons are behaving isothermally and that their inertia can be neglected. The ions, on the other hand, will be treated as cold. In that case I find, using (5.14),

$$p_e = \kappa T_e n_e = \kappa T_e N_0 \exp \frac{e\phi}{\kappa T_e},$$

$$P_e = \kappa T_e N_0. \tag{5.50}$$

For cold and non-streaming ions the density follows from (5.41) as

$$n_i = N_0 \left(1 - \frac{2e\phi}{m_i V^2}\right)^{-1/2}. \tag{5.51}$$

The corresponding Sagdeev potential is thus

$$\Psi(\phi) = \kappa T_e N_0 \left(1 - \exp\frac{e\phi}{\kappa T_e}\right) + N_0 m_i V^2 \left(1 - \sqrt{1 - \frac{2e\phi}{m_i V^2}}\right). \tag{5.52}$$

Since this is unbounded for $\phi \to -\infty$ and ceases to be defined for $\phi > m_i V^2/2e$, the solitons are compressive, with $\phi_m < m_i V^2/2e$ [Verheest and Hellberg 1997]. The underlying reason for the compressive character (in the electron density) is that the lighter/hotter species is negatively charged, so that an increase in ϕ corresponds to an increase in n_e. At the same time, the heavier/cooler species are positively charged, thus rendering the square root zero for some positive ϕ. Both together point to compressive solitons. The possible inclusion of cold charged dust influences the equilibrium densities and the linear phase velocities, as I explained when discussing linear ion-acoustic waves in dusty plasmas. Similar quantitative modifications occur for the nonlinear development of these [Popel and Yu 1995].

Had I be dealing with an antihydrogen plasma, composed of positrons and antiprotons, the reverse would be true. All this will have to be adapted for dusty plasmas, where the heavy negative dust grains will cause rarefactive solitons to occur, rarefactive with respect to the electrostatic potential. An example of this general conclusion can be found in the numerical treatment of nonlinear modes in dusty plasmas with Boltzmann ions and cold negative dust grains [Mamun *et al.* 1996a].

5.4.2 Reductive perturbation theory

Often one cannot deal with the equations in their full nonlinearity, not even in a co-moving frame. For want of anything better, a form of perturbation analysis is needed, in this case the method of stretched variables. This starts with stretching the independent variables as

$$\xi = \varepsilon^{1/2}(z - vt), \qquad \tau = \varepsilon^{3/2}t, \qquad (5.53)$$

where v is the linear phase velocity of the waves under study, to be determined later. Such a stretching is based on the consideration that the linear dispersion law (5.1) is even in k, so that corrections to the (linear) phase velocity ω/k will occur in powers of k^2, in the low-frequency, long-wavelength limit. In the traditional notation I then find that $\varepsilon^{1/2} \sim k$, and the slow time variations are induced by the nonlinearity of the system.

On the other hand and to be on the safe side, the dependent variables are developed as a series in $\varepsilon^{1/2}$,

$$n_\alpha = N_\alpha + \varepsilon^{1/2}n_{\alpha 1} + \varepsilon n_{\alpha 2} + \varepsilon^{3/2}n_{\alpha 3} + \dots,$$
$$p_\alpha = P_\alpha + \varepsilon^{1/2}p_{\alpha 1} + \varepsilon p_{\alpha 2} + \varepsilon^{3/2}p_{\alpha 3} + \dots,$$
$$u_\alpha = U_\alpha + \varepsilon u_{\alpha 1} + \varepsilon u_{\alpha 2} + \varepsilon^{3/2}u_{\alpha 3} + \dots,$$
$$\phi = \varepsilon^{1/2}\phi_1 + \varepsilon\phi_2 + \varepsilon^{3/2}\phi_{\alpha 1} + \dots \qquad (5.54)$$

The subscript on the terms in the expansions refers to the respective powers of $\varepsilon^{1/2}$, and the algebra will then determine how the expansions really go.

Substitute now (5.53) and (5.54) into the basic equations (5.32)–(5.34) and equate the different powers of $\varepsilon^{1/2}$. As will be seen, it is worth doing this carefully. To lowest significant order, here order ε, it is found that

$$n_{\alpha 1} = \frac{N_\alpha q_\alpha}{m_\alpha[(v - U_\alpha)^2 - c_{s\alpha}^2]}\phi_1,$$

$$p_{\alpha 1} = \frac{N_\alpha q_\alpha c_{s\alpha}^2}{(v - U_\alpha)^2 - c_{s\alpha}^2} \phi_1,$$

$$u_{\alpha 1} = \frac{(v - U_\alpha) q_\alpha}{m_\alpha [(v - U_\alpha)^2 - c_{s\alpha}^2]} \phi_1. \tag{5.55}$$

Using these results in Poisson's equation (5.35) to order $\varepsilon^{1/2}$ the linear dispersion law is recovered in this approach, namely

$$D(v) \equiv \sum_\alpha \frac{\omega_{p\alpha}^2}{(v - U_\alpha)^2 - c_{s\alpha}^2} = 0. \tag{5.56}$$

This corresponds to putting $\omega = vk$ in (5.1) and then taking the $k \to 0$ limit. In principle the new dispersion law determines the possibilities for v. A closer inspection of (5.56) shows that I need at least one highly mobile species (usually hot), in the sense that $(v - U_\alpha)^2 < c_{s\alpha}^2$, a role which traditionally has been played by the electrons. This species provides the pressure and restoring force for the waves. At the same time, at least one sluggish species (cooler or massive) is also necessary, defined here in the sense that $(v - U_\alpha)^2 > c_{s\alpha}^2$, so that there is inertia in the system to provide for the overshoot, yielding oscillations.

To order $\varepsilon^{3/2}$ I find from (5.32)–(5.34) that

$$n_{\alpha 2} = \frac{N_\alpha q_\alpha}{m_\alpha [(v - U_\alpha)^2 - c_{s\alpha}^2]} \phi_2$$
$$+ \frac{N_\alpha q_\alpha^2 [3(v - U_\alpha)^2 - c_{s\alpha}^2 + d_{s\alpha}^2]}{2m_\alpha^2 [(v - U_\alpha)^2 - c_{s\alpha}^2]^3} \phi_1^2,$$

$$p_{\alpha 2} = \frac{N_\alpha q_\alpha c_{s\alpha}^2}{(v - U_\alpha)^2 - c_{s\alpha}^2} \phi_2$$
$$+ \frac{N_\alpha q_\alpha^2 [(v - U_\alpha)^2 (3c_{s\alpha}^2 + d_{s\alpha}^2) - c_{s\alpha}^4]}{2m_\alpha [(v - U_\alpha)^2 - c_{s\alpha}^2]^3} \phi_1^2,$$

$$u_{\alpha 2} = \frac{(v - U_\alpha) q_\alpha}{m_\alpha [(v - U_\alpha)^2 - c_{s\alpha}^2]} \phi_2$$
$$+ \frac{(v - U_\alpha) q_\alpha^2 [(v - U_\alpha)^2 + c_{s\alpha}^2 + d_{s\alpha}^2]}{2m_\alpha^2 [(v - U_\alpha)^2 - c_{s\alpha}^2]^3} \phi_1^2. \tag{5.57}$$

Higher derivatives of the pressure are introduced through $m_\alpha d_{s\alpha}^2 = N_\alpha [d^2 p_\alpha / dn_\alpha^2]_{N_\alpha}$. Poisson's equation (5.35) to order ε yields

$$B(v) \phi_1^2 = 0, \tag{5.58}$$

$B(v)$ being defined (also for later purposes) as

$$B(v) = \sum_\alpha \frac{\omega_{p\alpha}^2 q_\alpha [3(v - U_\alpha)^2 - c_{s\alpha}^2 + d_{s\alpha}^2]}{2m_\alpha [(v - U_\alpha)^2 - c_{s\alpha}^2]^3}. \tag{5.59}$$

We have reached a bifurcation point, because either $B(v)$ or ϕ_1 has to vanish, unless $B(v)\phi_1^2$ is not exactly zero but so small that it has to be relegated to the next order. These different possibilities will be discussed in the next paragraphs.

5.4.3 Korteweg-de Vries-type expansions

In the generic case the plasma is not so special in its composition that v can be determined from (5.56) and at the same time annul $B(v)$. Hence $B(v) \neq 0$, and I have to take $\phi_1 = 0$, so that the expansion (5.54) of the potential and other dependent variables is in integer powers of ε rather than in $\varepsilon^{1/2}$, as I assumed initially. This leads to the usual Korteweg-de Vries (KdV) expansions, with the stretching (5.53) in $\varepsilon^{1/2}$ and $\varepsilon^{3/2}$, but the dependent variables going as integer powers of ε.

In that case (5.55) vanishes altogether and (5.57) then reduces to the form of (5.55), with all subscripts 1 replaced by 2. Continuing then to order ε^2 in Poisson's equation (5.35), I find a standard KdV equation

$$A(v)\frac{\partial \phi_2}{\partial \tau} + B(v)\phi_2\frac{\partial \phi_2}{\partial \xi} + \frac{1}{2}\frac{\partial^3 \phi_2}{\partial \xi^3} = 0, \tag{5.60}$$

with the new coefficient $A(v)$ given by

$$A(v) = \sum_\alpha \frac{\omega_{p\alpha}^2(v - U_\alpha)}{[(v - U_\alpha)^2 - c_{s\alpha}^2]^2}. \tag{5.61}$$

The single-soliton solution of this integrable evolution equation is

$$\phi_2(\xi, \tau) = \frac{3MA(v)}{B(v)}\mathrm{sech}^2\sqrt{\frac{MA(v)}{2}}\,(\xi - M\tau), \tag{5.62}$$

and there is a relation between the amplitude and the speed M of the soliton, in the co-moving frame. Moreover, $MA(v) > 0$ is required for such a solution to exist, and the sign of ϕ_2 is given by the sign of $B(v)$. Such solitons are termed compressive ($B(v) > 0$) or rarefactive ($B(v) < 0$), in the terminology of ion-acoustic solitons.

Returning to the classic example of ion-acoustic waves, the electrons are isothermal ($c_{se}^2 = \kappa T_e/m_e$, $d_{se} = 0$) and inertialess, while the ions are cold ($c_{si} = d_{si} = 0$), and there is no equilibrium streaming ($U_e = U_i = 0$). The dispersion law (5.56) then simply brings us back to the ion-acoustic phase velocity c_S introduced in (5.16). The coefficients (5.61) and (5.59) of the KdV equation (5.60) become

$$A = \frac{\omega_{pi}^2}{c_S^3}, \qquad B = \frac{e\omega_{pi}^2}{m_i c_S^4}, \tag{5.63}$$

and the KdV equation can be reduced to its most basic form,

$$\frac{\partial \phi_2}{\partial \tau} + \phi_2 \frac{\partial \phi_2}{\partial \xi} + \frac{1}{2} \frac{\partial^3 \phi_2}{\partial \xi^3} = 0, \tag{5.64}$$

by using nondimensional variables. Lengths are expressed in units of the electron Debye length λ_{De}, times in ion plasma periods ω_{pi}^{-1} and the electrostatic potential in units of $\kappa T_e/e$.

It is clear that the general KdV equation (5.60) can be cast in the same nondimensional form, so that the most important conclusion for the solutions is that they are solitons of a generic form, the same for different plasma compositions and for a wide range of pressure hypotheses.

5.4.4 Modified Korteweg-de Vries-type expansions

I now address plasmas where the composition is special enough so that v annuls both the dispersion law (5.56) and $B(v)$. This will certainly occur for plasmas with sufficient symmetry, as in pure electron-positron plasmas, or else when positive and negative ions are present in well determined, critical densities. In that case I need to continue with both expressions (5.55) and (5.57), and go one order higher in the expansions. When all this is substituted in Poisson's equation (5.35) to order $\varepsilon^{3/2}$, I obtain a modified Korteweg-de Vries (mKdV) equation

$$A(v)\frac{\partial \phi_1}{\partial \tau} + C(v)\phi_1^2 \frac{\partial \phi_1}{\partial \xi} + \frac{1}{2}\frac{\partial^3 \phi_1}{\partial \xi^3} = 0, \tag{5.65}$$

with yet another coefficient $C(v)$ given by

$$C(v) = \sum_\alpha \frac{\omega_{p\alpha}^2 q_\alpha^2 \mathcal{P}(v)}{4m_\alpha^2[(v - U_\alpha)^2 - c_{s\alpha}^2]^5}. \tag{5.66}$$

The polynomial

$$\mathcal{P}(v) = 15(v - U_\alpha)^4 - [4(v - U_\alpha)^2 - c_{s\alpha}^2][c_{s\alpha}^2 - 4d_{s\alpha}^2]$$
$$+ [(v - U_\alpha)^2 - c_{s\alpha}^2]e_{s\alpha}^2 + 3d_{s\alpha}^4 \tag{5.67}$$

contains the third derivatives of the pressure, formally defined through $m_\alpha e_{s\alpha}^2 = N_\alpha^2[d^3 p_\alpha/dn_\alpha^3]_{N_\alpha}$. The single-soliton solution of the mKdV equation (5.65) is

$$\phi_1(\xi, \tau) = \pm\sqrt{\frac{6MA(v)}{C(v)}}\operatorname{sech}\sqrt{2MA(v)}\,(\xi - M\tau). \tag{5.68}$$

Here $MA(v) > 0$ and $C(v) > 0$ are needed, and in mKdV structures, there is always symmetry between rarefactive and compressive solitons.

5.4.5 Higher-order expansions

In the extremely unlikely event that both $B(v)$ and $C(v)$ would vanish together for solutions v of the dispersion law (5.56), one would need to go a
step higher. This would lead to a higher-order KdV equation, which is no
longer integrable, as the KdV and mKdV equations are the only integrable
nonlinear equations of this class [Ablowitz and Clarkson 1991]. However,
for certain plasmas it can be shown generally that if $B(v) = 0$ for specific
density combinations, and using other constraints on the parameters (like
charge and current neutrality in equilibrium and the dispersion law), then
$C(v) > 0$ for these cases, and the higher-order equation has no meaning.
Specifically, Verheest [1988] has shown that in plasmas containing Boltzmann electrons and an arbitrary number of adiabatic ion species, although
one could for specific concentrations of negative ions make $B(v)$ vanish, it
then followed that $C(v) > 0$.

This general proof was completely overlooked by Das *et al.* [1997] in
their recent paper dealing with the Kadomtsev-Petviashvili (KP) equation,

$$\frac{\partial}{\partial \xi}\left[A(v)\frac{\partial \phi_2}{\partial \tau} + B(v)\phi_2\frac{\partial \phi_2}{\partial \xi} + \frac{1}{2}\frac{\partial^3 \phi_2}{\partial \xi^3}\right] + F(v)\left[\frac{\partial^2 \phi_2}{\partial \eta^2} + \frac{\partial^2 \phi_2}{\partial \vartheta^2}\right] = 0, \quad (5.69)$$

and higher order variants thereof, in a two-ion plasma with Boltzmann
electrons and cold ions. In these equations, η and ϑ are the coordinates
orthogonal to ξ. Although the KP and the mKP equation are more involved, the coefficients of the nonlinear terms are the same as those of the
corresponding KdV or mKdV equations, and hence the discussion about
the vanishing of these coefficients can literally be taken over.

Now Das *et al.* [1997] dreamt up higher and higher order KP equations, with the coefficients of the nonlinear terms vanishing successively.
These new evolution equations may be mathematically interesting enough
in themselves, but are without any relevance for plasma physics within
the limits of their own model! As long as one has isothermal or adiabatic
pressure laws, one cannot make both $B(v)$ and $C(v)$ vanish together for
a given plasma composition, no matter how many ion species are present.
And of course, as soon as $B(v) \neq 0$, one has the KdV equation and the
mKdV equation is then not a valid description of electrostatic modes in
such plasmas.

Similar remarks hold for the treatment by Tagare [1997] of solitons and
double layers in dusty plasmas with one cold dust species in the presence
of two-temperature Boltzmann distributed ions, all electrons having been
accreted on the dust grains. This is fully equivalent to the standard case
of ion-acoustic solitons or double layers in ordinary plasmas with cold ions
and two-temperature Boltzmann electrons, and thus cannot lead to qualitatively different nonlinear structures, being a mere translation exercise.

Of course, the signs of the charges are inverted, so that their compressive and rarefactive nature is interchanged, and there will be differences in the values of the charge-to-mass ratios, but these only affect the numerics.

One of the outstanding conclusions is that the behaviour of nonlinear electrostatic waves is governed by the KdV or mKdV equations, as soon as the pressures are purely a function of the corresponding densities. Since both these evolution equations are integrable, their solutions are expressed as well-behaved solitons, and in particular obey the interaction properties of the N-soliton solutions [Ablowitz and Clarkson 1991].

5.4.6   Expansion of the Sagdeev potential

There are now (at least) two methods to deal with the same physical problem, the Sagdeev and the reductive perturbation approaches. How are these related? Suppose that the problem is amenable to the Sagdeev potential method. Expansion of $n_\alpha(\phi)$ and $p_\alpha(\phi)$ in powers of ϕ, say

$$
\begin{aligned}
n_\alpha(\phi) &= N_\alpha + \nu_{\alpha 1}\phi + \nu_{\alpha 2}\phi^2 + \ldots, \\
p_\alpha(\phi) &= P_\alpha + \pi_{\alpha 1}\phi + \pi_{\alpha 2}\phi^2 + \ldots,
\end{aligned} \tag{5.70}
$$

and substitution of these in (5.34) and (5.38) allows the determination of the coefficients $\nu_{\alpha j}$ and $\pi_{\alpha j}$ to any desired order. These can in turn be used in the expansion of the Sagdeev potential (5.46), which eventually gives from (5.45) that

$$
\frac{1}{2}\left(\frac{d\phi}{d\zeta}\right)^2 + \frac{1}{2}D(V)\phi^2 + \frac{1}{3}B(V)\phi^3 + \frac{1}{6}C(V)\phi^4 + \ldots = 0. \tag{5.71}
$$

Lo and behold, the coefficients $D(V)$, $B(V)$ and $C(V)$ can immediately be obtained from the expressions in (5.56), (5.59) and (5.66), respectively, by replacing everywhere the phase velocity v of the linear modes in the system by the velocity V of the nonlinear structure. Of course, $D(V) \neq 0$, even though $D(v) = 0$. Taking the derivative of (5.71) twice with respect to ζ undoes the integrations performed to arrive at the formulation of the Sagdeev potential, and gives

$$
\frac{1}{2}D(V)\frac{\partial\phi}{\partial\zeta} + B(V)\phi\frac{\partial\phi}{\partial\zeta} + C(V)\phi^2\frac{\partial\phi}{\partial\zeta} + \frac{1}{2}\frac{\partial^3\phi}{\partial\zeta^3} = 0, \tag{5.72}
$$

which is very close to either the KdV equation (5.60) or the mKdV equation (5.65), according to which of the two terms with coefficients $B(V)$ and $C(V)$ one keeps. This general equation is obtained from the expansion of the Sagdeev potential up to 4th order. Similar results originate from the reductive perturbation method, if $B(v)\phi_1^2$ is taken, not strictly zero, but small, of the order of $C(v)\phi_1^3$.

Soliton solutions to (5.71) are typically of the form [Rao *et al.* 1990; Verheest 1992b]

$$\phi_{\mp}(\zeta) = \frac{3|D(V)|\text{sech}^2(\kappa\zeta)}{a_{\pm} - a_{\mp}\tanh^2(\kappa\zeta)}, \tag{5.73}$$

where

$$\kappa = \frac{\sqrt{|D(V)|}}{2},$$

$$a_{\pm} = B(V) \pm \sqrt{B(V)^2 - 3C(V)D(V)}, \tag{5.74}$$

and $D(V) < 0$. In the case of ion-acoustic modes this condition leads to solitary modes with supersonic velocities.

Before going on, I want to briefly discuss double layers, starting from (5.71). The general conditions (5.49) on the location of successive double roots for the potential require that

$$\frac{1}{2}D(V)\phi^2 + \frac{1}{3}B(V)\phi^3 + \frac{1}{6}C(V)\phi^4 = \frac{1}{6}C(V)\phi^2(\phi - \phi_m)^2. \tag{5.75}$$

One sees immediately that $C(V) < 0$, but in addition stringent relations between the coefficients must be obeyed, because

$$\phi_m = -\frac{B(V)}{C(V)}, \qquad B(V)^2 = 3C(V)D(V). \tag{5.76}$$

Not only should ϕ_m be small enough for the truncation of the expansion to order four to make sense, but the relations between the coefficients are usually next to impossible to fulfill [Baboolal *et al.* 1988; Verheest 1993]. This is seen more clearly if I rewrite the double-layer requirements as

$$B(V) = -\frac{3D(V)}{\phi_m}, \qquad C(V) = \frac{3D(V)}{\phi_m^2}. \tag{5.77}$$

The coefficients in the expansion tend to increase rapidly if ϕ_m is small, unless $D(V)$ were to be very small indeed. So in general, the expansion will converge too slowly in order to describe even weak double layers, as one would then neglect higher order terms comparable to the ones retained [Verheest and Hellberg 1997]. Similar remarks hold for mixed KdV equations with quadratic and cubic nonlinearities occurring together.

At the end of this section, the only term needing further attention is the first one in (5.72), because it has to be related to the time derivative present in (5.60) or (5.65). The idea of a slow time change inherent in the reductive perturbation approach means in a co-moving frame that

$$\zeta = \xi - M\tau = \varepsilon^{1/2}[z - (v + \varepsilon M)t]. \tag{5.78}$$

Both descriptions, the expansion of the Sagdeev potential or the reductive perturbation technique, can be reconciled by formally putting $\varepsilon = 1$ and $V = v + M$, and by using ϕ and $M = \mathcal{O}(\phi)$ or $M = \mathcal{O}(\phi^2)$ as expansions parameters. The derivatives in (5.60) or in (5.65) in the co-moving frame thus become

$$\frac{\partial}{\partial \xi} = \frac{\partial}{\partial \zeta}, \qquad A(v)\frac{\partial}{\partial \tau} = -MA(v)\frac{\partial}{\partial \zeta}. \qquad (5.79)$$

The latter is to be compared with the first term in (5.72). It is now rather straightforward to prove that

$$\frac{1}{2}D(V) = \frac{1}{2}D(v + M) \simeq -MA(v), \qquad (5.80)$$

up to first order in M, since $D(v) = 0$. Again I recover the picture of solitary modes with velocities that are very close to, but larger than the linear phase velocity, slightly supersonic in the case of ion-acoustic modes. For the coefficients $B(V)$ and $C(V)$ which are multiplied by higher powers of ϕ, I simply replace V to lowest order by v and thus obtain the desired connection between the two most commonly used theoretical descriptions of nonlinear electrostatic modes.

5.5 Nonlinear dusty plasma modes

5.5.1 Solitons

If I now adapt the general reasoning leading to the KdV equation to dusty plasmas, the simplest is to assume that both electrons and ions are isothermal and Boltzmann distributed, so that $c^2_{se,si} = \kappa T_{e,i}/m_{e,i}$ and $d_{se,si} = 0$. The dust grains will be treated as adiabatic ($p_d \sim n_d^\gamma$), and hence $d^2_{sd} = (\gamma - 1)c^2_{sd}$. The linear dispersion law then yields

$$v^2 = c^2_{da} + c^2_{sd} \qquad (5.81)$$

and shows the corrections to the dust-acoustic velocity due to some form of dust thermal effects. The coefficients of the KdV equation (5.60) are here

$$A(v) = \frac{\omega_{pd}^2\sqrt{c^2_{da} + c^2_{sd}}}{c^4_{da}},$$

$$B(v) = \frac{e}{2\kappa T_i\lambda^2_{Di}} - \frac{e}{2\kappa T_e\lambda^2_{De}} - \frac{\omega_{pd}^2 Z_d e[3c^2_{da} + (\gamma + 1)c^2_{sd}]}{2m_d c^6_{da}}. \qquad (5.82)$$

In different guises these results have been found by many authors. To be precise, in their seminal paper Rao *et al.* [1990] gave the nonlinear development of their novel dust-acoustic mode through a Boussinesq equation, of the generic form

$$\frac{A(v)}{c_{da}}\left(\frac{\partial}{\partial\tau}-c_{da}\frac{\partial}{\partial\xi}\right)\left(\frac{\partial}{\partial\tau}+c_{da}\frac{\partial}{\partial\xi}\right)\phi-\frac{\partial^4\phi}{\partial\xi^4}$$

$$-B(v)\frac{\partial^2}{\partial\xi^2}(\phi^2)-\frac{2C(v)}{3}\frac{\partial^2}{\partial\xi^2}(\phi^3)=0,\qquad(5.83)$$

obtained by heuristic methods. Next, unidirectional, slightly supersonic propagation is considered by putting $\partial/\partial\tau=-c_{da}\partial/\partial\xi$ and integrating once in ξ. In this way the generalized Korteweg-de Vries equation, with both quadratic and cubic nonlinearities is recovered,

$$A(v)\frac{\partial\phi}{\partial\tau}+B(v)\phi\frac{\partial\phi}{\partial\xi}+C(v)\phi^2\frac{\partial\phi}{\partial\xi}+\frac{1}{2}\frac{\partial^3\phi}{\partial\xi^3}=0.\qquad(5.84)$$

The solutions are compressive and rarefactive solitons, as discussed already. The rarefactive, potential-dip solutions are unique to the influence of the dust, as ordinary ion-acoustic waves propagate as localized potential humps only.

In an extension of the previous analysis, Verheest [1992b] considered a number of different cold dust fluids, in the presence of hot and cold electrons and of ions, all three described by Boltzmann distributions, much as done by Rao *et al.* [1990], except that they only considered one electron species and one ion species. Whereas rarefactive solutions can readily be found, compressive ones are not likely to exist within the limits imposed by the weakly nonlinear expansion techniques used. This shows once again that great care must be exercised when using perturbation techniques.

Bharuthram and Shukla [1992c] investigate large amplitude ion-acoustic solitons, by treating the dust in two different ways, once as a fixed background, and once fluid-like. Numerical criteria for finite solitons lead to maximum Mach numbers for the soliton speeds. Both amplitudes and speeds of compressive solitons are much larger than in the absence of dust. Incorporating the dynamics of the dust has a negligible effect on the compressive solitons, whereas the rarefactive soliton is very significantly affected. In addition, rarefactive solitons have a comparatively larger amplitude still, as indicated in Figure 5.7.

The nonlinear propagation of ion-acoustic modes is investigated by Mofiz *et al.* [1993], when the intergrain distance is large compared to the plasma Debye length. The plasma perturbations leave the dust essentially as it is, thus avoiding the complications due to dust charge variations. An interesting conclusion is that ion-acoustic solitons are found for small grain

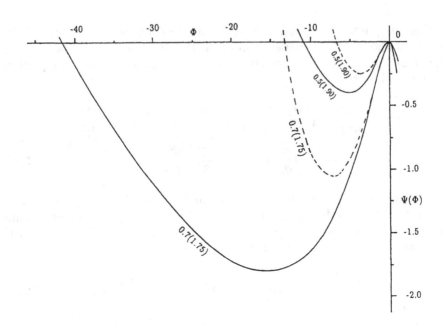

Figure 5.7. Sagdeev potential for rarefactive solitons. The curves are labeled by the normalized electron density N_e/N_0, with the corresponding Mach number M shown in parentheses. The continuous (broken) curves are for stationary (moving dust) particles. *Reprinted from Bharuthram and Shukla 1992c (Fig. 4 on p. 976), with permission of Elsevier Science*

charges, but increasing the grain charges render these solitons more unlikely and at large charges no solitons are possible.

Lakshmi and Bharuthram [1994] return to the problem of large amplitude, rarefactive dust-acoustic solitons in a plasma with Boltzmann distributed electrons, ion species at different temperatures and dust grains with constant charges. The main difference with the previous treatment by Bharuthram and Shukla [1992c] is in the description of the ions, which are treated as isothermal and inertialess, as was done already by Bharuthram and Shukla [1992a] in their description of double layers, showing no really qualitative differences with earlier results.

Lakshmi *et al.* [1997] improve on previous work by comparing a kinetic and a fluid treatment. Differences occur in the precise expressions for the coefficients of the KdV equation, having mostly to do with the temperature ratios T_d/T_e and T_i/T_e. No qualitative changes in the structure of the KdV equation itself crop up. This should not surprise us, because I tried to show how general and robust the KdV equation is.

On the other hand, Ma and Liu [1997] have treated the problem via the Sagdeev potential approach, as outlined in section 5.4.1. Although they purport to include dust charge fluctuations, the timescales considered are such that the latter are not important. The soliton behaviour then follows the expected rules governing the relation between amplitude, speed and width, verified numerically. Similar conclusions follow from the investigations of Singh and Rao [1997] who used an adiabatic equation of state for the charged dust. As is obvious from the derivation of the KdV equation, the pressure closures modify the coefficients in this equation, but cannot introduce qualitative departures from the generic structures. Here Singh and Rao find that the soliton amplitudes decrease with increasing the adiabatic exponent γ.

5.5.2   Double layers

The interest in double layers and vortices is that the former could play a large role in the acceleration of particles, much as occurs in the Earth's auroral regions where double layers are believed to be responsible for the observed energetic electrons and ions [Raadu 1989]. Furthermore, models for solar flares have been advanced in which the triggering mechanism is a double layer related phenomenon. Hence the question whether the intense dust jets of comet 1P/Halley could be explained this way. Another question is if dust gets trapped in double layers or better still in vortices and possibly transported over large distances.

Large-amplitude electrostatic double layers are treated by Bharuthram and Shukla [1992a]. As in many such papers, the equations have been written in the appropriate nondimensional units, relating to the dust, which makes quantitative comparisons between papers uneasy. Criteria for the existence of finite double layers are obtained numerically, as typified in Figure 5.8. As an extreme case, one could consider that most of the electrons have been accreted onto the dust grains, believed to be a acceptable first approximation for the F ring of Saturn. In that case, only compressive double layers are possible, supported by the ion non-isothermality, the dust providing the necessary inertia.

At the same time, Mace and Hellberg [1993] discussed the effects of ion inertia on the existence of dust-acoustic double layers. All species of the plasma are described as fluids, instead of having the lighter species Boltzmann distributed and effectively inertialess. The regimes in which dust-acoustic double layers can occur are found to be much more restricted than with Boltzmann distributions. Significantly, highly nonlinear double layers are ruled out, placing rather strict restrictions on the neglect of ion inertia. Interestingly, the strongest double layer profiles can be expected when most of the electrons have been absorbed by the dust grains. This disagrees with conclusions drawn by Bharuthram and Shukla [1992a] for a similar problem.

Verheest [1993] then tried to find out how well weak dust-acoustic dou-

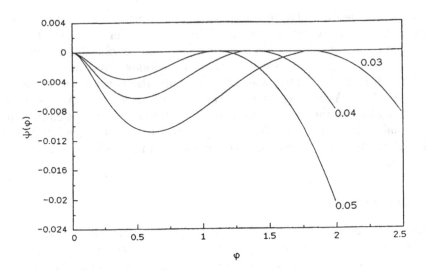

Figure 5.8. Typical form for the Sagdeev potential for $N_e = 0$ and $N_{i,cool}/N_{i,hot} = 0.11$. The parameter labeling the curves is the ratio of the cool to hot ion temperatures. *Reprinted from Bharuthram and Shukla 1992a (Fig. 1 on p. 468, with permission of Elsevier Science*

ble layers are described by modified Korteweg-de Vries equations, and more in particular whether the existence ranges correspond to what can be inferred from exact solutions. Computations were based on parameter values chosen, not so much on (astro)physical grounds, but so as to yield weak double layers if one numerically treats the full Sagdeev potential. Some of these weak double layers are not even recovered from the more approximate approaches, which give at best only order-of-estimate agreements. But one could be tempted to think that, if one found weak double layers from the expanded Sagdeev potential or from the mixed KdV equation and the value for the double-layer amplitude was thought to be consistent with the assumed smallness, then at least one would have a right indication. This is, however, also not always true.

Here the mixed KdV approach would let one assume that weak double layers existed, with an amplitude which was small enough to be considered fitting in the reductive perturbation framework. The full Sagdeev approach, alas, gives for the same parameter regimes double layers which have normalized amplitudes of order unity, too large to be called weak! Hence there are not only quantitative but, rather more important, also qualitative differences between the exact Sagdeev potential and the approximate methods, and the latter would predict the existence of weak double layers which could not be recovered from the exact Sagdeev potential as being weak. That on the other hand the approximate solutions would miss some of the truly weak double layers is less dangerous. Thus only very, very weak double layers might validly be described by the more approximate methods.

5.6 Oblique propagation

5.6.1 Upper and lower hybrid modes

I had already given the general dispersion law for electrostatic modes at oblique propagation in (4.81). Leaving the self-gravitational effects to Chapter 8, there remains to be discussed

$$\sum_\alpha \frac{\omega_{p\alpha}^2(\omega^2 - \Omega_\alpha^2 \cos^2\vartheta)}{\omega^2(\omega^2 - k^2 c_{s\alpha}^2 - \Omega_\alpha^2) + k^2 c_{s\alpha}^2 \Omega_\alpha^2 \cos^2\vartheta} = 1 \qquad (5.85)$$

analogous to (4.72) but more complicated.

Let me begin with the special case of strictly perpendicular propagation, when $\cos\vartheta = 0$. In a standard electron-ion plasma (5.85) becomes

$$\frac{\omega_{pe}^2}{\omega^2 - k^2 c_{se}^2 - \Omega_e^2} + \frac{\omega_{pi}^2}{\omega^2 - k^2 c_{si}^2 - \Omega_i^2} = 1. \qquad (5.86)$$

Without ion dynamics, when the latter form a neutralizing background, this has a high frequency limit

$$\omega^2 = \omega_{pe}^2 + \Omega_e^2 + k^2 c_{se}^2, \qquad (5.87)$$

leading to the upper hybrid mode. In the case of oblique propagation, this is modified to

$$\omega^2 \simeq \omega_{pe}^2 + \Omega_e^2 \sin^2\vartheta + k^2 c_{se}^2, \qquad (5.88)$$

under the standard assumption that $\Omega_e^2 \ll \omega_{pe}^2$ and for small k.

For lower frequency modes I need to include the ion dynamics as well. Once we understand how that works, I can generalize these modes to dusty plasmas, by going to the lower frequencies typical for the dust components. Again starting at strictly perpendicular propagation, I get for lower frequencies ($\Omega_i \ll \omega \ll |\Omega_e|$) that

$$\omega^2 \simeq |\Omega_e|\Omega_i + k^2 c_{ia}^2, \qquad (5.89)$$

giving now the lower hybrid mode, including an ion-acoustic velocity defined here through $c_{ia}^2 = c_{si}^2 + m_e c_{se}^2/m_i$, and $m_e \ll m_i$ has been used to simplify the description.

At really oblique propagation things become more tricky, because terms like $\Omega_i^2 \sin^2 \vartheta$ have to be compared with respect to $\Omega_e^2 \cos^2 \vartheta$, close to perpendicular propagation. Hence for the lowest frequency mode

$$\omega^2 \simeq \frac{\Omega_i^2 \cos^2 \vartheta}{\cos^2 \vartheta + m_e/m_i} + k^2 c_{ia}^2 \sin^2 \vartheta, \qquad (5.90)$$

provided $\cos^2 \vartheta \sim m_e/m_i$ or larger. Otherwise we are back to the lower hybrid mode with dispersion law (5.89). If $\cos^2 \vartheta \gg m_e/m_i$ then (5.90) starts as Ω_i^2 in the limit $k \to 0$.

5.6.2 Dust hybrid modes

In the light of what has been recalled in the previous section, I now bring charged dust into the description. Of course, at the higher frequencies the presence of the dust induces small differences, like in the treatment by Salimullah *et al.* [1992]. It is found that the combined presence of dust and magnetic field has a significant effect for perpendicular wave propagation, leading to an electrostatic ion Bernstein mode in a dusty plasma.

One can also include drifts between the species and obtain the related instabilities. Chow and Rosenberg [1995] thus investigate the electrostatic ion cyclotron instability, and show that the critical drift, needed to excite the instability, decreases as the negative dust charge density increases. As an example, the current system that connects Jupiter with its satellite Io is given. For further details of this system I refer *e.g.* to Grün *et al.* [1993,1998]. In the absence of dust measurements at Io, the detection of the electrostatic ion cyclotron instability would signal the presence of negative dust, as without it the mode would be stable under the conditions prevailing near Io. Afterwards Chow and Rosenberg [1996] came back to the same problem and compared theoretical and numerical analysis with experimental evidence, obtained by Barkan *et al.* [1995b]. The theoretical understanding agreed with experimental findings.

Whereas Rosenberg and Krall [1995] have studied a narrow cone of angles around perpendicular propagation ($k_\parallel \ll k_\perp$) and thus only studied the dust-modified two-stream instability, Bharuthram [1997] investigates much wider angles of propagation, so that also the dust-acoustic instability can be excited. A detailed numerical study is presented of the real frequencies and growth rates for both type of instabilities, from which it transpires that the transition from one type to the other can occur over a wide range of frequencies, depending on parameters. In particular, parameters were chosen as representative for the conditions in the E or G rings of Saturn. Growth rates increase with the particle drift speeds, as there is then more free energy available. A variation of the free electron density shows that the dust-acoustic instability is more easily excited in plasmas with fewer electrons, as in protostellar clouds. On the contrary, the growth rate of the dust-modified two-stream instability increases with the electron density, leaving less charges on the dust grains. In a similar vein, Amin [1996] studied quasi-perpendicular modes where the main modifications come from the difference between the electron and ion densities.

Once the simple modifications due to these density differences to various lower hybrid and electrostatic ion cyclotron modes had been studied, new lower frequency regimes were explored. Expliciting (5.85) for strictly perpendicular propagation in a three component dusty plasma as

$$\frac{\omega_{pe}^2}{\omega^2 - k^2 c_{se}^2 - \Omega_e^2} + \frac{\omega_{pi}^2}{\omega^2 - k^2 c_{si}^2 - \Omega_i^2} + \frac{\omega_{pd}^2}{\omega^2 - k^2 c_{sd}^2 - \Omega_d^2} = 1, \qquad (5.91)$$

I obtain for low frequencies (defined as $|\Omega_d| \ll \omega \ll \Omega_i \ll |\Omega_e|$) that

$$\omega^2 \simeq \frac{N_d Z_d}{N_i} |\Omega_d| \Omega_i + k^2 c_{da}^2. \qquad (5.92)$$

This is a new dust lower hybrid mode, including a suitably defined dust-acoustic velocity c_{da}. Also, the ordering $m_e \ll m_i \ll m_d$ has been used. The obvious name for the typical frequency thus obtained is dust lower hybrid, coined by Salimullah [1996], and defined through

$$\omega_{dlh}^2 = \frac{N_d Z_d |\Omega_d| \Omega_i}{N_i} = \frac{\omega_{pd}^2 V_{Ai}^2}{c^2}. \qquad (5.93)$$

The second way of writing ω_{dlh}^2 involves both the dust plasma frequency and the Alfvén velocity V_{Ai} in the (positive) ion fluid.

Again, at really oblique propagation things become even more tricky, because terms like $\Omega_d^2 \sin^2 \vartheta$ have to be compared to $\Omega_e^2 \cos^2 \vartheta$ and $\Omega_i^2 \cos^2 \vartheta$. Now I find for the lowest frequency mode that

$$\omega^2 \simeq \Omega_d^2 + k^2 c_{da}^2 \sin^2 \vartheta, \qquad (5.94)$$

in the limit $k \to 0$ and provided $\cos^2 \vartheta$ is not too small compared to the mass ratios m_e/m_d and m_i/m_d [D'Angelo 1990]. Recently, D'Angelo [1998] has extended the above ideas to current-driven electrostatic dust cyclotron instabilities in a collisional plasma, and applied them to laboratory studies and to certain wave modes in cometary tails. As the solar wind magnetic field lines are bent around the cometary nucleus, a magnetic field pile-up is produced together with rapid field-aligned flows. If the ion flow relative to the charged dust is rapid enough, the drift electrostatic dust instability is possible, and the critical velocities have to be 4 km/s or larger, which is of the order of the ion thermal velocities and certainly well below the solar wind speed.

5.6.3 Nonlinear developments

Nonlinear developments lead in general for oblique electrostatic modes to KdV equations, which I will derive in a slightly more general description in the next chapter, when electromagnetic waves are discussed. Examples of that equation are given by Kotsarenko *et al.* [1998b] for oblique dust-acoustic modes. Their results are applied to the creation of condensates in cometary tails and to Saturn's rings, where it provides an alternative explanation for spoke formation. Whereas in the model of Bliokh and Yaroshenko [1985] the spokes are interpreted through the excitation of space charge waves, Kotsarenko *et al.* [1998b] attribute the spokes to cylindrical solitons in an ion-dust plasma, where most of the electrons have been accreted onto the dust grains. Interestingly enough, both models give spatial sizes of the order of 1000 km, close to the observed sizes. Of course, plasma data are not so well constrained by observations yet, and thus a little manipulation or interpretation is always possible. The models differ in the lifetimes of the spoke structures, though, where Bliokh and Yaroshenko [1985] have a 1000 s or fractions of an hour, whereas the model of Kotsarenko *et al.* [1998b] predicts 10^5 s or a few days. Hence the debate is still open to what constitutes the most appropriate way to describe the spoke formation. In a similar vein, condensates in cometary tails are simulated with a spatial extent of the right order, also around 1000 km.

Further studies into off-parallel dust-acoustic solitons were given by Malik and Bharuthram [1998] for oblique propagation and by Malik *et al.* [1998] for strictly perpendicular propagation. The model is that of a two-ion plasma in the presence of electrons and negatively charged grains. This is the obvious extension of the classic ion-acoustic case, where the presence of two electron species at widely different temperatures not only permits the existence of rarefactive and compressive solitons, but could also lead to double layers in a suitably chosen parameter regime.

5.7 Collisional and kinetic effects

Most of what I detailed in the previous sections was written for a collisionless plasma, in a fluid description. The motivation was that I could thus point out the salient features of novel effects due to the presence of charged dust. However, many dusty plasmas contain a significant pressure of neutrals, so that the mean ion-neutral or dust-neutral collisional mean free paths become comparable to the typical wavelengths considered. Then the picture has to be refined, by either including explicitly some collisional effects through the appropriate drag forces, or even by going to a kinetic description where needed.

Rather than trying to give these modifications in detail, I refer to some recent papers dealing with such aspects. Kaw and Singh [1997] find a short wavelength branch to the dispersion law driven by ion drift and collisional ion-neutral momentum transfer. Nonlinearly this instability leads to the formation of KdV solitons. There is also a long wavelength mode, particularly relevant for high neutral pressure situations, driven by recombination effects on the surface of the dust particles. This instability nonlinearly favours the formation of envelope solitons in one dimensional descriptions, and to collapsing wave packets in more than one dimension. Of course, since the charge fluctuations will also involve damping or growth mechanisms, as will be seen in Chapter 7, there is a need to reconsider this and many related collisional problems by self-consistently including variable dust charges.

Related results were obtained by Rosenberg and Chow [1999] for different collisional regimes of the electrostatic dust cyclotron instability. While the dust-neutral collisions have a stabilizing influence, the ion-neutral collisions are destabilizing. This kinetic theory treatment corroborates the fluid description given by D'Angelo [1998].

A variation on that theme is the numerical simulation by Winske and Rosenberg [1998], where the presence of drifting neutrals maintains the initial relative drift between plasma and charged dust, until the unstable waves grow to large amplitudes and wave-particle interactions exceed the neutral collisions. As a result, stronger nonlinear effects, as manifested by enhanced fluctuations, larger amounts of plasma and dust heating, and a temporary reduction of the relative drift speed, can occur. Applications include dusty cometary atmospheres, specifically the region between the cometary nucleus and the ionopause, or noctilucent or polar mesospheric cloud regions at altitudes around 90 km.

CHAPTER 6

ELECTROMAGNETIC MODES

6.1 Parallel electromagnetic waves

We will, as I did in the previous chapter for electrostatic modes, first briefly recall some of the typical electromagnetic waves in ordinary plasmas, before embarking upon the modifications induced by the presence of charged dust. In the following sections, I start with parallel propagation and go then afterwards to oblique and perpendicular modes. In each case, linear modes will be reviewed before going to the nonlinear developments. For electromagnetic modes in magnetized plasmas, only the reductive perturbation scheme is indicated.

6.1.1 Dispersion law, resonances and cut-off frequencies

For wave propagation parallel to the external magnetic field $\mathbf{B}_0$ the dispersion law was derived in Chapter 4 as (4.67), or

$$\omega^2 = c^2 k^2 + \sum_\alpha \frac{\omega_{p\alpha}^2 (\omega - kU_\alpha)}{\omega - kU_\alpha \pm \Omega_\alpha}. \tag{6.1}$$

Since in the denominator the plus and minus signs refer to right and left hand circularly polarized (RHCP, LHCP) modes, respectively, these are called the R and the L modes in short.

Leaving out for the time being all equilibrium streaming, the dispersion law for the R mode in an electron-ion plasma is explicitly

$$\omega^2 = c^2 k^2 + \frac{\omega_{pe}^2 \omega}{\omega - |\Omega_e|} + \frac{\omega_{pi}^2 \omega}{\omega + \Omega_i}, \tag{6.2}$$

where it is remembered that the gyrofrequency Ω_e for the electrons has been defined with the sign of the charge included, as is also the case for negative ions or charged dust later on.

When studying these dispersion laws, typical frequencies occur whenever the refractive index $n = ck/\omega$ vanishes or becomes infinitely large. Cut-off frequencies correspond to $ck/\omega = 0$, at fixed ω for $k = 0$, whereas $ck/\omega \to +\infty$ gives the resonances, again at given ω when $k \to +\infty$. I can restrict myself to positive values for ω without loss of generality, because

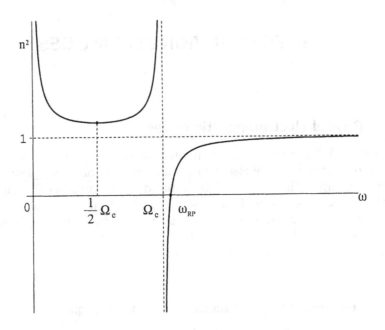

Figure 6.1. Typical dispersion curve giving n^2 as function of ω

negative frequencies mean that the notions of forward and backward propagating modes are interchanged, as are then also their senses of polarization. It is immediately seen that $|\Omega_e|$ is a resonance frequency, while

$$\omega_{RP} \simeq \frac{1}{2}\left(|\Omega_e| + \sqrt{\Omega_e^2 + 4\omega_{pe}^2}\right) \tag{6.3}$$

is the cut-off frequency. Because of the small electron-to-ion mass ratio, I have assumed that $\Omega_i \ll |\Omega_e|$ and $\omega_{pi} \ll \omega_{pe}$.

In a similar vein, the L mode with dispersion law

$$\omega^2 = c^2 k^2 + \frac{\omega_{pe}^2 \omega}{\omega + |\Omega_e|} + \frac{\omega_{pi}^2 \omega}{\omega - \Omega_i} \tag{6.4}$$

has a (positive) resonance in Ω_i and a cut-off frequency in

$$\omega_{LP} \simeq \frac{1}{2}\left(-|\Omega_e| + \sqrt{\Omega_e^2 + 4\omega_{pe}^2}\right). \tag{6.5}$$

However, in a two-ion plasma the L mode dispersion becomes

$$\omega^2 = c^2 k^2 + \frac{\omega_{pe}^2 \omega}{\omega + |\Omega_e|} + \frac{\omega_{p1}^2 \omega}{\omega - \Omega_1} + \frac{\omega_{p2}^2 \omega}{\omega - \Omega_2}, \tag{6.6}$$

where the subscripts 1 and 2 refer to the two ion species. There are now two resonances, in Ω_1 and Ω_2, and a new cut-off frequency appears in between the two at

$$\omega_{LR} \simeq x_1 \Omega_2 + x_2 \Omega_1. \tag{6.7}$$

Here $x_i = N_i q_i / N_e e$ (for $i = 1, 2$) are the relative ion charge densities, normalized with respect to the electron charge density, so that $x_1 + x_2 = 1$. There are corresponding changes when one of the ion species is negatively charged, and in addition, the above reasoning can easily be extended to plasmas with even more ion species.

When discussing (6.2) for various frequency regimes, I start at the highest frequencies, defined by $|\Omega_{e,i}| \ll \omega$, and see that both species are essentially unmagnetized and the dispersion law reduces to

$$\omega^2 \simeq c^2 k^2 + \sum_\alpha \omega_{p\alpha}^2. \tag{6.8}$$

This describes electromagnetic waves, modified because of their propagation through a plasma. I have written the result for a multispecies description, but the global plasma frequency is usually determined by the electrons. This might change when dusty plasmas are considered, depending on the level of electron depletion, as I discussed already when dealing with the respective plasma frequencies and their influences on electrostatic modes.

6.1.2 Whistler waves

When I now go to lower frequencies, it is the resonance frequencies that delineate the different regimes. Intermediate frequencies can be defined by $\Omega_i \ll \omega \ll |\Omega_e|$, so that the electrons are magnetized but the ions are not. This yields the whistler branch of the modes, with dispersion law

$$\frac{\omega}{|\Omega_e|} \simeq \frac{c^2 k^2}{\omega_{pe}^2} \tag{6.9}$$

in the Eckersley approximation [Booker 1984]. Including more traditional ion species does not really modify this result. However, as we shall see, the occurrence of charged dust opens up a new low-frequency regime below all ion gyrofrequencies and then additional whistler-like modes become possible.

6.1.3 Alfvén waves

Before going to dusty plasmas, I turn to the lowest possible frequencies, below all ion gyrofrequencies. I can easily be general here and work this out in a multispecies plasma description that can later be used also for dusty plasmas. This interesting low-frequency limit to (4.67) is normally defined by taking $\omega \ll |\Omega_\alpha|$, in the absence of streaming effects. However, as equilibrium streaming can generate interesting instabilities, I will define low frequencies in each of the species frame, i.e. all Doppler-shifted frequencies are small compared to the respective gyrofrequencies, $|\omega - kU_\alpha| \ll |\Omega_\alpha|$. This is thus not only a low-frequency, but also a long-wavelength limit.

Expanding (4.67) to second order in those small parameters yields

$$\omega^2 = c^2 k^2 - \sum_\alpha \frac{\omega_{p\alpha}^2 (\omega - kU_\alpha)^2}{\Omega_\alpha^2}. \tag{6.10}$$

The first-order terms

$$\sum_\alpha \frac{\omega_{p\alpha}^2 (\omega - kU_\alpha)}{\Omega_\alpha} = \frac{1}{\varepsilon_0 B_0} \sum_\alpha N_\alpha q_\alpha (\omega - kU_\alpha) = 0 \tag{6.11}$$

vanish because of total charge and current neutrality in equilibrium. Introducing the global Alfvén velocity V_A [Stix 1962] through

$$V_A^2 = B_0^2 \Big/ \mu_0 \sum_\alpha N_\alpha m_\alpha \tag{6.12}$$

allows us to rewrite (6.10) as

$$\omega^2 = c^2 k^2 - \frac{c^2}{V_A^2} \cdot \frac{\sum_\alpha N_\alpha m_\alpha (\omega - kU_\alpha)^2}{\sum_\alpha N_\alpha m_\alpha}. \tag{6.13}$$

It is useful to introduce mass averaged quantities as

$$\overline{f} = \sum_\alpha N_\alpha m_\alpha f_\alpha \Big/ \sum_\alpha N_\alpha m_\alpha \tag{6.14}$$

to finally obtain the dispersion law for Alfvén modes in streaming plasmas as

$$\left(\frac{\omega}{k} - \overline{U} \right)^2 + \frac{V_A^2}{c^2} \left(\frac{\omega}{k} \right)^2 = V_A^2 - \overline{(U - \overline{U})^2}. \tag{6.15}$$

In the absence of any streaming effects these low-frequency modes have phase velocities

$$\left(\frac{\omega}{k} \right)^2 = \frac{c^2 V_A^2}{c^2 + V_A^2} \simeq V_A^2, \tag{6.16}$$

since in most plasmas I can safely take $V_A \ll c$. The Alfvén velocity V_A is essentially determined by the mass density of the ions, and this will have important repercussions when I deal with dusty plasmas, because of the very heavy dust grains. In this limit, one could even treat the plasma in a magnetohydrodynamic (MHD) description [Shkarofsky *et al.* 1966], but this does not allow the proper inclusion of streaming or beam effects, and hence offers little advantage here.

Let us briefly mention the possibility of unstable waves, if there is enough kinetic energy in the streaming between the different plasma species. Such streaming is measured in normalized form by $W = \overline{(U - \overline{U})^2}$, and the instability criterion is

$$V_A^2 < W = \overline{(U - \overline{U})^2}. \qquad (6.17)$$

The mechanism is akin to a firehose instability, where a pressure anisotropy drives the instability, needing a sufficient excess of parallel over perpendicular pressure with respect to the ambient magnetic field. Here the normalized, net streaming energy along the field replaces the pressure anisotropy, and it needs a plasma with at least two ion species in order to be possible at all.

Since the instability is connected to the mass densities of the different plasma components, I consider a plasma where the streaming or beam component has a relative mass density σ, normalized to the total mass density of streaming and non-streaming components. The beam has velocity U in the frame of the stationary component. Then $W = \sigma(1 - \sigma)U^2$, and with the help of the reference Alfvén velocity V_{AR} in the stationary plasma component, the instability criterion simply becomes

$$V_{AR}^2 < \sigma U^2. \qquad (6.18)$$

This can be expressed in terms of the Alfvénic Mach number $M = U/V_{AR}$ of the flowing plasma component.

6.1.4 Cometary turbulence

An application of this nonresonant beam-plasma instability could explain part of the observed low-frequency turbulence in the neighbourhood of cometary nuclei, during the pickup process of cometary ions by the solar wind. Since I will later discuss what the modifications are when also charged cometary dust is present, it is useful to give a more detailed explanation of cometary ion pickup processes.

The concept of the mass loading of the solar wind by ionized molecules, such as water, from cometary atmospheres was introduced by Biermann *et al.* [1967]. This process is accompanied by instabilities, giving rise to MHD turbulence, and such strong turbulence around comets was discovered by satellite observations, in an extended region surrounding the comets (of a size greater than 10^6 km for comet Giacobini-Zinner) [Tsurutani and Smith 1986]. The turbulence is correlated with the presence of energetic heavy ions, which result from the ionization by the solar wind of cometary neutral molecules in the pickup process, and the wave spectrum exhibits a peak near the water group ion gyrofrequency, in the spacecraft (or cometary) frame [Tsurutani and Smith 1986; Thorne and Tsurutani 1987]. The waves making up the observed turbulence are in the magnetosonic (compressional Alfvén) mode [Tsurutani et al. 1987].

The plasma surrounding the comet is thus made up of solar wind ions (predominantly protons), electrons and energetic heavier ions, such as water, of cometary origin. The neutral molecules from the comet all have roughly the same velocity relative to the magnetic field of the solar wind, their outflow velocity from the cometary nucleus being about 1 km/s [Galeev 1991]. The cometary pickup ions, immediately following ionization, have a gyration velocity of $V_{c\perp} \simeq V_{SW} \sin \varphi$ around the magnetic field, and a drift velocity $V_{c\parallel} = V_{SW} \cos \varphi$ along the magnetic field is in the frame of the solar wind. Here V_{SW} is the solar wind velocity and φ the local angle between the solar wind velocity and the magnetic field [Winske et al. 1985; Thorne and Tsurutani 1987]. The ions thus form a combined gyrating ring and a field-aligned beam distribution, which has been shown to excite a number of instabilities [Wu and Davidson 1972; Winske et al. 1985; Sharma and Patel 1986; Lakhina and Verheest 1988; Tsurutani 1991; Brinca et al. 1993; Motschmann and Glassmeier 1993]. In the solar wind frame these cometary ions are viewed as a beam.

Although one could in principle consider both electrostatic and electromagnetic instabilities, the low-frequency electromagnetic fluctuations dominate. The higher-frequency instabilities have a large growth rate but saturate at very low levels of turbulence, and hence are not efficient in destroying the ion beams and in assimilating the beam ions into the main flow [Gary et al. 1984]. The MHD waves predicted to be excited have the largest growth rate for parallel propagation.

There are two distinct types of instability possible for the combined ring and beam distribution. The nonresonant firehose instability is a fluid instability arising from an imbalance in the effective parallel and perpendicular pressures. The resonant instability, on the other hand, is due to an interaction between the cometary ion and the electromagnetic wave that is left-hand polarized in the ion frame of reference, such that the Doppler-shifted wave frequency equals the ion gyrofrequency. The nonresonant instability requires either a large beam velocity or a high concentration of beam ions, conditions which are unlikely to occur except immediately cometward of the cometary bow shock ($\simeq 10^5$ km from comet Giacobini-Zinner) [Tsurutani et al. 1987]. Since waves are observed over a more extended region, the resonant instability is apparently the dominant process. The resonant instability can be for left hand or right hand polarized waves in the solar wind frame of reference, depending on the angle φ.

Nongyrotropic ion distribution functions, such as partially filled rings, are commonly observed, and also give rise to instabilities [Brinca *et al.* 1993; Motschmann and Glassmeier 1993]. The waves excited in the instabilities react back on the velocity distribution of cometary ions, and quasilinear theory has been used to predict the diffusion in velocity space of the cometary ions interacting resonantly with the waves [Galeev 1991]. A shell distribution in velocity space is formed rapidly from the initial ring distribution, within a few cometary ion gyroperiods. The theory of the shell distribution has been used to explain some features of the ion distribution observed at comet Halley by the Giotto spacecraft [Huddleston *et al.* 1993]. Dust grains may also be present in the plasma where the pickup ions are produced, provided some of this production occurs close to the comet in the Sunward direction, or downstream of the comet in the dust tail, but this I will discuss later under the pure applications to dusty plasmas.

When the nonresonant instability occurs, I find from (6.15) that the real part of the frequency and/or the phase velocity is determined as

$$\frac{\mathrm{Re}\,\omega}{k} = \frac{c^2\overline{U}}{c^2+V_A^2} \simeq \overline{U}, \qquad (6.19)$$

and indeed corresponds to the Doppler shift induced by the bulk motion of the plasma as a whole. In a frame co-moving with the combined plasma these are zero-frequency modes, with growth rate

$$\mathrm{Im}\,\omega = k\sqrt{\frac{c^2(W-V_A^2)}{c^2+V_A^2} + \frac{c^2V_A^2\overline{U}^2}{(c^2+V_A^2)^2}} \simeq k\sqrt{W-V_A^2}. \qquad (6.20)$$

I will come back to this in the next section, when modifications due to charged dust are introduced.

6.2 Dusty plasmas

Having recalled some of the electromagnetic wave modes at parallel propagation, I am now ready to discuss the modifications of these modes and new ones in dusty plasmas.

As highlighted in the previous section for ordinary plasmas, the mass is predominantly in the ions, and they determine the Alfvén speed V_A. However, for dusty plasmas the mass density could be much higher, even at small number densities, as the dust is so massive. There will be a consequent lowering of the Alfvén velocity, and at low phase velocities this is the main effect of including charged dust, if variations of the dust charges are not taken into account, but these are deferred to the next chapter.

One of the first to treat the effect of charged dust grains on the propagation and dissipation of Alfvén waves in interstellar clouds were Pilipp *et al.* [1987], as Alfvén waves are the slowest decaying modes, showing no nonlinear steepening. Although their model included collisional coupling between the different fluids, the dust grains were assumed to be either neutral or singly charged, without wave induced fluctuations in the charges themselves. The collisional coupling is of the standard form [Booker 1984], with an interchange term added in the equations of motion

$$\left(\frac{\partial}{\partial t} + u_{\alpha\parallel}\frac{\partial}{\partial z}\right)\mathbf{u}_{\alpha\perp} = \frac{q_\alpha}{m_\alpha}\mathbf{E}_\perp + \frac{q_\alpha}{m_\alpha}\left(\mathbf{u}_\alpha \times \mathbf{B}\right)_\perp + \mathbf{M}_\alpha, \qquad (6.21)$$

of the form

$$\mathbf{M}_\alpha = -\sum_\beta \nu_{\alpha\beta}(\mathbf{u}_{\alpha\perp} - \mathbf{u}_{\beta\perp}). \qquad (6.22)$$

The modes are transverse and so only the perpendicular components of velocities and fields are needed. The two dust populations (charged and neutral) interact through the conversion of ionized grains to neutral ones by recombination of ions on the grain surfaces, or on the contrary, by conversion of neutral to ionized grains by electron impact. The main effect of introducing charged dust is to reduce the minimum wavelengths and increase the maximum frequencies at which the waves remain coupled to the neutrals.

6.2.1 Dust whistler waves

In his general discussion of dust modes, Shukla [1992] also studied low-frequency electromagnetic modes, showing that modes purely driven by the dust occur, as well as modified shear Alfvén waves. To study some of these effects, I write the dispersion law (4.67) for a mixture of the usual multispecies plasma in the presence of one or more charged dust species:

$$\omega^2 = c^2 k^2 + \sum_{\text{plasma}} \frac{\omega_{pp}^2(\omega - kU_p)}{\omega - kU_p \pm \Omega_p} + \sum_{\text{dust}} \frac{\omega_{pd}^2(\omega - kU_d)}{\omega - kU_d \pm \Omega_d}. \qquad (6.23)$$

Suppose that the modes are low-frequency in the sense that $|\omega - kU_p| \ll |\Omega_p|$ for the plasma components. We have encountered this regime when discussing Alfvén waves or beam instabilities in the previous section, in the absence of charged dust. For those modes, all distinction between the RHCP and LHCP modes was lost, at the very low frequencies considered.

I will now show that the presence of the dust and the resulting imbalance between the electron and different ion densities restores the RHCP and LHCP character. Because the dust grains are so massive, I work in their frame of reference ($U_d = 0$) and at frequencies such that $\Omega_d \ll \omega$, where the dust is essentially non-magnetized. Expanding (6.23) to lowest significant order in the different small quantities yields

$$(\omega - k\overline{U})^2 \pm \omega\frac{V_{AR}^2}{c^2}\sum_{\text{dust}}\frac{\omega_{pd}^2}{\Omega_d} + k^2(W - V_{AR}^2) = 0. \qquad (6.24)$$

Here the mass averages in $\overline{U}$ and V_{AR} are computed for the pure plasma constituents, excluding the charged dust grains.

In the absence of equilibrium streaming between the plasma and the dust, and restricting ourselves to one average dust grain species, it follows that

$$\omega^2 \pm \frac{\omega_{pd}^2 V_{AR}^2}{c^2 \Omega_d} \omega - k^2 V_{AR}^2 = 0. \tag{6.25}$$

At the lowest frequencies this gives

$$\frac{\omega}{|\Omega_d|} \simeq \frac{c^2 k^2}{\omega_{pd}^2}. \tag{6.26}$$

This dust whistler mode [Verheest and Meuris 1995] is the generalization of the right-hand whistler mode in an electron plasma where the ions form a quasi-immobile neutralizing background, with dispersion law (6.9). Here, with $|\Omega_d| \ll \omega \ll |\Omega_\alpha|$, there is usually such a gap between $|\Omega_d|$ and the smallest of the standard ion gyrofrequencies that the dust can be considered as effectively immobile. Notice also that for the usual negative dust grains the dust whistler mode occurs for the LHCP mode, as the positive ions are then in the majority.

6.2.2 Charged dust in cometary plasmas

Another natural application of the dusty beam-plasma model introduced in this section is to low-frequency instabilities due to the pickup of cometary ions by the solar wind. We studied the gross features of the nonresonant instability in the previous section for pickup ions of the water group.

Dust grains may be present close to the comet in the Sunward direction, or downstream of the comet in the dust tail. The dust in the cometary environment can be charged due to radiation, and due to the flow of charged particles (electrons and ions) onto the grains from the background plasma. Even if the proportion of charge on the dust grains compared to that carried by free electrons is quite small, it can have a large effect on hydromagnetic shear and compressional Alfvén waves propagating at frequencies well below the ion gyrofrequency, as was shown by Pilipp et al. [1987] and Cramer and Vladimirov [1997] for waves in interstellar molecular clouds, where the dust to free electron charge ratio is of the order of 10^{-4}.

The character of the waves at frequencies less than the ion gyrofrequency is radically altered when a proportion of the negative charge resides on the dust grains [Cramer and Vladimirov 1996b], in analogy with the case of electromagnetic waves in plasmas in solids with unequal electron and hole numbers [Baynham and Boardman 1971]. With a negligibly small charge on the dust grains, the waves have the usual shear and compressional Alfvén wave properties. For a nonzero charge on the grains the RHCP mode experiences a cutoff due to the presence of the dust, while the LHCP mode is better described as a whistler or helicon wave extending to low frequencies, with the negatively charged dust grains playing the role of the heavy ion species, and the (relatively) light positive ions replacing the electrons [Mendis and Rosenberg 1994].

In discussing the effect of charged dust on the nonresonant firehose instabilities I will follow recent treatments given by Verheest and Meuris [1998] and Cramer *et al.* [1999]. In the latter paper there is also an investigation of the resonant instabilities, by using collisionless plasma kinetic theory of waves in a cold plasma with a ring-beam velocity distribution, but it would lead us to far to give the details of that. I will also neglect the effect of dust charging on the stability of these Alfvén waves, because those aspects will be discussed in the next Chapter 7. The linear and quasilinear firehose instabilities in dusty plasmas with anisotropic pressures with dust charging included were considered by Verheest and Shukla [1995] using a fluid model. The dust charging effect for a multibeam dusty plasma has also been considered by Reddy *et al.* [1996], and it was shown that drift-dissipative type modes can be excited.

Here I deal with a cometary plasma with dust, ions and electrons, and a solar wind plasma consisting of protons and electrons. With the indices s, c and d I will refer to solar wind particles (electrons and protons usually, but alpha particles could easily be included), to cometary plasma particles (electrons and protons and/or cometary water group ions assumed to be singly charged) and average cometary dust grains. A second subscript e or i refers to the corresponding electrons and ions. Working in the cometary frame means that for the parallel equilibrium drifts $U_{sw} = U$, where the beam velocity U is the projection of the solar wind velocity ($V_{sw} \sim 400$ km/s) on the direction of the interplanetary magnetic field. For the average Parker spiral magnetic field typically $U \sim 300$ km/s. Putting $U_{ci} = U_d = 0$ neglects the small parallel outflow velocities (~ 1 km/s) for cometary material.

I start with dispersion law (6.24), with now

$$\overline{U} = \frac{\rho_{se} + \rho_{si}}{\rho_{se} + \rho_{si} + \rho_{ce} + \rho_{ci}} U = \sigma U,$$

$$W = \frac{(\rho_{se} + \rho_{si})(\rho_{ce} + \rho_{ci})}{(\rho_{se} + \rho_{si} + \rho_{ce} + \rho_{ci})^2} U^2 = \sigma(1 - \sigma)U^2. \tag{6.27}$$

Here σ is the fractional mass density of the solar wind plasma. The reference Alfvén velocity V_{AR} as defined earlier by the summation of all plasma particles can be related to the solar wind Alfvén velocity V_{Asw} by

$$V_{AR}^2 = \sigma V_{Asw}^2. \tag{6.28}$$

No dust grains

When no dust grains ($N_d = 0$) are present, the dispersion relation is

$$(\omega - \sigma kU)^2 = \sigma k^2 V_{Asw}^2 [1 - (1 - \sigma)M^2], \tag{6.29}$$

with M denoting the Alfvénic Mach number $M = U/V_{Asw}$. Hence unstable modes occur whenever

$$\sigma < \sigma_{cr} = 1 - \frac{V_{Asw}^2}{U^2} < 1, \tag{6.30}$$

and they are given by

$$\omega = \sigma kU \pm ikU\sqrt{\sigma(\sigma_{cr} - \sigma)} \qquad \text{(for } \sigma < \sigma_{cr}). \tag{6.31}$$

When the fraction of the plasma mass density of solar wind origin is smaller than a threshold value, the instability is induced by a sufficient mass loading of the solar wind by cometary material.

Presence of charged dust

Now I allow for the presence of charged dust in the mixture of solar wind and cometary material, and assume for simplicity that it is negatively charged. It is afterwards straightforward to enlarge the discussion to positively charged dust, or even to a mixture of both. In (6.24) I introduce the proper definitions of $\overline{U}$ and W and get

$$(\omega - \sigma kU)^2 \mp \frac{\sigma \omega_{pd}^2 V_{Asw}^2}{c^2 |\Omega_d|} \omega - \sigma k^2 V_{Asw}^2 [1 - (1 - \sigma)M^2] = 0. \tag{6.32}$$

With the notation

$$A = \frac{\sigma \omega_{pd}^2 V_{Asw}^2}{c^2 |\Omega_d|} = \frac{N_d |q_d| B_0}{\rho_{se} + \rho_{si} + \rho_{ce} + \rho_{ci}} \simeq (1 - \sigma)\left(1 - \frac{N_{ce}}{N_{ci}}\right)\Omega_{ci}, \tag{6.33}$$

the dispersion law becomes

$$\omega^2 - [2\sigma kU \pm A]\omega + \sigma k^2 V_{Asw}^2 (M^2 - 1) = 0. \tag{6.34}$$

The discriminant of (6.34) can be written for the two different cases as

$$\begin{aligned}
\Delta &= 4\sigma U^2 (\sigma - \sigma_{cr})(k \pm k_1)(k \mp k_2), &\text{if } \sigma < \sigma_{cr}, \\
\Delta &= 4\sigma U^2 (\sigma - \sigma_{cr})(k \pm k_1)(k \pm k_2), &\text{if } \sigma > \sigma_{cr},
\end{aligned} \tag{6.35}$$

provided k_1 and k_2 are defined as

$$k_1 = \frac{A}{2U(\sqrt{\sigma} + \sqrt{\sigma_{cr}})\sqrt{\sigma}}, \qquad (6.36)$$

$$k_2 = \frac{A}{2U|\sqrt{\sigma} - \sqrt{\sigma_{cr}}|\sqrt{\sigma}}. \qquad (6.37)$$

The conclusions are as follows.

When in the absence of charged dust $\sigma \geq \sigma_{cr}$, the modes are stable because the mass loading is not strong enough. I see that due to the presence of (negatively) charged dust the discriminant Δ can be negative for the LHCP mode in the domain $k_1 < k < k_2$. On the other hand, when the modes are unstable in the absence of charged dust due to cometary pickup ($\sigma < \sigma_{cr}$), the RHCP mode is stabilized for $k < k_1$, whereas the LHCP mode becomes stable for $k < k_2$.

For positively charged dust I merely have to reverse the labels RHCP and LHCP in the above discussion.

Numerical evaluation

To estimate the relevance of the previous calculation, appropriate values should be put into the expressions (6.36) and (6.37). Consider a typical solar wind plasma (electrons and protons) with plasma density $N_{sw,p} = 6 \times 10^6$ m^{-3} ($\rho_{sw} = 10^{-20}$ kg m^{-3}), an energy of 10 eV and an ambient magnetic field of 9 nT. The solar wind velocity is given by $U = 400 \times 10^3$ m s^{-1} and with $V_{A,sw} \simeq 80 \times 10^3$ m s^{-1} I find that $M = 5$ and $\sigma_{cr} = 0.96$.

One expects the variation of the cometary ion density N_i to be similar to that of the neutral gas density. The ion mass of the cometary ion group ions is taken to be 16.8 times the proton mass. Observations of comet Halley by the Giotto spacecraft would indicate that N_i varies as d^{-2} outside the contact surface. At $d = 1.5 \times 10^8$ m, the ion density is given by 27×10^6 m^{-3} [Balsiger et al. 1986] or 10×10^6 m^{-3} [Mukai et al. 1986]. The parameter σ is therefore given as a function of d (in units of 10^9 m) by

$$\sigma = \frac{d^2}{D + d^2}, \qquad (6.38)$$

with $D = 0.756$ and $D = 2.04$ for Mukai et al. [1986] and Balsiger et al. [1986] respectively. This means that at a distance of 10^9 m, I can use values of $\sigma \simeq 0.3 - 0.6$. These values are smaller than σ_{cr}, and hence instabilities are expected.

For the dust density I use the measurement in the mass channels ranging from 10^{-13} to 10^{-20} kg for the VEGA spacecraft carried out by Vaisberg et al. [1986], who obtained a dust density of order 10^{-3} m^{-3} at a distance of 10^8 m. To fix the ideas, the dust is assumed to be charged to 7000 elementary charges, but the sign of the dust charge is not really relevant for the rest of this discussion.

When I put these numbers in (6.36) and (6.37), I get for an average value of $\sigma = 0.5$ that

$$k_1 = 5 \times 10^{-14} \text{ m}^{-1} \tag{6.39}$$

$$k_2 = 3 \times 10^{-13} \text{ m}^{-1}. \tag{6.40}$$

The wavelengths associated with k_1 and k_2 turn out to be very large, probably exceeding the dimensions of the cometary structures and influence spheres.

A first conclusion hence is that the presence of charged dust, even though theoretically able to significantly modify the streaming firehose-like instabilities and their ranges of occurrence, is in practice unlikely to be of any noticeable importance in cometary physics. Of course, this conclusion presently holds on the basis of the scarce data available from the few cometary missions flown up to now and might well have to be revised for different comets in the future, when more and accurate data become available [Verheest and Meuris 1998]. Secondly, however, the present study of the influence of charged dust has used the fact that the charge-to-mass ratios vastly differ from the values for ordinary ions, even water group ions, allowing a clear differentiation between the treatment of the heavy dust and the lighter ions. Some of the methodology might also be relevant for ordinary dust-free plasmas where there is a noticeable difference between heavy and light ions [Verheest *et al.* 1999b].

Finally, one would need to investigate also how other types of electromagnetic instabilities near comets are modified by the consideration of charged dust. The treatment of modes where the Doppler-shifted frequency for the cometary water group ions resonates with the gyrofrequency has been recently extended to include the possible effects of charged dust [Cramer *et al.* 1999]. These results indicate that indeed the charged dust could significantly alter previously obtained results.

6.3   Magnetosonic modes

6.3.1   Multispecies plasmas

After having given some examples of parallel electromagnetic modes affected by charged dust, I now turn to perpendicular propagation ($\vartheta = 90°$) and note from Chapter 4 that in the absence of self-gravitational effects and equilibrium drifts, the general dispersion law factorizes into the ordinary mode described by (4.82) and the extraordinary mode with

$$D_{yy}D_{zz} = D_{yz}^2, \tag{6.41}$$

with elements given from (4.77) as

$$D_{yy} = \omega^2 - c^2 k^2 - \sum_\alpha \frac{\omega_{p\alpha}^2 (\omega^2 - k^2 c_{s\alpha}^2)}{\omega^2 - k^2 c_{s\alpha}^2 - \Omega_\alpha^2},$$

$$D_{yz} = -\sum_\alpha \frac{\omega_{p\alpha}^2 \Omega_\alpha}{\omega^2 - k^2 c_{s\alpha}^2 - \Omega_\alpha^2},$$

$$D_{zz} = 1 - \sum_\alpha \frac{\omega_{p\alpha}^2}{\omega^2 - k^2 c_{s\alpha}^2 - \Omega_\alpha^2}. \tag{6.42}$$

For very small frequencies and long wavelengths, defined here in the same regime as for parallel modes, namely ω and $k c_{s\alpha} \ll |\Omega_\alpha|$, I develop the denominators in (6.42) to second order, use the mass averaged quantities introduced in (6.14), and suppose that $V_A \ll c$. The dispersion law for the low-frequency part of the X mode then become

$$\begin{vmatrix} \omega^2 - k^2 V_A^2 - k^2 \overline{c_s^2} & \dfrac{\overline{\omega^2 - k^2 c_s^2}}{\Omega_\alpha} \\[3mm] \dfrac{\overline{\omega^2 - k^2 c_s^2}}{\Omega_\alpha} & 1 + \dfrac{\overline{\omega^2 - k^2 c_s^2}}{\Omega_\alpha^2} \end{vmatrix} = 0, \tag{6.43}$$

and common factors c^2/V_A^2 have been divided out. To dominant order this is the dispersion law for the (fast) magnetosonic modes,

$$\omega^2 = k^2 V_{ms}^2 = k^2 (V_A^2 + C_s^2), \tag{6.44}$$

where the total Alfvén velocity V_A was introduced in (6.12), and a global thermal velocity C_s has been defined by

$$C_s^2 = \overline{c_s^2} = \sum_\alpha N_\alpha m_\alpha c_{s\alpha}^2 \bigg/ \sum_\alpha N_\alpha m_\alpha = \sum_\alpha P_\alpha \bigg/ \sum_\alpha N_\alpha m_\alpha \tag{6.45}$$

The magnetosonic velocity V_{ms}, given through $V_{ms}^2 = V_A^2 + C_s^2$, combines both the static magnetic field pressure and the thermal pressures of the different plasma constituents.

For the usual electron-proton plasma the electron pressure dominates in the numerator of C_s^2, whereas the mass density is that of the protons, so that C_s is essentially the ion-acoustic velocity c_{ia}.

6.3.2 Dusty plasmas

The generalization to dusty plasmas of the ideas exposed in the previous subsection is now obvious. If there is enough mass in the charged dust,

V_A is to be interpreted as the dust Alfvén velocity V_{Ad}, whereas $C_s \simeq c_{da}$, because the dust pressure is small compared to the electron and proton pressures. The dispersion law (6.44) then becomes

$$\omega^2 = k^2 V_{ms}^2 = k^2 (V_{Ad}^2 + c_{da}^2). \qquad (6.46)$$

Without external magnetic field, this reduces to the dispersion law for dust-acoustic modes, discussed in the previous chapter.

These ideas can be extended in various directions, either by looking at oblique propagation or by having higher frequencies, above the dust cyclotron frequency. In that case, either the dust alone is treated as unmagnetized ($|\Omega_d| \ll \omega \ll \Omega_i \ll |\Omega_e|$), or else both ions and dust are unmagnetized, but the electrons are magnetized ($|\Omega_d| \ll \Omega_i \ll \omega \ll |\Omega_e|$). In this context, various studies of Rao come to mind. First, hydromagnetic waves and shocks in dusty plasmas were studied [Rao 1993a]. Electron inertia is neglected, but ion and dust motion taken into account, at fixed dust charges. The modes considered are typical low-frequency ones such as Alfvén and (fast) magnetosonic waves, with suitable redefinitions. Shocks of magnetosonic waves driven by either upper-hybrid or O mode electromagnetic waves are shown to exist, allowing a transition in the dust flow velocity from sub- to supermagnetosonic values. Such transitions could occur when charged dust clouds act as obstacles to the plasma flow, as near comets.

Next comes a more systematic treatment [Rao 1993b], showing a new class of dust-magneto-acoustic waves, both of fast and slow type. The fast branch generalizes electrostatic dust-acoustic modes to magnetized plasmas, as I indicated at the beginning of this subsection. The wave frequencies for oblique modes are between the dust gyrofrequency (frozen-in dust) and the ion and electron gyrofrequencies. The departure from the frozen-in field approximation for the ordinary ions is given in a subsequent paper [Rao 1993c]. This is finally rounded off by a general overview of the different modes, including two new higher-frequency types of dust-magnetoacoustic waves, generalizations of dust-acoustic and dust-ion-acoustic modes when a static magnetic field is included [Rao 1995].

The conclusion of this series of papers is essentially that one can play around with different weightings of the electron, ion and dust pressures, in combination with ion and dust mass densities. According to what dominates, one finds for parallel propagation the transition from ion-acoustic to dust-acoustic modes, and for oblique propagation similar variations on the theme of lower-hybrid to dust lower-hybrid mode transitions, as I discussed in the previous chapter. At perpendicular propagation the modes are (fast and slow) magnetosonic, in various guises. The basic physics is the same, except that contributions of the different species might change, keeping in mind though that the dust is characterized by much lower eigenfrequencies.

6.4   Reductive perturbation for nonlinear waves

Having dealt with aspects of linear electromagnetic modes, both at parallel
and oblique directions of wave propagation, it is time to move on and treat
nonlinear modes at larger amplitudes. A survey of the existing literature
for ordinary plasmas shows that an exact description through a kind of
Sagdeev potential is not possible, when the magnetic part of the Lorentz
force is kept in the equations of motion. Then the reductive perturbation
techniques offer the most systematic way of obtaining information about
nonlinear electromagnetic mode behaviour.

And here you might think that, since we dwelled at length on the links
between the Sagdeev and the reductive perturbation methods in the previous chapter, that results from electrostatic waves could be taken over.
However, that is not the case, as there is a radical difference between parallel and really oblique wave propagation. For the former, the appropriate
nonlinear equation turns out to be of the derivative nonlinear Schrödinger
(DNLS) type, including the nonlinear Schrödinger (NLS) equation as a special case. At oblique propagation, one is led back to the KdV or mKdV
equations. Hence, care has to be taken and both cases distinguished accordingly.

6.4.1   Parallel propagation

At the linear level, parallel electromagnetic waves are purely transverse.
Contributions from the $\mathbf{u}_\alpha \times \mathbf{B}$ terms in the Lorentz force, however, will
nonlinearly generate longitudinal components, including density variations.
In addition, these same terms prevent the use of a Sagdeev-type approach
as for parallel electrostatic waves. Elements of singular or reductive perturbation analysis were given in the previous chapter, when dealing with
electrostatic modes. I have to adapt that now for vector equations and
quantities, and at the same time take another look at the stretching used
for the space and time coordinates.

To start with, it is useful to split all vector quantities in transverse and
longitudinal components,

$$\mathbf{a} = a_\parallel \mathbf{e}_z + \mathbf{a}_\perp. \tag{6.47}$$

The labels $\parallel$ and $\perp$ refer to the direction of wave propagation, for parallel
modes at the same time the direction of the external magnetic field. In the
basic set of equations

$$\frac{\partial n_\alpha}{\partial t} + \frac{\partial}{\partial z}(n_\alpha u_{\alpha\parallel}) = 0, \tag{6.48}$$

$$\mathbf{B} \cdot \frac{\partial \mathbf{u}_\alpha}{\partial t} + u_{\alpha\parallel}\mathbf{B} \cdot \frac{\partial \mathbf{u}_\alpha}{\partial z} = \frac{q_\alpha}{m_\alpha}\mathbf{E} \cdot \mathbf{B}, \tag{6.49}$$

$$\frac{\partial \mathbf{u}_{\alpha\perp}}{\partial t} + u_{\alpha\parallel}\frac{\partial \mathbf{u}_{\alpha\perp}}{\partial z} = \frac{q_\alpha}{m_\alpha}\left(\mathbf{E}_\perp + u_{\alpha\parallel}\mathbf{e}_z \times \mathbf{B}_\perp\right)$$
$$+ \Omega_\alpha \mathbf{u}_{\alpha\perp} \times \mathbf{e}_z, \tag{6.50}$$

I have replaced the parallel projection of the equations of motion by the projection of the total equation on the full magnetic field, static and wave parts together. This will turn out to be easier to handle for the perturbation analysis. I also need Maxwell's equations:

$$\mathbf{e}_z \times \frac{\partial \mathbf{E}_\perp}{\partial z} + \frac{\partial \mathbf{B}_\perp}{\partial t} = \mathbf{0},$$

$$-\varepsilon_0 \frac{\partial E_\parallel}{\partial t} = \sum_\alpha n_\alpha q_\alpha u_{\alpha\parallel},$$

$$\frac{1}{\mu_0}\mathbf{e}_z \times \frac{\partial \mathbf{B}_\perp}{\partial z} - \varepsilon_0 \frac{\partial \mathbf{E}_\perp}{\partial t} = \sum_\alpha n_\alpha q_\alpha \mathbf{u}_{\alpha\perp},$$

$$\varepsilon_0 \frac{\partial E_\parallel}{\partial z} = \sum_\alpha n_\alpha q_\alpha. \tag{6.51}$$

Contrary to what is usually done for Alfvén waves, I have retained in Maxwell's equations both the displacement current and the possibility of deviations from charge neutrality. Although one could argue that for most of the astrophysical and laboratory plasmas such corrections are small, retaining them presents no additional complications to get through the mathematical analysis.

In the usual stretching of coordinates and time [Mio et al. 1976; Mjølhus and Wyller 1986; Verheest 1990,1992a; Verheest and Buti 1992],

$$\xi = \varepsilon(z - Vt) \quad , \quad \tau = \varepsilon^2 t, \tag{6.52}$$

V is the linear wave velocity, to be determined later, and ε a small parameter measuring the weakness of the dispersion and of the nonlinear effects. This ordering differs from the Korteweg-de Vries (KdV) scaling for the simple reason that the corrections to the linear phase velocity behave differently. To see this, one goes back to the dispersion law, the solution of which gives $\omega(k)$. For low-frequency, long-wavelength modes this is expanded as

$$\omega(k) \simeq a_1 k + a_2 k^2 + a_3 k^3 + \dots \tag{6.53}$$

Here a_1 is the linear phase velocity ω/k for $k \to 0$, and the first nonzero correction determines the stretching. For electrostatic waves this is a_3, hence the KdV stretching (5.53), whereas for parallel electromagnetic modes a_2 is nonzero, so that (6.52) applies. As seen in the next section, this is not true for oblique or perpendicular modes, where again the KdV stretching must be used.

Density and parallel velocities have the following series expansions around a homogeneous equilibrium:

$$n_\alpha = N_\alpha + \varepsilon^{1/2} n_{\alpha 1} + \varepsilon n_{\alpha 2} + \varepsilon^{3/2} n_{\alpha 3} + \cdots,$$
$$u_{\alpha\|} = U_\alpha + \varepsilon^{1/2} u_{\alpha\|1} + \varepsilon u_{\alpha\|2} + \varepsilon^{3/2} u_{\alpha\|3} + \cdots, \qquad (6.54)$$

whereas all other variables go as e.g.

$$\mathbf{B}_\perp = \varepsilon^{1/2} \mathbf{B}_{\perp 1} + \varepsilon \mathbf{B}_{\perp 2} + \varepsilon^{3/2} \mathbf{B}_{\perp 3} + \cdots, \qquad (6.55)$$

I again expand all dependent variables consistently in powers of $\varepsilon^{1/2}$, as done in the previous chapter for electrostatic modes. The sequence of equations obtained by equating the coefficients of the various powers of $\varepsilon^{1/2}$ shows that Faraday's equation in (6.51) gives to orders 3/2 and 2 that

$$\mathbf{E}_{\perp 1} = V \mathbf{B}_{\perp 1} \times \mathbf{e}_z,$$
$$\mathbf{E}_{\perp 2} = V \mathbf{B}_{\perp 2} \times \mathbf{e}_z. \qquad (6.56)$$

From one of the equations of motion, namely (6.49), I find to lowest orders that

$$E_{\|1} = E_{\|2} = 0, \qquad (6.57)$$

and I can already see that charge neutrality is maintained up to order ε. Similarly, the perpendicular Ampère law in (6.51) indicates current neutrality, at least to order 1, whereas the parallel version shows current neutrality to order 3/2. These are the kind of arguments that have been used to neglect the displacement current from the outset, and impose charge neutrality. The perpendicular equations of motion first give

$$\mathbf{u}_{\alpha\perp 1} = (U_\alpha - V)\frac{\mathbf{B}_{\perp 1}}{B_0},$$
$$\mathbf{u}_{\alpha\perp 2} = (U_\alpha - V)\frac{\mathbf{B}_{\perp 2}}{B_0} + u_{\alpha\|1}\frac{\mathbf{B}_{\perp 1}}{B_0}. \qquad (6.58)$$

Next the continuity equations to order 2 show that

$$u_{\alpha\|1} = (V - U_\alpha)\frac{n_{\alpha 1}}{N_\alpha},$$
$$u_{\alpha\|2} = (V - U_\alpha)\left(\frac{n_{\alpha 2}}{N_\alpha} - \frac{n_{\alpha 1}^2}{N_\alpha^2}\right). \qquad (6.59)$$

The projection of the equations of motion onto the total magnetic field indicate to order 3/2, with the help of (6.49), that

$$(V - U_\alpha)^2 B_0 \frac{\partial n_{\alpha 1}}{\partial \xi} + \frac{N_\alpha q_\alpha}{m_\alpha}[\mathbf{E} \cdot \mathbf{B}]_{3/2} = 0. \qquad (6.60)$$

Here the notation $[\mathbf{E} \cdot \mathbf{B}]_{3/2}$ means that the corresponding quantity is evaluated at order 3/2. Multiplying (6.60) with $q_\alpha/(U_\alpha - V)^2$ and summing over all species yields, with the help of Poisson's equation in (6.51) to order 1/2, that

$$\sum_\alpha \frac{\omega_{p\alpha}^2}{(V - U_\alpha)^2}[\mathbf{E} \cdot \mathbf{B}]_{3/2} = 0. \tag{6.61}$$

Hence $[\mathbf{E} \cdot \mathbf{B}]_{3/2} = 0$, so that then $n_{\alpha 1}$ vanishes, making also $u_{\alpha \| 1}$ zero. In order to fully determine $n_{\alpha 2}$ and $u_{\alpha \| 2}$ I evaluate (6.49) to order 2 and get

$$(V - U_\alpha)^2 B_0 \left\{ \frac{1}{N_\alpha} \frac{\partial n_{\alpha 2}}{\partial \xi} - \frac{\mathbf{B}_{\perp 1}}{B_0} \cdot \frac{\partial \mathbf{B}_{\perp 1}}{\partial \xi} \right\} + \frac{q_\alpha}{m_\alpha}[\mathbf{E} \cdot \mathbf{B}]_2 = 0. \tag{6.62}$$

A reasoning similar to what was done before at lower order, in (6.60), leads to the vanishing of $[\mathbf{E} \cdot \mathbf{B}]_2$ and to

$$n_{\alpha 2} = \frac{1}{2} N_\alpha \frac{B_{\perp 1}^2}{B_0^2},$$

$$u_{\alpha \| 2} = \frac{1}{2}(V - U_\alpha) \frac{B_{\perp 1}^2}{B_0^2}. \tag{6.63}$$

To this order, the plasma still is charge neutral, and indeed carries no parallel current. The wave phenomena are too slow, and the plasma can adjust easily. From the perpendicular equations of motion to order 3/2 I get that

$$\mathbf{u}_{\alpha \perp 3} = \frac{\mathbf{E}_{\perp 3}}{B_0} \times \mathbf{e}_z + U_\alpha \frac{\mathbf{B}_{\perp 3}}{B_0} + \frac{1}{2}(V - U_\alpha)\frac{B_{\perp 1}^2}{B_0^3}\mathbf{B}_{\perp 1}$$

$$+ \frac{(V - U_\alpha)^2}{\Omega_\alpha B_0}\mathbf{e}_z \times \frac{\partial \mathbf{B}_{\perp 1}}{\partial \xi}. \tag{6.64}$$

Rather than go on like this to higher orders, I will determine a global equation by multiplying (6.50) by $N_\alpha m_\alpha$, summing the result over all species and eliminating the sums containing $N_\alpha q_\alpha \ldots$ with the help of Maxwell's equations in (6.51). This procedure gives

$$\sum_\alpha n_\alpha m_\alpha \frac{\partial \mathbf{u}_{\alpha \perp}}{\partial t} + \sum_\alpha n_\alpha m_\alpha u_{\alpha \|} \frac{\partial \mathbf{u}_{\alpha \perp}}{\partial z}$$

$$= \varepsilon_0 \frac{\partial E_\|}{\partial z}\mathbf{E}_\perp - \varepsilon_0 \frac{\partial E_\|}{\partial t}\mathbf{e}_z \times \mathbf{B}_\perp + \frac{B_0}{\mu_0}\frac{\partial \mathbf{B}_\perp}{\partial z} + \varepsilon_0 B_0 \mathbf{e}_z \times \frac{\partial \mathbf{E}_\perp}{\partial t}, \tag{6.65}$$

and to order 3/2 this yields the dispersion law

$$\sum_\alpha N_\alpha m_\alpha (V - U_\alpha)^2 = \frac{B_0^2}{\mu_0}\left(1 - \frac{V^2}{c^2}\right) \tag{6.66}$$

determining V. In the absence of equilibrium streaming between the plasma components this shows that

$$V^2 = \frac{c^2 V_A^2}{c^2 + V_A^2} \simeq V_A^2, \qquad (6.67)$$

in other words the linear phase velocity is close to the Alfvén velocity for the plasma as whole, as defined before.

To order 2 (6.65) does not learn anything new. Inserting all the known expressions and eliminating $\mathbf{B}_{\perp 3}$ and $\mathbf{E}_{\perp 3}$ with the help of Faraday's law to order 5/2 finally gives the desired nonlinear evolution equation

$$A\frac{\partial \mathbf{B}_{\perp 1}}{\partial \tau} + \frac{C}{4}\frac{\partial}{\partial \xi}\left(B_{\perp 1}^2 \mathbf{B}_{\perp 1}\right) + D\mathbf{e}_z \times \frac{\partial^2 \mathbf{B}_{\perp 1}}{\partial \xi^2} = 0. \qquad (6.68)$$

This is the DNLS equation, with coefficients

$$A = \sum_\alpha N_\alpha m_\alpha (V - U_\alpha) + \frac{B_0^2 V}{\mu_0 c^2},$$

$$C = \frac{1}{\mu_0}\left(1 - \frac{V^2}{c^2}\right),$$

$$D = \sum_\alpha \frac{N_\alpha m_\alpha (V - U_\alpha)^3}{2\Omega_\alpha}. \qquad (6.69)$$

In the absence of equilibrium streaming these reduce to

$$A = \frac{B_0^2}{V},$$

$$C = 1 - \frac{V^2}{c^2},$$

$$D = \frac{V^3 \mu_0}{2}\sum_\alpha \frac{N_\alpha m_\alpha}{\Omega_\alpha}, \qquad (6.70)$$

if common factors are eliminated from the equation. By projecting (6.68) on two orthogonal directions like the x and y axes, and combining B_{x1} and B_{y1} into a single complex variable $\varphi = B_{x1} + iB_{y1}$, one gets the complex scalar representative of the DNLS equation, written as

$$A\frac{\partial \varphi}{\partial \tau} + \frac{C}{4}\frac{\partial}{\partial \xi}\left(|\varphi|^2 \varphi\right) + iD\frac{\partial^2 \varphi}{\partial \xi^2} = 0. \qquad (6.71)$$

The typical solutions of (6.68) or (6.71) are envelope solitons of the form

$$|\mathbf{B}_{\perp 1}|^2 = |\varphi|^2 = \pm\frac{8AM}{C}\left(\sqrt{2}\cosh\frac{AM}{D}(\xi - M\tau) \mp 1\right)^{-1}, \qquad (6.72)$$

where M represents the velocity in the co-moving frame. The $\pm$ signs are not related to circular polarization, but stem from the requirement that $|\varphi|^2$ has to be positive, with corresponding restrictions on M [Verheest and Buti 1992].

The NLS equation itself is obtained for long-wavelength modulation of the envelope of a plane wave, by putting

$$\varphi(\xi,\tau) = \Phi(\zeta,\tau)\exp i\kappa\left(\xi + \frac{D}{A}\kappa\tau\right), \qquad (6.73)$$

where $\zeta = \xi + 2D\kappa\tau/A$. By approximating the derivative in the cubic nonlinearity by its limit for large κ, a NLS equation results, of generic form

$$iA\frac{\partial\Phi}{\partial\tau} = \frac{\kappa C}{4}|\Phi|^2\Phi + D\frac{\partial^2\Phi}{\partial\zeta^2}. \qquad (6.74)$$

It has typical soliton solutions [Ablowitz and Clarkson 1991]

$$|\Phi|^2 = \frac{8D\gamma^2}{kC}\mathrm{sech}^2\gamma(\zeta - M\tau). \qquad (6.75)$$

All the preceding results are fully general as far as the plasma composition is concerned, and have been applied in various guises to different problems involving multispecies plasmas, with or without streaming effects. A DNLS equation similar to (6.68) has been derived by Vladimirov and Cramer [1996], where the dust is considered as a heavy quasi-immobile background, but interesting differences arise because of the imbalance between electron and ion densities.

Buti [1997] then applied the general DNLS equation (6.68) to explore the effect of really massive and heavily charged dust grains on chaotic Alfvénic systems. Interestingly, because of the dust grains, chaos is not only reduced but disappears even at very low dust grain densities, as small as 10^{-8} of the proton density. The computations were done for typical parameters in the rings of Saturn and in cometary cases. Chaos is reduced simply due to inertial stabilization caused by the heavy dust grains. Similar conclusions had already been drawn for impurity ions, but the dust grains are much more massive, hence the very low densities needed.

6.4.2 Oblique and perpendicular modes

For simplicity at oblique propagation, I take a cold plasma model and look at waves or structures propagating along the z axis, the static magnetic field then being taken as $\mathbf{B}_0 = B_0(\sin\vartheta\,\mathbf{e}_x + \cos\vartheta\,\mathbf{e}_z)$ [Verheest and Meuris 1996c]. I will not repeat the basic equations, but first go to the corrections to the linear phase velocity. For parallel low-frequency electromagnetic waves the phase velocity has a linear correction in k [Mio et al. 1976], which I discussed and used in the preceding subsection. However, for really oblique modes the correction is quadratic [Kakutani et al. 1968], and inducing the classical KdV stretching (5.53).

Although I will not give the details here, because this becomes very repetitive, one could proceed very systematically by trying to start for low-frequency waves from nonzero $\mathbf{E}_1$ and $\mathbf{B}_1$, where the expansion is in powers of $\varepsilon^{1/2}$. Then, after straightforward but tedious algebra, one would arrive at a true bifurcation point in third order, such that

$$B_{x1} \sin \vartheta = 0. \tag{6.76}$$

In other words, for really oblique propagation ($\vartheta \neq 0$ and $\sin \vartheta$ finite), $B_{x1} = 0$ is imposed, and as a consequence all first-order variables vanish. The expansion thus starts at order ε and now follows the usual KdV ordering. In addition, $u_{\alpha y}$, B_y, E_x and E_z will only start at order $\varepsilon^{3/2}$ [Verheest and Meuris 1996c]. That polarization is singled out, for which the wave magnetic field is to lowest order perpendicular to the direction of wave propagation, in the plane spanned by the wave vector and the static field. The linear wave electric field is perpendicular to that plane. For the special case of perpendicular wave propagation, this polarization corresponds to the low-frequency branch of the extraordinary mode [Meuris and Verheest 1996b].

To lowest nonzero order I find in the ordering chosen that

$$n_{\alpha 2} = \frac{N_\alpha B_{\perp 0}}{B_0^2} B_{x2},$$

$$u_{\alpha x2} = -\frac{V_A B_{\parallel 0}}{B_0^2} B_{x2},$$

$$u_{\alpha z2} = \frac{V_A B_{\perp 0}}{B_0^2} B_{x2},$$

$$u_{\alpha y3} = \frac{V_A^2}{\Omega_\alpha B_0} \frac{\partial B_{x2}}{\partial \xi} + \frac{E_{z3}}{B_{\perp 0}}. \tag{6.77}$$

The vanishing of Ampère's law to lowest nontrivial (third) order again yields the phase velocity as V_A, because thermal and streaming effects have been omitted. The global equation (6.65) is then also fulfilled to this order, whereas to order ε^2 it shows that

$$\sum_\alpha N_\alpha m_\alpha u_{\alpha y3} = \frac{1}{\mu_0 B_{\perp 0}} \left(\frac{B_{\parallel 0}^2}{V^2} - \frac{B_0^2}{c^2} \right) E_{z3}. \tag{6.78}$$

Together with the expression for $u_{\alpha y3}$ this yields

$$E_{z3} = -\frac{V_A^4}{c^2 \sin^2 \vartheta} \sum_\alpha \frac{\omega_{p\alpha}^2}{\Omega_\alpha^3} \frac{\partial B_{x2}}{\partial \xi}, \tag{6.79}$$

allowing me to rework $u_{\alpha y3}$ as

$$u_{\alpha y3} = \frac{V_A^2}{B_0} \left(\frac{1}{\Omega_\alpha} - \frac{V_A^2}{c^2 \sin^2 \vartheta} \sum_\beta \frac{\omega_{p\beta}^2}{\Omega_\beta^3} \right) \frac{\partial B_{x2}}{\partial \xi}. \tag{6.80}$$

To order $\varepsilon^{5/2}$ the KdV equation is obtained,

$$A \frac{\partial B_{x2}}{\partial \tau} + C B_{x2} \frac{\partial B_{x2}}{\partial \xi} + D \frac{\partial^3 B_{x2}}{\partial \xi^3} = 0, \tag{6.81}$$

with coefficients

$$
\begin{aligned}
A &= \frac{2c^4}{V_A^7}, \\
C &= \frac{3c^2 \sin \vartheta}{B_0 V_A^4} \sum_\alpha \frac{\omega_{p\alpha}^2}{\Omega_\alpha^2}, \\
D &= \left(1 + \sum_\alpha \frac{\omega_{p\alpha}^2}{\Omega_\alpha^2} \right) \sum_\alpha \frac{\omega_{p\alpha}^2}{\Omega_\alpha^4} - \frac{1}{\sin^2 \vartheta} \left(\sum_\alpha \frac{\omega_{p\alpha}^2}{\Omega_\alpha^3} \right)^2.
\end{aligned}
\tag{6.82}
$$

This encompasses results for oblique modes in a standard hydrogen plasma [Kakutani *et al.* 1968], and for perpendicular modes in bi-ion plasmas [Toida and Ohsawa 1994]. Whereas quasi-neutrality indeed follows to the orders needed, the displacement current could have a non negligible influence, both on the definition of the Alfvén velocity V_A and of the coefficients (6.82) in the KdV equation (6.81).

The remarkable property is that the nonlinearity is strongest at strictly perpendicular propagation. It vanishes if one tries to take the parallel limit, but that is no surprise and merely reminds us of the different stretching needed there. Furthermore, as pointed out already by Kakutani *et al.* [1968], the coefficient of the dispersive term is usually negative for most of the range of oblique propagation, positive for strictly perpendicular modes and changes sign close to perpendicular propagation. The vanishing of the dispersive term at the crossover angle is an indication of wave steepening, and the solitons become shocks [Meuris 1998]. Moreover, D blows up if one tries to take the parallel limit, except for special plasmas.

I now briefly discuss the behaviour of the coefficient D for specific plasmas, and determine in particular the critical angle ϑ_0 where D changes sign. Assuming traditionally that $V_A \ll c$, I find that in a classical hydrogen plasma

$$\sin^2 \vartheta_0 \simeq 1 - \frac{m_e}{m_i}, \tag{6.83}$$

with the usual small mass ratio. This gives a critical angle $\vartheta = 88°$, very close to perpendicular propagation. On the other hand, in an electron-positron plasma ($m_e = m_p$) I see that

$$D = \left(1 + \frac{2\omega_p^2}{\Omega^2}\right) \frac{2\omega_p^2}{\Omega^4} \tag{6.84}$$

is always positive. Multi-ion plasmas, including dusty plasmas, lie in between these two extremes.

A relevant question is then of whether in other plasmas a similar positive dispersion as in electron-positron plasmas could occur. That requires the vanishing of part of D, such that

$$\sum_\alpha \frac{\omega_{p\alpha}^2}{\Omega_\alpha^3} = 0. \tag{6.85}$$

In a plasma composed of electrons and different kinds of positive ions this cannot be, but if there is at least one species of negative ions, the condition can be fulfilled at critical densities. Thinking of course of dusty plasmas, I get from (6.85) that

$$N_d \simeq \frac{Z_d N_i m_i^2}{m_d^2}. \tag{6.86}$$

Except for the special cases of electron-positron plasmas (where also several other waves decouple [Verheest and Lakhina 1996]) or plasmas with negative ions at critical densities, the critical angle is rather close to 90°. Hence the dispersive term has a negative coefficient D for most of the range of oblique waves, and turns positive in a narrow range around perpendicular propagation. Whenever that happens, the solitons solutions to (6.81)

$$B_{x2} = \frac{3MA}{C}\text{sech}^2\left[\frac{1}{2}\sqrt{\frac{A|M|}{|D|}}(\xi - M\tau)\right] \tag{6.87}$$

will change from subalfvénic ($M < 0$), rarefactive ($B_{x2} < 0$) to super-alfvénic ($M > 0$), compressive ($B_{x2} > 0$) in nature, because A and C are positive, and then both M and B_{x2} have the sign of D.

CHAPTER 7

FLUCTUATING DUST CHARGES

7.1 Charge fluctuations

In the previous chapters I have discussed many different modes in dusty plasmas, but always keeping the dust charges fixed. The need has now come to look at the specific effects induced because dust charges are not really independent of the variations in the plasma potentials. Hence comes, even in fluid theory, the first of the crucial differences between ordinary multispecies and dusty plasmas. All the modes we have encountered in the two previous chapters will influence the charging mechanism, and the resulting self-consistent feedback will lead to interesting and sometimes totally unexpected changes. These are totally outside what is covered in standard plasma physics textbooks and monographs.

Indeed, as we have seen in Chapter 2, the charging of the grains depends on the local plasma characteristics. If a wave disturbs these characteristics, the charging of the grains is affected and the dust grain charge is modified, with a resulting feedback on the wave mode. Simplest to deal with are the parallel electrostatic modes in an unmagnetized plasma. However, for magnetized dusty plasmas changes in dust grain gyro motion and gyroresonances may occur. Let me stress, however, that the description of the latter modifications requires a self-consistent analytic charging theory for magnetized plasmas, which is not (yet) available. One may even wonder whether such a consistent picture will ever be at hand, because (analytic) probe theory has not advanced in this direction for many decades, and much of what is used in dust grain charging is borrowed from that.

7.1.1 History

Before I dive into possible mathematical descriptions, I recall some of the recent developments when dealing with the problems caused by fluctuating dust charges. This is novel territory, compared to standard plasma physics textbook treatment. Once dust-acoustic modes became popular, the dusty plasma community soon realized that at some stage fluctuating dust charges needed to be taken into account. The main question was, how?

Progress was made by several groups simultaneously. The first to include variations in the dust grain charge due to collective modes in the plasma were Melandsø et al. [1993a], who analyzed the damping of the dust-acoustic mode. Their treatment, however, was not really self-consistent, as the issue of sink and source terms in the electron and ion continuity equations is sidestepped by assuming these species to be Boltzmann distributed. We will see that this neglects some essential elements for both small and long wavelengths [Meuris and Verheest 1996a]. Nevertheless, they introduced a charging frequency, the inverse of the charging time of a dust particle. The damping of the wave is then due to the delay in the charging of the dust particles, and in later papers by other authors referred to as Tromsø damping.

At the same time and totally independently, Varma *et al.* [1993] analyzed also Langmuir and dust ion-acoustic modes, but again without using source/sink terms in the continuity equations. Electrostatic oscillations and instabilities were studied by treating the dust charge as an additional dynamical variable. Dust charge fluctuations give rise to two purely damped modes, in addition to causing a collisionless damping of existing normal modes, much as in the recombination process in partially ionized plasmas. Furthermore, some interesting electrostatic instabilities arise in the presence of an equilibrium drift of the charged particles. At the end, a brief allusion is made to the incorporation of loss terms in the ion and electron momentum equations, although no explicit results are presented.

Several papers then follow and embroider on these ideas. Jana *et al.* [1993] discuss charge fluctuations in response to oscillations in the plasma currents flowing to the grains, but also not fully self-consistently. Dissipative and instability mechanisms for ion and dust waves in the plasma lead to interesting applications in laboratory and astrophysical situations. One of these is that there could be anomalous transport of grains due to a turbulent spectrum of waves. Interestingly enough, but then in a totally different field, related phenomena are described by quantum chromodynamics in quark-gluon plasmas [Kaw 1992].

The first fully self-consistent approaches were carried out by Bhatt and Pandey [1994], later refined by Meuris and Verheest [1996a] and Bhatt [1997]. Losses in electron and ion number densities are represented by Bhatt and Pandey [1994] through two attachment frequencies, as will discussed in the next section. Charge variations occur, due to changes not only in electron and ion currents flowing to the grains, but also in electron and ion densities themselves. Various approximations are given for dust- and ion-acoustic modes, as well as for the streaming instability. The results are consistent with other earlier papers and lead to mode damping.

Phase differences between the dust charge and wave potential variations can lead to strong damping of electrostatic modes. At sufficiently low frequencies, the phase shift is small, and hence there is little damping, even though the grain charge fluctuations could be important, as very little energy is exchanged between waves and grains. At the highest frequencies, although the phase shift is large, there is hardly any damping as the charging or discharging of the dust cannot follow! It is at the middle frequencies, of the order of the charging frequencies, that the strongest damping occurs.

Besides Tromsø damping, D'Angelo [1994] included another mechanism, that of creation damping due to the continuous injection of fresh ions in the plasma, to replace those that are lost on the dust grains. Such fresh ions do not share initially in the wave motion of the existing ones and hence lower the average momentum of the ion population. Creation damping had been considered much earlier by D'Angelo [1967] in discussing recombination instabilities in ordinary plasmas. Creation damping is usually dominant, except for such low-frequency motions that the grain dynamics becomes important. Then inertia is provided by the dust grains, and ion momentum matters little. Similar results occur with the ionization instability [D'Angelo 1997] in a system where neutral gas is ionized by a constant flux of energetic electrons, so that the ion-acoustic wave has increased frequency and growth rate.

Similar self-consistent descriptions were given by Ma and Yu [1994a], when the dust is an immobile background but still fluctuating in charge. The associated damping could easily surpass Landau damping, as shown from a proper kinetic treatment. One of the drawbacks of the description, as with so many similar ones which rely on standard probe models for the charging of the dust, is that it corresponds to a thermodynamically open system in which the dust grains can absorb electrons and ions indefinitely. Although total charge is conserved in this way, and the dust charge could reach a stationary value, the total number of electrons and ions is not conserved, and an outside source for these particles must exist. High-frequency Langmuir waves become unstable when the ions and electrons stream with respect to the dust, but on rather long timescales [Ma and Yu 1994b]. This resembles the ionization instability, and because of the open nature of the system it is not clear how the instability saturates.

Since the attachment frequencies are very similar to collision frequencies, as will be seen, the treatments discussed below could easily be extended to collisional drag between plasma and/or dust and neutral particles, or even between the plasma particles themselves [Shukla et al. 1997b]. That would give opposite effects on the ions and the negative dust grains, leading to the possibility that the ion drag force exceeds the restoring electrostatic force, so that dust-acoustic modes can grow [Ivlev et al. 1999]. The mathematical details are similar to what will be discussed below from the perspective of charge fluctuations, so there is no need here to specify all these additional details.

Because of the importance and novelty of dust charge fluctuations, I will give in the next section a systematic treatment along the lines of Bhatt and Pandey [1994], starting with a minimum of hypotheses [Meuris and Verheest 1996a]. This provides a general frame in which to judge the relevance of the different assumptions made by other authors. In tune with the available literature, I will devote quite some pages to parallel electrostatic modes, before turning later to electromagnetic and oblique modes.

7.1.2 Continuity equations and attachment frequencies

For the continuity equations (4.42),

$$\frac{\partial n_\alpha}{\partial t} + \nabla \cdot (n_\alpha \mathbf{u}_\alpha) = S_\alpha. \tag{7.1}$$

the basic assumption now will be that the number densities of the plasma species proper are not conserved, because fluctuations in the dust grain charges liberate or capture electrons and other plasma species, and hence influence their number densities. Extra source/sink terms are needed, and I have to try and determine some form for these. Before going into that, I assume the source/sink terms S_α to vanish in equilibrium for all species. This is tantamount to saying that in equilibrium as much plasma particles are created as are collected by dust grains per time unit. This creation process can be a consequence of the emission of plasma particles by dust grains, by charge exchange at the grain's surface and/or ionization of neutrals in the plasma. It is worth recalling here what we remarked in Chapter 4, that we do not really know yet whether a true stationary state exists in a closed system.

For the dust grains, I will omit sink/source terms. The argument for that is that the dust number density is not really affected by the dust loosing or picking up some charges. Coalescing or breaking up of the dust thus not been incorporated, and indeed has not been treated so far. So the continuity equation for the dust remains as usual

$$\frac{\partial n_d}{\partial t} + \nabla \cdot (n_d \mathbf{u}_d) = 0. \tag{7.2}$$

Now I turn to (4.46), expressing conservation of charge in the plasma as a whole, and separate ordinary plasma from dust species, to get

$$\frac{\partial}{\partial t} \left(\sum_\alpha n_\alpha q_\alpha + n_d q_d \right) + \nabla \cdot \left(\sum_\alpha n_\alpha q_\alpha \mathbf{u}_\alpha + n_d q_d \mathbf{u}_d \right) = 0. \tag{7.3}$$

This can be rewritten with the help of the continuity equations as

$$n_d \left(\frac{\partial}{\partial t} + u_d \cdot \nabla \right) q_d = - \sum_\alpha q_\alpha S_\alpha. \tag{7.4}$$

In all these equations I restrict the summations over α strictly to the usual plasma species, and for simplicity only discuss one average dust grain constituent. On the other hand the charge fluctuations are given by

$$\frac{dq_d}{dt} = \left(\frac{\partial}{\partial t} + u_d \cdot \nabla \right) q_d = \sum_\alpha I_\alpha(n_\alpha, q_d), \tag{7.5}$$

where I_α stands for the charging current that involves species α. When I combine (7.5) and (7.4), I get

$$- \sum_\alpha q_\alpha S_\alpha = \sum_\alpha n_d I_\alpha(n_\alpha, q_d). \qquad (7.6)$$

No specific assumptions were needed so far, but to proceed further I will assume that there exists a homogeneous equilibrium state, for which the total charging current vanishes

$$\sum_\alpha I_{\alpha 0} = 0, \qquad (7.7)$$

and that the system subsequently remains close to it. In that case one can Taylor expand (7.6) to obtain to linear order that

$$- \sum_\alpha q_\alpha S_\alpha \simeq N_d \sum_\alpha \frac{\partial I_\alpha}{\partial n_\alpha}\bigg|_{(N_\alpha, Q_d)} \delta n_\alpha + N_d \delta q_d \sum_\alpha \frac{\partial I_\alpha}{\partial q_d}\bigg|_{(N_\alpha, Q_d)}$$
$$+ \delta n_d \sum_\alpha I_{\alpha 0}. \qquad (7.8)$$

However, the last term vanishes because of (7.7). Inspired by this result, I will introduce a form for S_α, vanishing at equilibrium for all species as stated before,

$$S_\alpha = - \nu_\alpha \delta n_\alpha - \mu_\alpha \delta q_d, \qquad (7.9)$$

and keeping in mind that this expression is only valid close to equilibrium. Thus two categories of attachment or charging frequencies ν_α and μ_α are introduced naturally, and these will be discussed at various places further on. With such a form for S_α I have two separate consequences, namely that

$$\sum_\alpha q_\alpha \nu_\alpha \delta n_\alpha = N_d \sum_\alpha \frac{\partial I_\alpha}{\partial n_\alpha}\bigg|_{(N_\alpha, Q_d)} \delta n_\alpha,$$
$$\delta q_d \sum_\alpha q_\alpha \mu_\alpha = N_d \delta q_d \sum_\alpha \frac{\partial I_\alpha}{\partial q_d}\bigg|_{(N_\alpha, Q_d)}. \qquad (7.10)$$

This reasoning holds for any number of plasma species, provided I define ν_α and μ_α by

$$\nu_\alpha = \frac{N_d}{q_\alpha} \frac{\partial I_\alpha}{\partial n_\alpha}\bigg|_{(N_\alpha, Q_d)},$$
$$\mu_\alpha = \frac{N_d}{q_\alpha} \frac{\partial I_\alpha}{\partial q_d}\bigg|_{(N_\alpha, Q_d)}. \qquad (7.11)$$

Before these are discussed in more detail, I want to briefly deal with the momentum equations.

7.1.3 Momentum equations

Also the momentum equations are altered by the presence of dust and related problems occur. The equations of motion (4.43) are, without self-gravitational effects, given by

$$\left(\frac{\partial}{\partial t} + \mathbf{u}_\alpha \cdot \nabla\right)\mathbf{u}_\alpha + \frac{1}{n_\alpha m_\alpha}\nabla \cdot \mathsf{p}_\alpha = \frac{q_\alpha}{m_\alpha}(\mathbf{E} + \mathbf{u}_\alpha \times \mathbf{B}) + \mathbf{M}_\alpha,$$

$$\left(\frac{\partial}{\partial t} + \mathbf{u}_d \cdot \nabla\right)\mathbf{u}_d + \frac{1}{n_d m_d}\nabla \cdot \mathsf{p}_d = \frac{q_d}{m_d}(\mathbf{E} + \mathbf{u}_d \times \mathbf{B}) + \mathbf{M}_d. \quad (7.12)$$

For the specific form of the momentum exchanges due to the charging process, we have been inspired by the linearized expressions usually written for ordinary collisional exchanges [Booker 1984]. In principal this would mean that

$$\mathbf{M}_\alpha = \delta\left\{\gamma_{\alpha d}(\mathbf{u}_\alpha - \mathbf{u}_d)\right\},$$

$$\mathbf{M}_d = \delta\left\{\sum_\alpha \gamma_{d\alpha}(\mathbf{u}_d - \mathbf{u}_\alpha)\right\}. \quad (7.13)$$

The coefficients $\gamma_{\alpha d}$ denote the frequency characterizing the rate of capture of plasma particles by dust particles, and they should be computed from an appropriate and self-consisting theory, as outlined further in this chapter. I will assume for simplicity that these frequencies are constant, unlike the model of Bhatt and Pandey [1994], where the variation of $\gamma_{\alpha d}$ is taken into account. This means that the momentum exchanges will be used in the form

$$\mathbf{M}_\alpha = \gamma_{\alpha d}\delta(\mathbf{u}_\alpha - \mathbf{u}_d),$$

$$\mathbf{M}_d = \sum_\alpha \gamma_{d\alpha}\delta(\mathbf{u}_d - \mathbf{u}_\alpha). \quad (7.14)$$

Note that for a non-drifting plasma both models correspond.

As will be seen in the next section, upon linearizing and Fourier transforming the relevant equations with source/sink terms, one finds that there occur complex quantities, whereas in the absence of charge fluctuations everything was real. The upshot is that wave damping or unstable growth becomes the rule, if one analyzes the dispersion law. Before doing that, however, I briefly look at the charging frequencies.

7.1.4 Charging frequencies

The calculation for the different dust related parameters ν_α, μ_α, $\gamma_{d\alpha}$ and $\gamma_{\alpha d}$ was given by Meuris and Verheest [1996a], assuming that the charging of the grains takes place solely due to the attachment of electrons and ions onto the dust grains, and that other effects of secondary charging, like radiation, can be ignored. Perturbations in the charging currents as well as in the dust charges are assumed to be small, so that I can consider the different charging frequencies to be essentially independent of these perturbations.

Let me repeat some arguments from Chapter 2 in a slightly more general context. Consider spherical dust grains with radius a. The current on such a dust particle is in orbit-limited theory for an unmagnetized plasma given by

$$I_\alpha(\Gamma_\alpha) = n_\alpha q_\alpha \int_{|\mathbf{v}|=\eta\sqrt{\Gamma_\alpha}}^{|\mathbf{v}|=\infty} v\sigma_\alpha f_\alpha(\mathbf{v})d^3\mathbf{v}, \tag{7.15}$$

where the subscript α stands for electrons ($\alpha = e$) or ions ($\alpha = i$), $\mathbf{v}$ is the relative velocity between a plasma particle and a dust grain, with $v = |\mathbf{v}|$, while $f_\alpha(\mathbf{v})$ is the equilibrium distribution function for species α, which can be a Maxwellian or shifted Maxwellian, but its form need not yet be specified. Then σ_α is the charging cross section introduced in (2.7),

$$\sigma_\alpha = \pi a^2 \left(1 - \frac{\Gamma_\alpha}{v^2}\right), \tag{7.16}$$

while $\eta = 1$ is taken for negative particles, because only sufficiently fast negative particles can charge the dust grains. For positive particles, $\eta = 0$. Γ_α was introduced to make the notation easier and stands for

$$\Gamma_\alpha = \frac{2q_\alpha(V - V_p)}{m_\alpha} = \frac{2q_\alpha q_d}{4\pi\varepsilon_0 a m_\alpha}. \tag{7.17}$$

Note that the use of this orbit-limited theory implies that the dust grains are considered as point particles, and hence $a \ll \lambda_D$. The charging currents, due to the collisions of plasma particles of species α with dust grains, can be written as

$$I_\alpha = n_\alpha[I_1(\Gamma_\alpha) - \Gamma_\alpha I_2(\Gamma_\alpha)], \tag{7.18}$$

with

$$I_1(\Gamma_\alpha) = q_\alpha \pi a^2 \int_{|v|=\eta\sqrt{\Gamma_\alpha}}^{|v|=\infty} v f_\alpha(\mathbf{v})d^3\mathbf{v},$$

$$I_2(\Gamma_\alpha) = q_\alpha \pi a^2 \int_{|v|=\eta\sqrt{\Gamma_\alpha}}^{|v|=\infty} \frac{f_\alpha(\mathbf{v})}{v}d^3\mathbf{v}. \tag{7.19}$$

This becomes to first order

$$I_{\alpha 0} + \delta I_\alpha = (N_\alpha + \delta n_\alpha)\left[I_1(\Gamma_{\alpha 0}) - \Gamma_{\alpha 0}I_2(\Gamma_{\alpha 0})\right]$$
$$+ N_\alpha \left[\frac{dI_1}{d\Gamma}(\Gamma_{\alpha 0}) - \Gamma_{\alpha 0}\frac{dI_2}{d\Gamma}(\Gamma_{\alpha 0}) - I_2(\Gamma_{\alpha 0})\right]\delta\Gamma_\alpha. \tag{7.20}$$

It is readily verified, however, with the help of the definitions (7.19) that

$$\frac{dI_1}{d\Gamma}(\Gamma_{\alpha 0}) - \Gamma_{\alpha 0}\frac{dI_2}{d\Gamma}(\Gamma_{\alpha 0}) = 0. \tag{7.21}$$

Since charge fluctuations are governed by

$$\frac{dq_d}{dt} = \sum_\alpha I_\alpha, \tag{7.22}$$

this is up to linear order

$$\frac{d}{dt}\delta q_d = \sum_\alpha \frac{I_{\alpha 0}}{N_\alpha}\delta n_\alpha - \frac{2}{4\pi\varepsilon_0 a}\sum_\alpha \frac{q_\alpha}{m_\alpha} \cdot \frac{I_{\alpha 0}I_2(\Gamma_{\alpha 0})}{I_1(\Gamma_{\alpha 0}) - \Gamma_{\alpha 0}I_2(\Gamma_{\alpha 0})}\delta q_d. \tag{7.23}$$

When I compare this result with (7.4) and (7.9), I find to lowest order that

$$\nu_\alpha = \frac{N_d I_{\alpha 0}}{q_\alpha N_\alpha},$$

$$\mu_\alpha = -\frac{2N_d}{4\pi\varepsilon_0 a m_\alpha} \cdot \frac{I_{\alpha 0}I_2(\Gamma_{\alpha 0})}{I_1(\Gamma_{\alpha 0}) - \Gamma_{\alpha 0}I_2(\Gamma_{\alpha 0})}. \tag{7.24}$$

Furthermore, the frequency $\gamma_{\alpha d}$ characterizing the rate of capture of plasma particles by dust particles can be calculated to lowest order as

$$\gamma_{\alpha d} = N_d \int_{|v|=\eta\sqrt{\Gamma_\alpha}}^{|v|=\infty} v\sigma_\alpha f_\alpha(\mathbf{v})d^3\mathbf{v} = \frac{N_d I_{\alpha 0}}{N_\alpha q_\alpha}, \tag{7.25}$$

and it is readily verified that $\gamma_{\alpha d} = \nu_\alpha$. Assuming conservation of momentum between any two species, as true collisions would do, then yields an expression for $\gamma_{d\alpha}$ as well, namely

$$\gamma_{d\alpha} = \gamma_{\alpha d}\frac{N_\alpha m_\alpha}{N_d m_d}. \tag{7.26}$$

To find expressions for μ_α is not so simple. Hence, a shifted Maxwellian distribution function is assumed

$$f_\alpha(\mathbf{v}) = \left(\frac{m_\alpha}{2\pi\kappa T_\alpha}\right)^{\frac{3}{2}} \exp\left[-\frac{m(\mathbf{v} - \mathbf{U}_\alpha)^2}{2\kappa T_\alpha}\right], \tag{7.27}$$

and used to compute the ratios I_1/I_2 for electrons and protons. Define

$$b_\alpha = \frac{\eta\sqrt{\Gamma_{\alpha 0}}}{\sqrt{2}c_{s\alpha}} \quad \text{and} \quad w_\alpha = \frac{U_\alpha}{\sqrt{2}c_{s\alpha}}, \tag{7.28}$$

to calculate the ratio in a more compact form. For electrons ($\eta = 1$) the thermal velocity is usually much higher than the drift velocity in the dust frame, hence the limit $w_e \to 0$ is taken, which results in

$$\left[\frac{I_1}{I_2}\right]_e = \frac{2\kappa T_e}{m_e} + \Gamma_{e0}. \tag{7.29}$$

On the other hand for the ions the limit $b_i \to 0$ is more appropriate, so that

$$\left[\frac{I_1}{I_2}\right]_i = \frac{2\kappa T_i}{m_i}\left(\frac{1}{2} + w_i^2 + \frac{w_i \exp(-w_i^2)}{\sqrt{\pi}\mathrm{Erf}(w_i)}\right). \tag{7.30}$$

This becomes

$$\left[\frac{I_1}{I_2}\right]_i = \frac{2\kappa T_i}{m_i} \tag{7.31}$$

for driftless ions ($w_i \to 0$). In this way the remaining attachment frequencies

$$\mu_e = \frac{N_d|I_{e0}|}{4\pi\varepsilon_0 a\kappa T_e},$$

$$\mu_i = -\frac{N_d|I_{e0}|}{4\pi\varepsilon_0 a\kappa T_i - eQ_d} = -\frac{N_d|I_{e0}|}{4\pi\varepsilon_0 a(\kappa T_i - eV_0)} \tag{7.32}$$

are obtained from (7.11), assuming a fairly simple model. In these expressions, as seen in more detail in Chapter 2, a stands for the radius and V_0 for the equilibrium surface potential of a typical grain.

7.2 Parallel electrostatic waves with variable dust charges

For waves in plasmas with particles of fixed charge, recalled in Chapter 5, fluctuations in charge density are in reality fluctuations in number density. However, for the dust there are two effects, the fluctuations in number density and in the charge itself. And now we have to tackle the self-consistent description in its glory, or rather, intricacies!

7.2.1 Dispersion law in general

For a start, repeat the continuity equations

$$\frac{\partial n_e}{\partial t} + \frac{\partial}{\partial z}(n_e u_e) = -\nu_e \delta n_e - \mu_e \delta q_d,$$

$$\frac{\partial n_i}{\partial t} + \frac{\partial}{\partial z}(n_i u_i) = -\nu_i \delta n_i - \mu_i \delta q_d,$$

$$\frac{\partial n_d}{\partial t} + \frac{\partial}{\partial z}(n_d u_d) = 0, \tag{7.33}$$

and the equations of motions

$$\left(\frac{\partial}{\partial t} + u_e\frac{\partial}{\partial z}\right)u_e + \frac{c_{se}^2}{n_e}\frac{\partial n_e}{\partial z} = \frac{e}{m_e}\frac{\partial\varphi}{\partial z} - \gamma_{ed}\delta(u_e - u_d),$$

$$\left(\frac{\partial}{\partial t} + u_i\frac{\partial}{\partial z}\right)u_i + \frac{c_{si}^2}{n_i}\frac{\partial n_i}{\partial z} = -\frac{e}{m_i}\frac{\partial\varphi}{\partial z} - \gamma_{id}\delta(u_i - u_d),$$

$$\left(\frac{\partial}{\partial t} + u_d\frac{\partial}{\partial z}\right)u_d = -\frac{q_d}{m_d}\frac{\partial\varphi}{\partial z} - (\gamma_{de} + \gamma_{di})\delta u_d$$
$$+ \gamma_{de}\delta u_e + \gamma_{di}\delta u_i, \qquad (7.34)$$

explicitly for a three-component dusty plasma. The fluid equations are supplemented by Poisson's equation

$$\varepsilon_0\frac{\partial^2\varphi}{\partial z^2} - en_e + en_i + n_d q_d = 0, \qquad (7.35)$$

together with the conservation of total charge

$$\frac{\partial}{\partial t}(en_i - en_e + n_d q_d) + \frac{\partial}{\partial z}(en_i u_i - en_e u_e + n_d q_d u_d) = 0. \qquad (7.36)$$

Before going on, it is instructive to use the continuity equations (7.33) with source/sink terms and obtain

$$N_d\frac{d}{dt}\delta q_d + e(\mu_e - \mu_i)\delta q_d = e(\nu_i\delta n_i - \nu_e\delta n_e). \qquad (7.37)$$

This shows that there is a natural decay rate for the dust charge dynamics, driven by the perturbations in the attachment frequencies of electrons and ions [Bhatt and Pandey 1994].

Now I linearize and Fourier transform the relevant equations, as done several times in the previous chapters. However, since the plasma and dust equations are coupled by the charge exchange terms, I want to proceed very cautiously. With the help of some compact notations,

$$\begin{array}{ll} \omega_{\nu e} = \omega - kU_e + i\nu_e, & \omega_{\gamma e} = \omega - kU_e + i\gamma_{ed}, \\ \omega_{\nu i} = \omega - kU_i + i\nu_i, & \omega_{\gamma i} = \omega - kU_i + i\gamma_{id}, \\ \omega_d = \omega - kU_d, & \omega_{\gamma d} = \omega - kU_d + i\gamma_{de} + i\gamma_{di}, \end{array} \qquad (7.38)$$

the electron and ion equations yield

$$\delta n_e = \frac{kN_e}{\omega_{\nu e}}\delta u_e - i\frac{\mu_e}{\omega_{\nu e}}\delta q_d,$$

$$\delta u_e = \frac{kc_{se}^2}{N_e\omega_{\gamma e}}\delta n_e + \frac{kq_e}{m_e\omega_{\gamma e}}\delta\varphi + i\frac{\gamma_{ed}}{\omega_{\gamma e}}\delta u_d,$$

$$\delta n_i = \frac{kN_i}{\omega_{\nu i}}\delta u_i - i\frac{\mu_i}{\omega_{\nu i}}\delta q_d,$$

$$\delta u_i = \frac{kc_{si}^2}{N_i\omega_{\gamma i}}\delta n_i + \frac{kq_i}{m_i\omega_{\gamma i}}\delta\varphi + i\frac{\gamma_{id}}{\omega_{\gamma i}}\delta u_d. \qquad (7.39)$$

Solving for the electron variables leads to

$$\delta n_e = \frac{\varepsilon_0 k^2\omega_{pe}^2}{q_e(\omega_{\nu e}\omega_{\gamma e} - k^2c_{se}^2)}\delta\varphi + i\frac{kN_e\gamma_{ed}}{\omega_{\nu e}\omega_{\gamma e} - k^2c_{se}^2}\delta u_d$$
$$- i\frac{\mu_e\omega_{\gamma e}}{\omega_{\nu e}\omega_{\gamma e} - k^2c_{se}^2}\delta q_d,$$

$$\delta u_e = \frac{kq_e\omega_{\nu e}}{m_e(\omega_{\nu e}\omega_{\gamma e} - k^2c_{se}^2)}\delta\varphi + i\frac{\gamma_{ed}\omega_{\nu e}}{\omega_{\nu e}\omega_{\gamma e} - k^2c_{se}^2}\delta u_d$$
$$- i\frac{kc_{se}^2\mu_e}{N_e(\omega_{\nu e}\omega_{\gamma e} - k^2c_{se}^2)}\delta q_d, \qquad (7.40)$$

with analogous expressions for the ion density and velocity. For the dust grains, however, I have that

$$\delta n_d = \frac{kN_d}{\omega_d}\delta u_d,$$

$$\delta u_d = \frac{kQ_d}{m_d\omega_{\gamma d}}\delta\varphi + i\frac{\gamma_{de}}{\omega_{\gamma d}}\delta u_e + i\frac{\gamma_{di}}{\omega_{\gamma d}}\delta u_i. \qquad (7.41)$$

Inserting these results into the linearized and Fourier transformed equation of Poisson (7.35) gives us a global expression of the form

$$A\delta\varphi + B\delta u_d + C\delta q_d = 0, \qquad (7.42)$$

with

$$A = \varepsilon_0 k^2\left\{\frac{\omega_{pe}^2}{\omega_{\nu e}\omega_{\gamma e} - k^2c_{se}^2} + \frac{\omega_{pi}^2}{\omega_{\nu i}\omega_{\gamma i} - k^2c_{si}^2} - 1\right\},$$

$$B = \frac{kN_dQ_d}{\omega_d} + i\frac{kN_eq_e\gamma_{ed}}{\omega_{\nu e}\omega_{\gamma e} - k^2c_{se}^2} + i\frac{kN_iq_i\gamma_{id}}{\omega_{\nu i}\omega_{\gamma i} - k^2c_{si}^2},$$

$$C = N_d - i\frac{\mu_eq_e\omega_{\gamma e}}{\omega_{\nu e}\omega_{\gamma e} - k^2c_{se}^2} - i\frac{\mu_iq_i\omega_{\gamma i}}{\omega_{\nu i}\omega_{\gamma i} - k^2c_{si}^2}. \qquad (7.43)$$

Similarly, the global conservation of charge (7.36) gives that

$$D\delta\varphi + E\delta u_d + F\delta q_d = 0, \qquad (7.44)$$

with

$$D = \varepsilon_0 k^2 \left\{ \frac{\nu_e \omega_{pe}^2}{\omega_{\nu e}\omega_{\gamma e} - k^2 c_{se}^2} + \frac{\nu_i \omega_{pi}^2}{\omega_{\nu i}\omega_{\gamma i} - k^2 c_{si}^2} \right\},$$

$$E = ik \frac{N_e q_e \nu_e \gamma_{ed}}{\omega_{\nu e}\omega_{\gamma e} - k^2 c_{se}^2} + ik \frac{N_i q_i \nu_i \gamma_{id}}{\omega_{\nu i}\omega_{\gamma i} - k^2 c_{si}^2},$$

$$F = iN_d \omega_d + q_e \mu_e + q_i \mu_i - i \frac{q_e \nu_e \mu_e \omega_{\gamma e}}{\omega_{\nu e}\omega_{\gamma e} - k^2 c_{se}^2}$$

$$- i \frac{q_i \nu_i \mu_i \omega_{\gamma i}}{\omega_{\nu i}\omega_{\gamma i} - k^2 c_{si}^2}. \tag{7.45}$$

Finally, there remains to insert the electron velocity from (7.40) and similarly the ion velocity into the dust velocity in (7.41) to get

$$G\delta\varphi + H\delta u_d + J\delta q_d = 0, \tag{7.46}$$

with

$$G = \frac{kQ_d}{m_d} + ik \frac{q_e \omega_{\nu e} \gamma_{de}}{m_e(\omega_{\nu e}\omega_{\gamma e} - k^2 c_{se}^2)} + ik \frac{q_i \omega_{\nu i} \gamma_{di}}{m_i(\omega_{\nu i}\omega_{\gamma i} - k^2 c_{si}^2)},$$

$$H = -\omega_{\gamma d} - \frac{\omega_{\nu e} \gamma_{de} \gamma_{ed}}{\omega_{\nu e}\omega_{\gamma e} - k^2 c_{se}^2} - \frac{\omega_{\nu i} \gamma_{di} \gamma_{id}}{\omega_{\nu i}\omega_{\gamma i} - k^2 c_{si}^2},$$

$$J = \frac{kc_{se}^2 \mu_e \gamma_{de}}{N_e(\omega_{\nu e}\omega_{\gamma e} - k^2 c_{se}^2)} + \frac{kc_{si}^2 \mu_i \gamma_{di}}{N_i(\omega_{\nu i}\omega_{\gamma i} - k^2 c_{si}^2)}. \tag{7.47}$$

The dispersion law is then

$$\begin{vmatrix} A & B & C \\ D & E & F \\ G & H & J \end{vmatrix} = 0, \tag{7.48}$$

and clearly very complicated. For the following discussions, I work in the dust frame of reference, so that $U_d = 0$, without loss of generality.

If constant dust charges are assumed, all coefficients μ_e, μ_i, ν_e, ν_i, γ_{de}, γ_{di}, γ_{ed} and γ_{id} have to be set equal to zero, and we are back to Chapter 5. In the limit of negligible electron and proton inertia (giving Boltzmann distributions for their densities) the dispersion law for dust-acoustic modes is recovered, as given by Rao et al. [1990] and Rao and Shukla [1994]. For variable dust charges, but electrons and ions that still are Boltzmann distributed, our dispersion law includes results given by Melandsø et al. [1993a], subsequently also found by Bhatt and Pandey [1994], up some misprints. These cases are discussed below.

7.2.2 No dust dynamics

For high-frequency modes, I can assume $m_d \to \infty$ to a good approximation and thereby neglect the dust dynamics altogether. The dispersion law (7.48) then reduces to

$$\begin{vmatrix} A & C \\ D & F \end{vmatrix} = 0 \qquad (7.49)$$

or more explicitly

$$\left(\frac{\omega_{pe}^2}{\mathcal{L}_e} + \frac{\omega_{pi}^2}{\mathcal{L}_i} - 1\right)\left(i\omega - \alpha_e - \alpha_i + i\frac{\alpha_e \nu_e \omega_{\gamma e}}{\mathcal{L}_e} + i\frac{\alpha_i \nu_i \omega_{\gamma i}}{\mathcal{L}_i}\right)$$
$$- \left(\frac{\nu_e \omega_{pe}^2}{\mathcal{L}_e} + \frac{\nu_i \omega_{pi}^2}{\mathcal{L}_i}\right)\left(1 + i\frac{\alpha_e \omega_{\gamma e}}{\mathcal{L}_e} + i\frac{\alpha_i \omega_{\gamma i}}{\mathcal{L}_i}\right) = 0, \qquad (7.50)$$

with

$$\mathcal{L}_\alpha = (\omega - kU_\alpha + i\nu_\alpha)(\omega - kU_\alpha + i\gamma_{\alpha d}) - k^2 c_{s\alpha}^2, \qquad (7.51)$$

and

$$\alpha_e = \frac{e\mu_e}{N_d}, \qquad \alpha_i = -\frac{e\mu_i}{N_d}. \qquad (7.52)$$

The dispersion law is a polynomial of degree 5 in ω. When there are no equilibrium drifts, it has the structure

$$\omega^5 + ia_1\omega^4 + a_2\omega^3 + ia_3\omega^2 + a_4\omega + ia_5 = 0, \qquad (7.53)$$

with real coefficients a_i. This equation describes two damped Langmuir and two damped dust-ion-acoustic modes, all with dispersion laws of the form $\omega = \pm\omega_1 - i\omega_2$, coupled to a purely imaginary mode ($\omega = -i\omega_3$). For the dust-acoustic modes that occur at lower frequencies, however, the dynamics of the dust grains must be included and in that case I need to analyze the dispersion law (7.48) in full. All these dispersion laws can be solved numerically, of course, but it is instructive to find analytic approximations for the different branches. In that way I can also compare various results from the literature.

7.2.3 Langmuir modes

The Langmuir modes that I briefly recalled in Chapter 5 are excited when in a small region of a neutral plasma some particles are displaced from their equilibrium position. A charge imbalance is thereby created, and the associated electrostatic fields will pull the particles back, but particle inertia makes them overshoot their equilibrium position. This results in an oscillatory motion around the equilibrium position at a frequency ω_{pa}, usually maintained by the lightest particles, the electrons. In the case of a dusty plasma these Langmuir oscillations couple to the charge relaxation mechanism and become damped, as I now show.

For simplicity, I assume that the electrons and ions are Maxwellian so that $\gamma_{ed} = \nu_e$ and $\gamma_{id} = \nu_i$. When ν_e, ν_i, α_e, α_i, kU_e, kU_i and kc_{si} are considered as small frequencies, compared to the plasma frequencies, I can solve the dispersion law (7.50) and get to lowest order

$$\left(\frac{\omega_{pe}^2}{\mathcal{L}_e} + \frac{\omega_{pi}^2}{\mathcal{L}_i} - 1\right)(i\omega - \alpha_e - \alpha_i) - \frac{\nu_e \omega_{pe}^2}{\mathcal{L}_e} - \frac{\nu_i \omega_{pi}^2}{\mathcal{L}_i} \simeq 0. \qquad (7.54)$$

The solution is of the form

$$\omega = \omega_o - i\nu_e \frac{2\omega_o^2 - \omega_{pe}^2 - 2\omega_{pi}^2}{2\mathcal{N}} - i\nu_i \frac{\omega_o^2 - \omega_{pe}^2 - k^2 c_{se}^2}{2\mathcal{N}}$$
$$+ kU_e \frac{\omega_o^2 - \omega_{pi}^2}{\mathcal{N}} + kU_i \frac{\omega_o^2 - \omega_{pe}^2 - k^2 c_{se}^2}{\mathcal{N}}, \qquad (7.55)$$

where

$$\mathcal{N} = 2\omega_o^2 - \omega_{pe}^2 - \omega_{pi}^2 - k^2 c_{se}^2 \qquad (7.56)$$

and ω_o is a solution of the zeroth order dispersion law

$$\left(\omega_o^2 - \omega_{pe}^2 - \omega_{pi}^2\right)\left(\omega_o^2 - k^2 c_{se}^2\right) = \omega_{pe}^2 k^2 c_{se}^2. \qquad (7.57)$$

Restrict the analysis to solutions with a frequency close to the total plasma frequency, so that $\omega_o \geq \omega_p$ and $\omega_o > kc_{se}$. The latter restriction naturally occurs because in a fluid treatment the phase velocity should be larger than the thermal velocity. For most three-component dusty plasmas $\omega_{pe} \simeq \omega_p$, but when all electrons reside on the dust grains $\omega_{pe} \to 0$.

When the zeroth order relation (7.57) is rewritten as

$$\omega_o^2 - \omega_{pe}^2 - k^2 c_{se}^2 = \omega_{pi}^2 \left(1 - \frac{k^2 c_{se}^2}{\omega_o^2}\right), \qquad (7.58)$$

I see that $\mathcal{N}$ and the coefficients of ν_e and ν_i in (7.55) are all positive, so that there indeed is damping.

This damping mechanism may be understood as follows [Bhatt 1997]. Fluctuations in the electrostatic potential will occur because of
   (1) electron density fluctuations, as in usual plasmas, but also
   (2) electron density variations due to the capture/release of electrons by the charged dust grains, and finally
   (3) dust charge fluctuations proper.

When only (1) and (3) are considered (no sink/source terms in the electron and ion continuity equations nor in a Vlasov approach were I to work in a kinetic description), the modes become unstable [Ma and Yu 1994b; Li *et al.* 1994], as the stabilizing effect of (2) does not occur. In a fully self-consistent description, however, the effect of (2) plays a crucial role. Indeed, the damping rate due to (2) is twice as large as the growth rate due to (1) and (3), and therefore damping occurs. This damping is a clear consequence of the coupling between the Langmuir waves and the charge relaxation mechanisms.

For very small thermal effects the dispersion law can be approximated as

$$\omega = \omega_p - i\frac{\nu_e\omega_{pe}^2 + \nu_i\omega_{pi}^2}{2\omega_{pe}^2 + 2\omega_{pi}^2} + \frac{kU_e\omega_{pe}^2 + kU_i\omega_{pi}^2}{2\omega_{pe}^2 + 2\omega_{pi}^2}. \tag{7.59}$$

The ion terms in this dispersion law are usually an order m_e/m_i smaller than the electron terms, and hence I can safely neglect them. In order to evaluate the importance of this charge exchange damping, I should compare this with the rates obtained from a kinetic theory, which predicts an additional Landau damping [Stix 1992]. Without going into details, it can be shown that there exists a domain $k_{cr1} \le k \le k_{cr2}$ for which classical Landau damping exceeds charge exchange damping.

With hindsight, it is easy to sketch in general terms what happens. I could have found (7.59) without the small ion terms by simply assuming that not only the dust grains, but also the ions are part of the static neutralizing background, as they are in the classic derivation of Langmuir modes. In this picture, only the dust charge variation and the electron density variations sustain the oscillations, while the ions do contribute to the equilibrium charge of the grains. The system can then be described by the continuity and momentum equations for the electrons alone, the charge variation equation and Poisson's equation. Provided I leave all equilibrium streaming out when deriving the dispersion laws, I recover the result derived by Bhatt [1997].

However, for a drifting plasma the result obtained by Bhatt [1997] is different because variations of the attachment frequencies were included in the model. When $\omega_{pi} \ll \omega_{pe}$ and equal streaming of the electrons and the ions with respect to the dust background is assumed ($U_e = U_i = U$), (7.50) can be solved for small α_e, α_i, ν_e, ν_i and kc_{si} as

$$\omega = kU \pm \sqrt{\omega_{pe}^2 + k^2 c_{se}^2} - i\frac{\nu_e}{2}\frac{\omega_{pe}^2 + 2k^2 c_{se}^2 \pm 2kU\sqrt{\omega_{pe}^2 + k^2 c_{se}^2}}{\omega_{pe}^2 + k^2 c_{se}^2 \pm kU\sqrt{\omega_{pe}^2 + k^2 c_{se}^2}}. \tag{7.60}$$

Taking for simplicity $kU > 0$, the mode for which $\omega \simeq kU + \sqrt{\omega_{pe}^2 + k^2 c_{se}^2}$ is damped, while a weak instability can occur for the mode with $\omega \simeq$

$kU - \sqrt{\omega_{pe}^2 + k^2 c_{se}^2}$, provided

$$\frac{\omega_{pe}^2 + 2k^2 c_{se}^2}{2\sqrt{\omega_{pe}^2 + k^2 c_{se}^2}} < kU < \sqrt{\omega_{pe}^2 + k^2 c_{se}^2}. \tag{7.61}$$

At small thermal effects this range is approximately given by

$$\frac{1}{2} + \frac{3k^2 c_{se}^2}{4\omega_{pe}^2} < \frac{kU}{\omega_{pe}} < 1 + \frac{k^2 c_{se}^2}{2\omega_{pe}^2}. \tag{7.62}$$

For $kU < 0$ the conclusions are just reversed. The instability is clearly a consequence of the coupling between the charging process and the streaming.

7.2.4 Dust-ion-acoustic modes

At lower frequencies, sound waves driven by long-range electrostatic forces are modified when the motion of the more massive ions is also taken into account. Such ion-acoustic modes are generated in the frequency range for which

$$k^2 c_{se}^2 \gg k^2 c_{se}^2 \omega_{pi}^2/(\omega_{pe}^2 + k^2 c_{se}^2) + k^2 c_{si}^2 = \omega_{dia}^2. \tag{7.63}$$

In the presence of charged dust, the imbalance between the electron and ion densities will make the dust-ion-acoustic frequency ω_{dia} deviate from the ordinary ion-acoustic frequency ω_{ia}.

Again, treat the charging frequencies ν_e, ν_i, α_e and α_i as small, this time compared to ω_{dia}, and then the dispersion law (7.50) is to lowest order in these small quantities

$$\omega = \omega_o - i\frac{k^2 \lambda_{De}^2 \omega_{pi}^2}{2\omega_o(\omega_o - kU_i)(1 + k^2 \lambda_{De}^2)^2}\nu_e$$
$$- i\left\{1 - \frac{k^2 \lambda_{De}^2 \omega_{pi}^2}{2\omega_o(\omega_o - kU_i)(1 + k^2 \lambda_{De}^2)}\right\}\nu_i. \tag{7.64}$$

Here

$$\omega_o = kU_i \pm \sqrt{\frac{k^2 \lambda_{De}^2 \omega_{pi}^2}{1 + k^2 \lambda_{De}^2} + k^2 c_{si}^2} = kU_i \pm \omega_{dia} \tag{7.65}$$

is Doppler-shifted from the dust-ion-acoustic frequency.

For a non-drifting plasma, (7.64) becomes

$$\omega = \omega_o - i\nu_e \frac{\lambda_{De}^2}{2(1 + k^2 \lambda_{De}^2)(\lambda_{De}^2 + \lambda_{Di}^2 + k^2 \lambda_{Di}^2 \lambda_{De}^2)}$$
$$- i\nu_i \frac{\lambda_{De}^2 + 2\lambda_{Di}^2 + 2k^2 \lambda_{De}^2 \lambda_{Di}^2}{2(\lambda_{De}^2 + \lambda_{Di}^2 + k^2 \lambda_{Di}^2 \lambda_{De}^2)}. \tag{7.66}$$

This clearly indicates damping due to the coupling of the dust-ion-acoustic mode with the grain charging mechanism.

This dispersion law could further be simplified for an isothermal plasma, for which $\lambda_{De}^2/\lambda_{Di}^2 = \nu_e/\nu_i$. At low dust concentrations $N_e \simeq N_i$, both conditions together give

$$\omega = \omega_o - i\nu_e\frac{(1+k^2\lambda_{De}^2)^{-1} + N + 2N^2(1+\lambda_{De}^2)}{2[1+N(1+k^2\lambda_{De}^2)]}, \tag{7.67}$$

if $N = N_e/N_i$ is the fractional electron density. Both for very short and very long wavelengths, the imaginary part of the frequency becomes $\mathrm{Im}\,\omega = -\nu_e \simeq -\nu_i$, valid as long as $\omega_o \gg \nu_e$ and $k \gg \nu_e/(\omega_{pi}\lambda_{De})$. A minimum damping of $(\sqrt{2}-0.5)\nu_e = 0.91\nu_e$ occurs for $k\lambda_{De} = \sqrt[4]{2}$.

There is also a new dust-ion-acoustic drift driven instability which can be analyzed using equation (7.64). When $kc_{si} \ll \omega_o$, the dispersion relation is simplified to

$$\omega = \omega_o - i\frac{\omega_{dia}}{2(1+k^2\lambda_{De}^2)(\omega_{dia}\pm kU_i)}\nu_e - i\frac{\omega_{dia}\pm 2kU_i}{2(\omega_{dia}\pm 2kU_i)}\nu_i, \tag{7.68}$$

and even further when $k\lambda_{De} \ll 1$, so that

$$\omega = kU_i \pm k\lambda_{De}\omega_{pi} - i\frac{\lambda_{De}\omega_{pi}(\nu_e+\nu_i)\pm 2U_i\nu_i}{2(\lambda_{De}\omega_{pi}\pm U_i)}. \tag{7.69}$$

It is readily verified that the mode for which $\mathrm{Re}\,\omega = kU_i + \omega_{dia}$ is always damped, while the $\mathrm{Re}\,\omega = kU_i - \omega_{dia}$ mode can become unstable [Bhatt and Pandey 1994] and grow in a very narrow streaming regime given by

$$\lambda_{De}\omega_{pi} < U_i < \left(1+\frac{N_i}{N_e}\right)\frac{\lambda_{De}\omega_{pi}}{2}. \tag{7.70}$$

However, the damping for high η values is finite for our approach, while a infinitely strong damping is encountered in the approach of Bhatt and Pandey [1994].

7.2.5   Dust-acoustic modes

For still lower phase velocities, the dust dynamics can no longer be neglected, and the dust participates in the wave motion. These dust-acoustic modes occur in the range $kc_{se} \gg kc_{si} \gg |\omega - kU_a| \gg kc_{sd}$. When I use the expressions for the different coefficients and take the limit for $m_e \to 0$ and $m_i \to 0$ in the full dispersion law (7.48), I find the result of Melandsø et al. [1993a], which can be written as

$$(\alpha - i\omega)\left(A^2\omega_{pd}^2 - \omega^2\right) = A^2\beta\omega^2, \tag{7.71}$$

with the definitions

$$
\alpha = \alpha_e + \alpha_i = \frac{a}{\sqrt{2\pi}} \left[\frac{\omega_{pi}}{\lambda_{Di}} + \frac{\omega_{pe}}{\lambda_{De}} \exp \frac{eV_0}{\kappa T_e} \right],
$$

$$
\beta = \frac{\nu_e}{k^2 \lambda_{De}^2} + \frac{\nu_i}{k^2 \lambda_{Di}^2} = \frac{f}{k^2 \lambda_D^2} \cdot \frac{a}{\sqrt{2\pi}} \left[\frac{\omega_{pi}}{\lambda_{Di}} \left(1 - \frac{eV_0}{\kappa T_i} \right) + \frac{\omega_{pe}}{\lambda_{De}} \exp \frac{eV_0}{\kappa T_e} \right],
$$

$$
A^2 = \frac{k^2 \lambda_D^2}{1 + k^2 \lambda_D^2}. \tag{7.72}
$$

In these expressions, α and $\beta k^2 \lambda_D^2 / f$ are sometimes referred to as the charging frequencies, λ_D is of course the global plasma Debye length, and the dimensionless parameter

$$
f = 4\pi a N_d \lambda_D^2 = 4\pi \left(N_d \lambda_D^3 \right) \left(\frac{a}{\lambda_D} \right), \tag{7.73}
$$

essentially represents the number of dust grains in a Debye sphere weighted by the grain size expressed in units of the Debye length. This parameter was first introduced by Melandsø et al. [1993a], as a measure of how dense the dusty plasma would be, but recently linked in more physical terms to the concept of fugacity by Rao [1999]. I will come back to it at the end of this section.

As (7.71) clearly indicates, there will be damping because the dust-acoustic mode couples with the zero frequency mode due to the charging mechanism. This dispersion law (7.71) is only valid when all the finite mass effects of the plasma particles can be neglected. As alluded to already before, the derivation by Melandsø et al. [1993a] uses Boltzmann distributions for the plasma particles, and therefore completely sidesteps the issue of sink and source terms in continuity and momentum equations.

These terms become important at both ends of the wavelength domain, because for these k-values the damping rate will be less than γ_{de} and γ_{di} in (7.71), but equal to their sum in a self-consistent approach. To see this, go back to (7.48) in the frequency domain for the dust-acoustic mode but retain ion and electron inertial effects. Again supposing that the charging frequencies ν_e, ν_i, α_e, α_i, γ_{ed}, γ_{id}, γ_{de} and γ_{di} are small compared to ω, I find from (7.48) that

$$
\omega = \pm A\omega_{pd} - i\nu_e \frac{A^2}{2k^2 \lambda_{De}^2} - i\gamma_{ed} \frac{A^2 m_e Q_d}{2k^2 \lambda_{De}^2 m_d q_e} - i\frac{\gamma_{de}}{2} \left\{ 1 + \frac{A^2 N_d Q_d}{k^2 \lambda_{De}^2 N_e q_e} \right\}
$$

$$
- i\nu_i \frac{A^2}{2k^2 \lambda_{Di}^2} - i\gamma_{id} \frac{A^2 m_i Q_d}{2k^2 \lambda_{Di}^2 m_d q_i} - i\frac{\gamma_{di}}{2} \left\{ 1 + \frac{A^2 N_d Q_d}{k^2 \lambda_{Di}^2 N_i q_i} \right\}. \tag{7.74}
$$

It is easily seen that for large k values the frequency becomes

$$
\omega = \pm A\omega_{pd} - i\frac{\gamma_{de} + \gamma_{di}}{2}, \tag{7.75}
$$

and the damping is dominated by the momentum exchanges [Meuris and Verheest 1996a].

For Maxwellian electrons and ions, and restricting the dust charging only to primary charging, (7.74) becomes

$$
\omega = \pm A\omega_{pd} - i\frac{\nu_e}{2}\left\{\frac{N_e m_e}{N_d m_d} + \frac{A^2}{k^2\lambda_{De}^2}\left(1 + 2\frac{Q_d m_e}{q_e m_d}\right)\right\}
$$
$$
- i\frac{\nu_i}{2}\left\{\frac{N_i m_i}{N_d m_d} + \frac{A^2}{k^2\lambda_{Di}^2}\left(1 + 2\frac{Q_d m_i}{q_i m_d}\right)\right\}. \tag{7.76}
$$

For intermediate k-values (7.71) adequately describes the dust-acoustic mode, even for high dust densities. From the general structure of this dispersion law we see that the solutions are of the form

$$
\omega = \pm\omega_1 - i\omega_2, \qquad\qquad \omega = -i\omega_3. \tag{7.77}
$$

There is always one purely imaginary mode. However, (7.71) can have three purely imaginary roots for ω, and in that case the dust-acoustic mode disappears as such. This can occur in a finite interval for k, and the marginal k-values are those for which (7.71) has a double and a simple purely imaginary root. The k-values in question have to be determined from

$$
4\alpha(\alpha + A^2\beta)^3 + A^2(8\alpha^2 - 20A^2\alpha\beta - A^4\beta^2)\omega_{pd}^2 + 4A^4\omega_{pd}^4 = 0. \tag{7.78}
$$

Finally, the discussion about the dust-acoustic mode can be rounded of by looking at the opposite limit, when $\omega \ll \alpha \ll A^2\beta$. This would seem valid in laboratory experiments [Thompson et al. 1997], and then (7.71) reduces to [Melandsø et al. 1993a; Rao 1999]

$$
\frac{\omega^2}{k^2} = \frac{C_{da}^2}{1 + k^2\lambda_D^2 + f\Delta}. \tag{7.79}
$$

Here

$$
\Delta = \frac{\beta k^2 \lambda_D^2}{f\alpha} = \frac{\dfrac{\omega_{pi}}{\lambda_{Di}}\left(1 - \dfrac{eV_0}{\kappa T_i}\right) + \dfrac{\omega_{pe}}{\lambda_{De}}\exp\dfrac{eV_0}{\kappa T_e}}{\dfrac{\omega_{pi}}{\lambda_{Di}} + \dfrac{\omega_{pe}}{\lambda_{De}}\exp\dfrac{eV_0}{\kappa T_e}} \tag{7.80}
$$

is a parameter of order unity under those conditions. The dense limit is defined by $k^2\lambda_D^2 \ll f\Delta$ and $1 \ll f\Delta$, so that f is a more natural way of characterizing dense dusty plasmas than the Havnes parameter P, given in (2.42), which is too dependent on the dust temperature. Here f can be introduced even for cold dusty plasmas, as we did, and in the dense limit (7.79) takes the form

$$
\omega^2 = \frac{\omega_{pd}^2\alpha}{\beta} = \frac{k^2 C_{da}^2}{f\Delta}. \tag{7.81}
$$

For a further discussion of this dust-Coulomb wave, a normal mode of dense dusty plasmas arising essentially due to grain charge fluctuations, which are proportional to dust number density perturbations, and existing in a frequency regime which is much lower than that of the usual dust-acoustic waves, I refer to Rao [1999].

7.2.6 Zero-frequency modes

The charging mechanism itself gives rise to a zero-frequency damped mode. When (7.48) or for that matter (7.71) are solved to lowest order in the small quantities ν_e, ν_i, α_e, α_i, γ_{de}, γ_{di}, γ_{ed} and γ_{id}, and the frequency ω itself is included in the small frequencies, I get for small wavelengths

$$\omega = -i(\alpha_e + \alpha_i). \tag{7.82}$$

This corresponds to the charge relaxation frequency. Analogous results have been obtained by Das *et al.* [1996], with or without streaming between the electrons and the ions, in a cold plasma approach which, however, renders the charging of the dust grains rather questionable. Indeed, even in the simplest orbit limited theory the equilibrium charge Q_d on a dust grain is determined from

$$N_i c_{si} \left(1 - \frac{eQ_d}{4\pi\varepsilon_0 a\kappa T_i}\right) = N_e c_{se} \exp\left[\frac{eQ_d}{4\pi\varepsilon_0 a\kappa T_e}\right]. \tag{7.83}$$

For long wavelengths, the situation is more complex, because the mode couples to the other modes, as explained before, and no tractable analytical expressions exist.

7.2.7 Case studies: Rings of Saturn

The validity of the analysis and previous dispersion laws can be verified by solving the full dispersion law (7.48) numerically. This equation is a polynomial with 17 solutions! In order to obtain some physical insight, several approximations were made in the previous subsections. Mostly the various attachment frequencies were considered small compared to the wave frequencies. However, as seen from the discussion of the dust-acoustic mode in denser dusty plasmas, the laboratory experiments indicate that the charging is very rapid, and so charge fluctuations would give rise to some jitter on the dispersion laws. At the typical low frequencies characterizing the dust you would not have to worry unduly then about such high-frequency noise. Nevertheless, there is evidence from space observations that in certain parts of the solar system the attachment frequencies are small, not large, and so small indeed that the type of analysis presented in the preceding subsections is a valid picture.

Hence I will quote some results obtained by Meuris [1997b] in his PhD thesis. The parameters were derived by assuming that only primary charging is important, which is not really true for the rings of Saturn, but to a first approximation gives a fair idea of the order of magnitude of the different frequencies involved. The parameters used are listed in Table 7.1.

		E ring	Spokes
N_d	(m^{-3})	10^2	10^6
N_i	(m^{-3})	10^7	10^7
T_e	(K)	5×10^5	5×10^5
T_i	(K)	5×10^5	5×10^5
λ_{De}	(m)	23.3	101
λ_{Di}	(m)	15.4	15.4
λ_D	(m)	12.9	15.3
a	(m)	10^{-6}	10^{-6}
N		0.438	0.0233
χ		-1.87	-0.000326
$N_d a^2 \lambda_D$		1.2810^{-9}	15.2×10^{-6}
$N_d a \lambda_D^2$		0.017	233

Table 7.1. Data used in the case studies [Meuris 1997b]

E ring of Saturn: Dust in plasma

Neglecting all relative drifts and also the magnetic field, the different attachment frequencies are computed as

$$\nu_e = \gamma_{ed} = 2.11 \times 10^{-4} \text{ s}^{-1}$$
$$\nu_i = \gamma_{id} = 0.92 \times 10^{-4} \text{ s}^{-1}$$
$$\alpha_e = 3.09 \times 10^{-4} \text{ s}^{-1}$$
$$\alpha_i = 1.08 \times 10^{-4} \text{ s}^{-1}$$
$$\gamma_{de} = 2.01 \times 10^{-15} \text{ s}^{-1}$$
$$\gamma_{di} = 3.69 \times 10^{-12} \text{ s}^{-1}$$
$$\omega_p = 118 \times 10^3 \text{s}^{-1} \tag{7.84}$$

It turns out that the Langmuir modes are perfectly described by (7.59) for the k-domain considered. The Langmuir damping is larger than the charge exchange damping for k values outside the domain [0.01, 1.60] in normalized units. Because Landau damping decays exponentially for both smaller and larger wavelengths, I may safely neglect it outside this interval.

The dust-ion-acoustic modes, on the other hand, are described by (7.67) because ω_{dia} is much larger than all the small charging and attachment frequencies.

Finally, the dust-acoustic mode does not disappear in the k-interval considered. As long as $A\omega_{pd}$ is an order of magnitude larger than the small frequencies, (7.74) is a very good approximation. For larger wavelengths, (7.71) is valid. Note that this dispersion law is not valid for higher k values, because $\gamma_{d\alpha}$ effects are coming into play.

Spokes in the B ring: Dusty plasma

Here the different charging frequencies are, under the same assumptions as in the previous subsection,

$$\nu_e = 13.8 \ s^{-1}$$
$$\nu_i = 0.322 \ s^{-1}$$
$$\alpha_e = 1.07 \times 10^{-4} \ s^{-1}$$
$$\alpha_i = 1.07 \times 10^{-4} \ s^{-1}$$
$$\gamma_{de} = 7.00 \times 10^{-16} \ s^{-1}$$
$$\gamma_{di} = 1.29 \times 10^{-12} \ s^{-1}. \tag{7.85}$$

Again, the Langmuir modes are perfectly described by (7.59) for the k-domain considered. The Langmuir damping is at least an order of magnitude stronger than the classical Landau damping, so that the latter can be neglected.

The dust-ion-acoustic modes are given by (7.67) as long as $\omega_{dia} \gg \nu_e$. When this is no longer the case, the mode disappears, and the situation becomes too complicated for a simple analytical treatment.

For the dust-acoustic mode, the situation is as follows. In the normalized k-interval [0.187, 0.411], the dust-acoustic mode vanishes because of a mode bifurcation and three purely damped modes appear. Wavelengths shorter than this critical interval are in the small wavelength regime, for which $\text{Re}\omega = \omega_{pd}$. For longer wavelengths, (7.79) is recovered for the real part of the mode, until a k-value is reached at which the mode disappears. The dust-acoustic damping can be described by (7.74) for smaller wavelengths, but not for very small wavelengths [Meuris and Verheest 1996a].

7.2.8 Implications for other electrostatic modes

Although early results [Ma and Yu 1994b; Li et al. 1994] seemed to indicate that the presence of dust grain fluctuations would make the Langmuir mode unstable, a self-consistent approach shows the opposite to be true. Langmuir modes are damped by the presence of dust grains, as also recently shown by Bhatt [1997], using rather ad hoc assumptions.

From the self-consistent approach outlined in this section it also follows that the dust-ion-acoustic modes are damped due to their coupling to the charging mechanism. The damping rate is given by (7.66).

The standard dispersion law, first derived by Melandsø et al. [1993a] to describe the coupling between the dust-acoustic mode and the charging mechanism, is not adequate at very small and long wavelengths. In that case a fully consistent approach is needed that includes the sink/source terms in the continuity and momentum equations. When (7.71) is valid, three purely imaginary modes could exist, because the coupling between the dust-acoustic mode and the charging process leads to the disappearance of the dust-acoustic mode in a small k-interval.

7.3 Kinetic theory and nonlinear developments

7.3.1 Kinetic treatments

Melandsø et al. [1993b] have given a kinetic model for dust-acoustic waves applied to planetary rings, which can have thicknesses of up to 100 km. They include both Landau and Tromsø damping or growth, in addition to velocity differences between dust and plasma, as dealt with by Rosenberg [1993] who used constant dust charges, however. As in their previous paper [Melandsø et al. 1993a], the electron and ion Vlasov equations do not contain sink or source terms. The instability criterion of this dust-acoustic mode is applied to planetary ring systems, leading in the most likely case to mode damping. For certain parameter regimes there is competition between Landau and Tromsø damping. Resonances with planetary satellites or corotational perturbations would lead to clumping of the rings.

Also Li et al. [1994] looked at longitudinal waves in plasmas where the dust forms an immobile background but has variable charges. There is damping for ion-acoustic and electrostatic ion cyclotron waves, whereas high-frequency Langmuir waves grow. The damping or growth rates are proportional to the dust charging frequency. One would also have to include friction between charged and neutral dust for certain planets, and later also magnetic effects on the charging of the grains.

An interesting new development comes from Aslaksen and Havnes [1994], who assume the dust grains to be of one size but with a discrete distribution of electric charges, in other words as heavy ions with a large number of ionization levels. The dust charge thus becomes an additional phase space variable, and in Chapter 9 some other examples will be given of dealing with a spectrum of dust sizes and masses. Classically, microscopic velocity fluctuations around an average result in pressure terms in the macroscopic fluid equations. Similarly, fluctuations in dust charges will give additional terms. The absorption frequency of electrons and ions by the dust is comparable to and often even larger than the frequency of typical dust-acoustic modes. Hence, transitions between the dust ionization levels are of great importance, and described by a Balescu-Lenard collision term. Sink or source terms in the momentum equations were omitted, in order not to burden the computations too much. Analytical and numerical estimates are given for the relaxation times both towards a Maxwellian distribution and towards an equilibrium distribution for the ionization levels. Summing over the different ionization levels gives a hierarchy of charge-moment equations for dust density function, and yields estimates of the importance of terms coming from the ionization distribution.

As an aside, in plasmas with large dust grains the coupling between the grains becomes very strong and might lead to crystallization of the plasma, as observed in recent experiments [Thomas et al. 1994]. The work discussed here is one of the first to tackle the problem of fluctuating dust charges at the microscopic level, and while far from complete gives already estimates of what can happen. In particular for space plasmas, there will be need to extend the treatment to include the interplay between a nonzero electromagnetic field and the distribution of the dust grains.

Finally in this section, Vladimirov [1994] investigates the propagation of electromagnetic and Langmuir waves, taking the effect of capture of plasma electrons and ions into account, extending thereby the recent kinetic theory of Tsytovich and Havnes [1993]. The new wave damping leads to a lowering of the frequencies.

7.3.2 Electrostatic solitons

Turning now to papers which include fluctuating grain charges at the nonlinear level, Rao and Shukla [1994] extended their original work [Rao et al. 1990] on nonlinear dust-acoustic waves. For small but finite amplitudes the waves are shown to be governed by a driven Boussinesq-like nonlinear equation coupled to the charge fluctuation equation, similar to what was obtained by Varma et al. [1993]

$$\frac{\partial^2 \varphi}{\partial t^2} - c_{da}^2 \frac{\partial^2 \varphi}{\partial z^2} - \frac{A}{2}\frac{\partial^2 \varphi^2}{\partial z^2} - \frac{c_{da}^4}{\omega_{pd}^2}\frac{\partial^4 \varphi}{\partial z^4} = -\frac{m_d c_{da}^3}{Q_d^2}\frac{\partial^2}{\partial z \partial t}\delta q_d. \qquad (7.86)$$

Here A is a rather complicated coefficient involving all the parameters of the electron, ion and dust fluids. For unidirectional propagation this driven Boussinesq equation reduces to a Korteweg-de Vries equation with a source term. The picture given here is not fully self-consistent, as the dust charges fluctuate but there are no corresponding sink or source terms in the plasma continuity equations. Localized solutions are found to be damped due to the fluctuations in the dust charges.

7.4 Electromagnetic waves with variable dust charges

Charge fluctuation instabilities for electromagnetic waves occur because of momentum exchange when plasma electrons and ions are captured or released by the dust. Damping of different modes is usually found, as for the electrostatic modes. However, equilibrium drifts between the dust grains and the plasma particles can sometimes balance the charging losses, especially for Alfvén waves. Verheest and Meuris [1995] describe the plasma species as moving in a static background of dust particles with variable charges. Use is made of momentum exchanges modelled on what happens for ordinary collisions (6.22) or (7.14), and this leads to a kind of damped whistler wave.

Full dust dynamics have been included in the description by Reddy *et al.* [1996], of which I recall the principal elements. First come the continuity equations in the model of Bhatt and Pandey [1994], as we saw used for electrostatic modes,

$$\frac{\partial n_\alpha}{\partial t} + \frac{\partial}{\partial z}(n_\alpha u_{\alpha\parallel}) = -\nu_\alpha \delta n_\alpha - \mu_\alpha \delta q_d, \tag{7.87}$$

and the full equations of motion in a cold plasma,

$$\frac{\partial \mathbf{u}_\alpha}{\partial t} + u_{\alpha\parallel}\frac{\partial \mathbf{u}_\alpha}{\partial z} = \frac{q_\alpha}{m_\alpha}[\mathbf{E} + \mathbf{u}_\alpha \times \mathbf{B}] - \sum_\beta \gamma_{\alpha\beta}(\delta\mathbf{u}_\alpha - \delta\mathbf{u}_\beta). \tag{7.88}$$

Momentum loss/gain terms are included, and for the electrons and ions the summation over β is restricted to interfacing with the dust, hence terms in γ_{ed} or γ_{id} only. On the other hand, for the dust there will be terms both with γ_{de} and γ_{di}. The description is then closed by Maxwell's equations (6.51), given in the previous chapter.

To first order I obtain for the transverse equations of motion that

$$(\omega - kU_\alpha)\mathbf{u}_{\alpha 1} + i\Omega_\alpha \mathbf{e}_z \times \mathbf{u}_{\alpha 1} + i\sum_\beta \gamma_{\alpha\beta}(\mathbf{u}_{\alpha 1} - \mathbf{u}_{\beta 1})$$

$$= i\frac{Q_\alpha(\omega - kU_\alpha)}{\omega m_\alpha}\mathbf{E}_1, \tag{7.89}$$

and these are all coupled. Progress is possible for circularly polarized, low-frequency and long-wavelength modes such that

$$\left|\frac{\gamma_{\alpha\beta}}{\Omega_\alpha}\right| \sim \left|\frac{\omega - kU_\alpha}{\Omega_\alpha}\right| \ll 1. \tag{7.90}$$

In this vein, attachment frequency need not be small compared to the wave frequencies, but cannot be much larger. Up to second order in these small quantities the electron and ion velocities are

$$\mathbf{u}_{e1} = \frac{\omega - kU_e}{\omega B_0}\mathbf{E}_1 \times \mathbf{e}_z - m_e\frac{i(\omega - kU_e)^2 + \gamma_{ed}k(U_e - U_d)}{\omega q_e B_0^2}\mathbf{E}_1,$$

$$\mathbf{u}_{i1} = \frac{\omega - kU_i}{\omega B_0}\mathbf{E}_1 \times \mathbf{e}_z - m_i\frac{i(\omega - kU_i)^2 + \gamma_{id}k(U_i - U_d)}{\omega q_i B_0^2}\mathbf{E}_1, \tag{7.91}$$

and similarly for the dust velocity

$$
\begin{aligned}
\mathbf{u}_{d1} = {} & \frac{\omega - kU_d}{\omega B_0} \mathbf{E}_1 \times \mathbf{e}_z \\
& - m_d \frac{i(\omega - kU_d)^2 + k\gamma_{de}(U_d - U_e) + k\gamma_{di}(U_d - U_i)}{\omega Q_d B_0^2} \mathbf{E}_1.
\end{aligned} \quad (7.92)
$$

In this way the linear dispersion law is obtained as

$$
\left(\omega - k\overline{U}\right)^2 - k^2(V_A^2 - W) + ik\Gamma = 0. \quad (7.93)
$$

Here $\overline{U}$ is the bulk mass velocity of the plasma, including the charged dust, and W refers to the mass averaged relative kinetic energy in the parallel flows, which I encountered already in the previous chapter when discussing beam-plasma instabilities in general. Here it is

$$
W = \frac{\rho_{e0}\rho_{i0}(U_e - U_i)^2 + \rho_{e0}\rho_{d0}(U_e - U_d)^2 + \rho_{i0}\rho_{d0}(U_i - U_d)^2}{(\rho_{e0} + \rho_{i0} + \rho_{d0})^2}, \quad (7.94)
$$

with $\rho_{\alpha 0} = m_\alpha N_\alpha$. A totally new term is due to the combined charging and streaming effects, characterized by

$$
\Gamma = \frac{(\rho_{e0}\gamma_{ed} - \rho_{d0}\gamma_{de})(U_d - U_e) + (\rho_{i0}\gamma_{id} - \rho_{d0}\gamma_{di})(U_d - U_i)}{\rho_{e0} + \rho_{i0} + \rho_{d0}}. \quad (7.95)
$$

Fluctuations in the dust charges themselves drop out of the linear dispersion law (7.93) because, like density fluctuations, dust charge fluctuations give rise to first order current perturbations in the parallel direction, which do not couple to the transverse waves considered here.

The solution of (7.93) is given by

$$
\omega = \left[k\overline{U} \mp \left(\frac{(A^2 + k^2\Gamma^2)^{\frac{1}{2}} + A}{2} \right)^{\frac{1}{2}} \right] \pm i \left(\frac{(A^2 + k^2\Gamma^2)^{\frac{1}{2}} - A}{2} \right)^{\frac{1}{2}}, \quad (7.96)
$$

where $A = k^2(V_A^2 - W)$. This will be discussed in two different limits.

Case $\Gamma = 0$

If the charging effects conserve momentum between any two species, as true collisional effects usually do, then

$$
\rho_{e0}\gamma_{ed} = \rho_{d0}\gamma_{de}, \qquad \rho_{i0}\gamma_{id} = \rho_{d0}\gamma_{di}, \quad (7.97)
$$

and lo and behold, Γ vanishes exactly. If that is the case and $W > V_A^2$, A becomes negative and (7.93) yields the dispersion law

$$\omega = k\overline{U} + ik\sqrt{W - V_A^2}, \tag{7.98}$$

that I encountered already in (6.15). For $Zm_e/m_d \ll m_e/m_i \ll 1$ the bulk expressions are simplified to

$$\overline{U} = \frac{U_i + \sigma U_d}{1+\sigma}, \qquad W = \frac{\sigma(U_i - U_d)^2}{(1+\sigma)^2}, \tag{7.99}$$

and

$$V_A^2 = \frac{V_{Ai}^2}{1+\sigma}, \tag{7.100}$$

where V_{Ai} is the Alfvén velocity with respect to the mass of the ion beams and $\sigma = \rho_{d0}/\rho_{i0}$ is the ratio of the mass densities of dust and ions. The condition $W > V_A^2$ tells us that instability occurs provided the relative Alfvén Mach number $M = |U_i - U_d|/V_{Ai} > M_{cr}$, where the critical Mach number is given by $M_{cr} = (1 + 1/\sigma)^{1/2}$.

Coming back to the general conditions of validity of the dispersion law, with Ω_d being by far the smallest of all the gyrofrequencies in absolute value and the frequency becoming complex, I need for the moduli that

$$|\omega - kU_d|^2 = |\mathrm{Re}\,\omega - kU_d|^2 + |\mathrm{Im}\,\omega|^2 \ll \Omega_d^2. \tag{7.101}$$

Working this out gives the allowable range of wave numbers as

$$k \ll \sqrt{\frac{\sigma}{M^2-1}} \cdot \frac{|\Omega_d| M_{cr}}{V_{Ai}}. \tag{7.102}$$

At the same time, one can compute that the growth rate will peak for a given Mach number, corresponding to relative densities obeying

$$\sigma_m = \frac{M^2+1}{M^2-1} > \frac{1}{M^2-1} = \sigma_{cr}. \tag{7.103}$$

Here σ_{cr} stands for the minimum value needed to obtain instability. It is now simple to see that

$$|\mathrm{Re}\,\omega - kU_d| = |\mathrm{Im}\,\omega| = kV_{Ai}\frac{M^2-1}{2M}, \tag{7.104}$$

at the maximum growth rate for a given M. For most of the practical applications, however, σ is a rather large number and therefore (7.104) is not of much help in the case of dusty plasmas. From (7.98) I obtain in the dust frame approximate expressions for large σ as:

$$\mathrm{Re}\,\omega \simeq \frac{kV_{Ai}M}{\sigma},$$

$$\mathrm{Im}\,\omega = \frac{kV_{Ai}}{M_{cr}^2}\sqrt{\frac{M^2 - M_{cr}^2}{\sigma}}. \qquad (7.105)$$

It is readily seen that $\mathrm{Im}\,\omega$ increases with M and decreases with σ. Conversely, the allowed k increases with σ, so that finally, at given M, the growth rates and the wavelengths $\lambda = 2\pi/k$ of the structures both decrease with σ.

Once k obeys the conditions imposed by the dust, it is usually the case that conditions for the ions and the electrons

$$|\mathrm{Re}\,\omega - kU_{i,e}|^2 + |\mathrm{Im}\,\omega|^2 \ll \Omega_{i,e}^2 \qquad (7.106)$$

are also satisfied. Hence the biggest constraint upon the wave numbers is the one coming from the very small dust gyrofrequencies. These conclusions have been borne out by some numerical computations [Reddy et al. 1996].

Case $\Gamma \neq 0$

When the charging and discharging effects do not conserve momentum between any two species, then $\Gamma \neq 0$. Unlike true collisions where momentum is exchanged by changes in the velocities of the colliding particles, here momentum is exchanged by shifting charges from the electron and ion fluids to the dust and vice versa. Hence it would not seem necessary for the dust to behave so as to make Γ vanish exactly, but more complicated three-way processes might be involved. As a zeroth order approximation I consider all charging frequencies to be constant. For $\Gamma \neq 0$, I find that for the case of $M > M_{cr}$, the effect of Γ is negligible on the modes considered here [Reddy et al. 1996]. However, the effect of finite Γ is significant for $W \leq V_A^2$ or $M \leq M_{cr}$, which I discuss below. For simplicity I shall consider again $Zm_e/m_d \ll m_e/m_i \ll 1$, and now also $\gamma_{di} \simeq \gamma_{de} \simeq 0$, so that Γ is approximately

$$\Gamma = \frac{\gamma_{id}(U_d - U_i)}{1 + \sigma}. \qquad (7.107)$$

For $W < V_A^2$ or $M < M_{cr}$, and in the limit $A^2 \gg k^2\Gamma^2$, (7.96) simplifies to

$$\omega = \frac{kV_{Ai}}{M_{cr}^2}\left(\frac{M}{\sigma} + \frac{U_d M_{cr}^2}{V_{Ai}} - \sqrt{\frac{M_{cr}^2 - M^2}{\sigma}}\right) + i\frac{\gamma_{id}M}{2\sqrt{\sigma(M_{cr}^2 - M^2)}}. \qquad (7.108)$$

The different assumptions made in deriving the above equation demand that the following inequalities are satisfied,

$$\frac{M_{cr}^2 M}{M_{cr}^2 - M^2} \cdot \frac{\gamma_{id}}{|\Omega_d|} \ll \frac{k V_{Ai}}{|\Omega_d|} \ll M_{cr}^2 \sqrt{\frac{\sigma}{M_{cr}^2 - M^2}}. \tag{7.109}$$

The condition on the growth rate is then automatically fulfilled. Numerical computations [Reddy et al. 1996] indicate that the growth rates are nearly constant, whereas the real frequencies increase with increasing wave number, or with an increase in γ_{id}.

For the modes to grow, both $M \neq 0$ and $\gamma_{id} \neq 0$ are necessary. These properties are similar to those of drift dissipative modes where density gradients give rise to negative energy modes which grow due to collisional dissipation. Here the role of a density gradient drift is replaced by the relative streaming between ion and dust beams. Both type of modes could generate spatial structures over a wide range of hundreds to tens of million kilometers in dusty cometary tails. As in Chapter 6, the influence of charged dust could lead to modifications on very large scales.

7.5   Nonlinear electromagnetic waves with variable dust charges

Electromagnetic waves will be considered in the regime where they are not much affected by the fact that dust charges fluctuate, at least at the linear level. The rationale for this computational strategy is quite obvious. Indeed, were charge and wave fluctuations to occur on comparable timescales, then the effects of variable charges would show up already in the linear description and lead there to wave damping. In that case, a possible nonlinear behaviour is not really relevant.

The nonlinear behaviour of electromagnetic modes in dusty plasmas with variable dust charges is relatively new [Verheest and Meuris 1996a] and little studied, in contrast to electrostatic modes for which much more nonlinear results are available. I will follow an analogous procedure as in the previous chapter, with modifications where necessary to incorporate fluctuating dust charges [Verheest and Meuris 1996b]. Besides the basic equations (7.88) and Maxwell's equations (6.51), I shall also need a global equation. Compared to the previous chapter, (6.65) is now modified to

$$\sum_\alpha n_\alpha m_\alpha \frac{\partial \mathbf{u}_{\alpha\perp}}{\partial t} + \sum_\alpha n_\alpha m_\alpha u_{\alpha\parallel} \frac{\partial \mathbf{u}_{\alpha\perp}}{\partial z}$$
$$= \varepsilon_0 \frac{\partial E_\parallel}{\partial z} \mathbf{E}_\perp - \varepsilon_0 \frac{\partial E_\parallel}{\partial t} \mathbf{e}_z \times \mathbf{B}_\perp + \frac{B_0}{\mu_0} \frac{\partial \mathbf{B}_\perp}{\partial z} + \varepsilon_0 B_0 \mathbf{e}_z \times \frac{\partial \mathbf{E}_\perp}{\partial t}$$
$$- \sum_\alpha \sum_\beta n_\alpha m_\alpha \gamma_{\alpha\beta} (\delta \mathbf{u}_{\alpha\perp} - \delta \mathbf{u}_{\beta\perp}). \tag{7.110}$$

The stretching of coordinates and time remains given in (6.52).

Although in the previous chapter all possible terms in Maxwell's equations were included, the algebra told us that the displacement current and the deviations from electrical neutrality did not play any role at all, for the weakly nonlinear description we were considering. This was to be expected on physical grounds for low-frequency phenomena, and there is thus no need to be so cautious here. Anyway, the derivation will be already complicated enough by the presence of the sink/source terms in the momentum equations. This means that the expansion of the dependent variables in (6.54) and (6.55) is modified for parallel quantities to

$$
\begin{aligned}
n_\alpha &= N_\alpha + \varepsilon n_{\alpha 2} + \cdots, \\
q_\alpha &= Q_\alpha + \varepsilon q_{\alpha 2} + \cdots, \\
u_{\alpha\|} &= U_\alpha + \varepsilon u_{\alpha\|2} + \cdots, \\
E_\| &= \varepsilon E_{\|2} + \cdots,
\end{aligned}
\tag{7.111}
$$

and for the transverse variables to

$$
\begin{aligned}
\mathbf{u}_{\alpha\perp} &= \varepsilon^{1/2} \mathbf{u}_{\alpha\perp 1} + \varepsilon^{3/2} \mathbf{u}_{\alpha\perp 3} + \cdots, \\
\mathbf{B}_\perp &= \varepsilon^{1/2} \mathbf{B}_{\perp 1} + \varepsilon^{3/2} \mathbf{B}_{\perp 3} + \cdots, \\
\mathbf{E}_\perp &= \varepsilon^{1/2} \mathbf{E}_{\perp 1} + \varepsilon^{3/2} \mathbf{E}_{\perp 3} + \cdots
\end{aligned}
\tag{7.112}
$$

As seen, the variable dust charges are also expanded.

Substitution of the stretching (6.52) and the perturbation expansions (7.111) and (7.112) into the basic equations again gives a sequence of equations. To orders 3/2 and 5/2 the global equation (7.110) gives

$$
\sum_\alpha N_\alpha m_\alpha (U_\alpha - V) \frac{\partial \mathbf{u}_{\alpha\perp 1}}{\partial \xi} = \frac{B_0}{\mu_0} \frac{\partial \mathbf{B}_{\perp 1}}{\partial \xi},
\tag{7.113}
$$

and

$$
\begin{aligned}
&\sum_\alpha N_\alpha m_\alpha \frac{\partial \mathbf{u}_{\alpha\perp 1}}{\partial \tau} + \sum_\alpha N_\alpha m_\alpha (U_\alpha - V) \frac{\partial \mathbf{u}_{\alpha\perp 3}}{\partial \xi} \\
&+ \sum_\alpha m_\alpha \left(n_{\alpha 2}(U_\alpha - V) + N_\alpha u_{\alpha\|2} \right) \frac{\partial \mathbf{u}_{\alpha\perp 1}}{\partial z} \\
&= \frac{B_0}{\mu_0} \frac{\partial \mathbf{B}_{\perp 3}}{\partial \xi} - \sum_\alpha \sum_\beta N_\alpha m_\alpha \gamma_{\alpha\beta} (\mathbf{u}_{\alpha\perp 1} - \mathbf{u}_{\beta\perp 1}),
\end{aligned}
\tag{7.114}
$$

so that the dependent variables only need to be determined to third order. Furthermore, the coefficients related to the different aspects of the fluctuating dust charges like ν_α, μ_α and $\gamma_{\alpha\beta}$ are supposed to be small, so that the linear waves can be considered as stable and the influence of these coefficients only comes into play at the nonlinear level, governed by slow time scales,

$$\nu_\alpha \sim \mu_\alpha \sim \gamma_{\alpha\beta} \sim \frac{\partial}{\partial\tau}. \tag{7.115}$$

Because of this ordering, the results to lowest orders are reminiscent to what was found in the previous chapter and not repeated here. I can now derive from (7.113) the dispersion law as

$$\sum_\alpha N_\alpha m_\alpha (V - U_\alpha)^2 = \frac{B_0^2}{\mu_0}, \tag{7.116}$$

giving V_A if there are no drifts. At the next order I find that

$$\mathbf{u}_{\alpha\perp3} = \frac{U_\alpha}{B_0}\mathbf{B}_{\perp3} + \frac{u_{\alpha\|2}}{B_0}\mathbf{B}_{\perp1} + \frac{(V-U_\alpha)^2}{B_0\Omega_\alpha}\mathbf{e}_z \times \frac{\partial\mathbf{B}_{\perp1}}{\partial\xi} + \frac{1}{B_0}\mathbf{E}_{\perp3} \times \mathbf{e}_z. \tag{7.117}$$

With the help of this I find from a combination of the parallel equations and charge neutrality that

$$n_{\alpha2} = \frac{N_\alpha}{2B_0^2}B_{\perp1}^2 - \frac{\omega_{p\alpha}^2 N_d q_{d2}}{Q_\alpha(V-U_\alpha)^2}\bigg/ \sum_\beta \frac{\omega_{p\beta}^2}{(V-U_\beta)^2},$$

$$u_{\alpha\|2} = \frac{V-U_\alpha}{2B_0^2}B_{\perp1}^2 - \frac{\omega_{p\alpha}^2 N_d q_{d2}}{N_\alpha Q_\alpha(V-U_\alpha)}\bigg/ \sum_\beta \frac{\omega_{p\beta}^2}{(V-U_\beta)^2}. \tag{7.118}$$

The interplay between the nonlinear effects and the dust charging is clear, coupling density and dust charge variations at this level. Nevertheless, it turns out that direct influences of fluctuating dust charges disappear from the final result, when I insert all known expressions in (7.114) and eliminate $\mathbf{B}_{\perp3}$ and $\mathbf{E}_{\perp3}$ with the help of Faraday's law to order 5/2. This yields a nonlinear evolution equation

$$A\frac{\partial\mathbf{B}_{\perp1}}{\partial\tau} + \frac{1}{4\mu_0}\frac{\partial}{\partial\xi}\left(B_{\perp1}^2\mathbf{B}_{\perp1}\right) + C\mathbf{e}_z \times \frac{\partial^2\mathbf{B}_{\perp1}}{\partial\xi^2} + D\mathbf{B}_{\perp1} = 0, \tag{7.119}$$

the coefficients of which are

$$A = \sum_\alpha N_\alpha m_\alpha (V - U_\alpha), \tag{7.120}$$

$$C = \sum_\alpha \frac{N_\alpha m_\alpha (V-U_\alpha)^3}{2\Omega_\alpha}, \tag{7.121}$$

$$D = \frac{1}{2}\sum_\alpha\sum_\beta [N_\alpha m_\alpha \gamma_{\alpha\beta} - N_\beta m_\beta \gamma_{\beta\alpha}] U_\beta. \tag{7.122}$$

From the vector form (7.119) I can derive the usual scalar form of the extended derivative nonlinear Schrödinger equation by projection and combination of $\phi = B_{x1} + iB_{y1}$ as

$$A\frac{\partial \phi}{\partial \tau} + \frac{1}{4\mu_0}\frac{\partial}{\partial \xi}\left(|\phi|^2\phi\right) + iC\frac{\partial^2 \phi}{\partial \xi^2} + D\phi = 0. \qquad (7.123)$$

If the charging effects conserve momentum between any two species, as true collisional effects, then

$$N_\alpha m_\alpha \gamma_{\alpha\beta} = N_\beta m_\beta \gamma_{\beta\alpha}, \qquad (7.124)$$

and D vanishes exactly. This also occurs when there are no equilibrium drifts of the different plasma species along the external magnetic field. In both cases all reference to fluctuating dust charges is lost at the order considered and I am left discussing the ordinary DNLS equation in multispecies plasma, with its attendant soliton solutions. Here I can fall back on existing multispecies descriptions [Verheest 1990,1992; Verheest and Buti 1992; Deconinck *et al.* 1993a,b], given in the previous chapter.

I note from these investigations that the inclusion of pressure and temperature effects vastly complicates the coefficients of the DNLS equations, without altering the structural form of the equation. Hence, similar conclusions will also hold for the DNLS-like equation (7.119) when the source term does not vanish.

When the charging and discharging effects do not conserve momentum between any two species in the sense of (7.124), and there are different streaming velocities, then $D \neq 0$. In that case one is reduced to a discussion of a DNLS-like equation (7.119) with a source term. If I integrate (7.119) over all ξ, as is normally done to investigate conservation laws of the ordinary DNLS equation, it is found that

$$A\frac{\partial}{\partial \tau}\int_{-\infty}^{+\infty}\phi\, d\xi + \left[\frac{1}{4\mu_0}|\phi|^2\phi + iC\frac{\partial \phi}{\partial \xi}\right]_{\xi=-\infty}^{\xi=+\infty} + D\int_{-\infty}^{+\infty}\phi\, d\xi = 0. \qquad (7.125)$$

Since all ϕ vanish at infinity for solitary structures, I conclude that

$$\int_{-\infty}^{+\infty}\phi\, d\xi = \left[\int_{-\infty}^{+\infty}\phi\, d\xi\right]_{\tau=0}\exp\left\{-\frac{D}{A}\tau\right\}. \qquad (7.126)$$

There is no stable solution possible, and certainly no conserved densities, in the traditional sense of soliton theory.

I have given a discussion of nonlinear electromagnetic modes in dusty plasmas, when the variability of the dust charges is taken into account, by introducing in the fluid equations appropriate coefficients to account for the inherent density and momentum exchanges. I have looked in particular at the intermediate frequency case where the linear waves are not affected, but the dust charging influences the slower nonlinear development. This approach leads to a derivative nonlinear Schrödinger equation with a source term. In general no stable solitary waves are possible, showing the damping mechanisms to be still at work, even if shuffled under the rug for the linear description. In the absence of equilibrium streaming, stable envelope solitons are possible. Although I took a cold plasma description for simplicity, I do not expect that the inclusion of temperature effects will qualitatively modify our conclusions.

One outstanding problem remains as usual, that of more accurate descriptions of the charging mechanisms for the dust, which so far are too reliant on the probe model. While maybe a good first approximation for electrostatic modes, the influence of static magnetic fields needs to be investigated, as a prerequisite for a proper treatment of electromagnetic modes. This was bypassed here by the use of phenomenological attachment frequencies, that are very *ad hoc* indeed.

7.6   Oblique and perpendicular modes

Few results are available for oblique or perpendicular wave propagation in magnetized plasmas with fluctuating dust charges, as if the complexities deterred investigation. In line with the discussions given in this chapter, when electrostatic modes propagate perpendicular to an ambient magnetic field, and when the dust dynamics can be neglected, the dispersion law (7.48) remains formally the same, provided one uses

$$\omega_{\gamma\alpha} = \omega + i\gamma_{\alpha d} - \frac{\Omega_\alpha^2}{\omega + i\gamma_{\alpha d}} \tag{7.127}$$

to describe the resonances of the extraordinary mode. For that mode, the energy is divided in an electrostatic and an electromagnetic part. The upper-hybrid wave is a purely electrostatic wave and can be seen as the limit of the extraordinary wave when approaching the resonance ($k \to \infty$), as the electromagnetic energy is converted into electrostatic waves. To have some idea what is happening here, assume again that ν_e, ν_i, α_e, α_i, γ_{ed}, γ_{id} and ω_{pi} are small frequencies in a cold, non-drifting plasma, take (7.127) into account and obtain the dispersion law for the upper-hybrid waves

$$\omega^2 = \omega_{pe}^2 + \Omega_e^2 - i\gamma_{ed}\frac{\omega_{pe}^2 + 2\Omega_e^2}{2\omega_{pe}^2 + 2\Omega_e^2}. \tag{7.128}$$

The damping is mediated through the momentum exchanges between the electron fluid and the quasi-immobile dust background.

More sophisticated is the kinetic theory given by De Juli and Schneider [1998] for magnetized plasmas with variable charges on the grains. The latter are treated as immobile, so that waves at the typical dust frequencies are also here not considered. The dielectric response of the plasma is modified in two different ways. On the one hand, the dielectric tensor now contains additional terms connected with the variability of the dust charges. This results in damping or amplification, which can compete with Landau damping, as pointed out already on various occasions. In addition, new terms arise due to the charging process itself. These findings are applied to the dispersion of the perpendicular magnetosonic wave. Usually this is thought of as an undamped mode, but here the dust grains cause absorption through the attachment frequencies, computed from the kinetic model equations.

Finally, dust dynamics are included for oblique electrostatic modes by Tripathi and Sharma [1996a]. Their model is slightly inconsistent, because the plasma is assumed cold, also for the electrons and ions, that normally need some temperature before grain charging can occur. Otherwise, sink/source terms have been included in the continuity and momentum equations, and these hence lead to rather formidable expressions for the elements of the dispersion tensor. The dispersion law is then discussed for various frequency regimes, when compared to the gyrofrequencies. At the lowest frequencies ($\omega \sim |\Omega_d| \ll \Omega_i \ll |\Omega_e|$—) there seems to be little damping for perpendicularly propagating dust-cyclotron waves. For higher frequencies, when $|\Omega_d| \ll \omega \sim \Omega_i \ll |\Omega_e|$— or higher, one gets in the range of the lower-hybrid frequencies with some charge fluctuation damping. Restrictions on the results come also from the assumption that the grain sizes are small compared to the electron Larmor radius. It is difficult to gauge how well this is obeyed in all dusty space plasmas.

CHAPTER 8

SELF-GRAVITATION

8.1 Janus faces of Jeans instabilities

For certain dusty plasmas, containing rather heavy charged grains, is was hypothesized that the intergrain gravitational force could become of the order of the intergrain electrostatic force, and this has prompted a revival of the interest in Jeans instabilities and self-gravitational effects, reconsidered in a novel context. As I had the occasion of pointing out in Chapter 3, cosmic dust is a well-known and common constituent of the interstellar medium. It may be charged or neutral depending on the nearby sources of radiation like stars and/or the presence of charged particles. Such cosmic clouds of dust can be very large with widely varying grain diameters, and, in the main, it is these dust clouds that I shall keep in mind here. Before we can go on and address the issue of how charged dust affects the stability or instability, however, I have to recall some notions from neutral gases and in particular, the ambiguities connected with the proper description of gravitational phenomena in such systems.

8.1.1 Jeans (in)stabilities

The Jeans instability of a self-gravitating cloud is probably the most basic instability in gravitating systems, known in the astrophysical literature for almost a century, ever since Jeans obtained in 1902 the instability criterion for harmonic waves whose wavelength exceeds some critical value. His study was based upon the assumption that the unperturbed gaseous cloud is initially uniform [Jeans 1929]. The instability thus derived is non-oscillatory and purely growing, a macro instability corresponding to gravitational collapse of the gas or plasma. Besides the possible importance for star formation, Alfvén [1981] already expressed the idea that the solar system was formed out of a dusty plasma. Although this instability has now been abandoned as a possible explanation for Solar System formation, it is still recognized as of fundamental importance in modern cosmology.

As mentioned in Chapter 4 while discussing basic stationary states in dusty plasmas, Newtonian gravitation in extended mass systems precludes truly homogeneous equilibria. Nevertheless, true disciples of Jeans skirt around this difficulty by considering small wavelength, local perturbations, small compared to the inhomogeneity scale lengths. While in itself plausible, this procedure can in most cases not be tested for internal consistency,

185

because real knowledge about the equilibrium is lacking. Small wonder that the severe limitations of such an approach were already pointed out in the case of purely neutral gases [Spitzer 1978; Boss 1987; Čadež 1990; Vranješ and Čadež 1990].

Such an at present unavoidable dichotomy will affect the structure of this chapter. After recalling some elements of what happens in neutral gases, I will start in the next section with an example of a plasma where the computations can be done explicitly, both for the stationary as for the perturbed state, showing that the system is stable [Čadež *et al.* 1999]. The remaining sections after that, however, will be devoted to a review of recent work involving Jeans instabilities in dusty plasmas, and here the "Jeans swindle" will be used... The present state of affairs does not allow anything more consistent, and this will typically remain so, until more equilibria are worked out in complex systems as dusty plasmas.

8.1.2   Physics of the classic Jeans instability

In order to set the scene, I will briefly recall some of the fundamental ideas related to the ordinary Jeans instability in neutral gases — pertinent to stellar or galactic systems. The discussion is later extended to plasmas containing charged dust grains, in which other mechanisms can occur that resist gravitational collapse. Usually, the restrictions which follow from possible boundary effects in real, bounded plasmas are only taken into account when essential to the understanding of the problems studied. However, cosmic dust clouds or dusty plasmas on galactic scales and the associated instability lengths are so vast that their sizes have to be kept in mind before jumping to conclusions.

To brush a simple picture of the Jeans instability in a neutral gas, consider a nearly spatially uniform distribution slightly overdense region of radius L and density $N_d m_d$. The overdense region will collapse if the random velocity v_{rms} of the particles, due to their thermal motions, or due to the presence of collisional sound waves, for example, is insufficient to carry them out of the dense region before collapse can occur. The collapse timescale τ_c can be estimated by considering a test particle initially at rest on the edge of the dense sphere. The gravitational acceleration experienced by this particle is just

$$a = \frac{GM}{L^2} = \tfrac{4}{3}\pi G N_d m_d L, \tag{8.1}$$

where $M = (4/3)\pi L^3 \cdot N_d m_d$ has been used. The test particle will oscillate around the center of the density enhancement, and its motion is described by

$$r = L \cos\left(t\sqrt{\tfrac{4}{3}\pi G N_d m_d}\right). \tag{8.2}$$

Hence, the time to traverse the density enhancement is half the period, so that

$$\tau_c = \left(\frac{3\pi}{4GN_d m_d}\right)^{1/2}. \tag{8.3}$$

On the other hand, the timescale τ_e for particles with random speed v_{rms} to escape the overdense region is of the order $\tau_e = 2L/v_{rms}$. It is evident that τ_c is independent of L, while τ_e increases linearly with L. Consequently, small overdense regions, i.e. those that have $\tau_e < \tau_c$, are stable, while large regions where $\tau_e > \tau_c$ are unstable.

Strictly speaking, the results obtained so far hold for a test particle only, like in the gravitational escape of atoms and molecules from stellar atmospheres. Nevertheless, it has been interpreted as applying to the whole system, and the critical radius where collapse of extended regions is just possible can be estimated by setting $\tau_e = \tau_c$, yielding the Jeans length

$$L_J = \frac{\pi v_{rms}\sqrt{3}}{2\omega_J} \sim \frac{v_{rms}}{\omega_J}. \tag{8.4}$$

In this picture, only those overdense regions that have $L > L_J$ are subject to the Jeans instability. In a neutral gas v_{rms} is the thermal speed c_{sd} of an average particle, and then L_J takes the explicit form (leaving out numerical factors of order unity)

$$L_J \equiv \frac{c_{sd}}{\omega_J} = \left(\frac{\kappa T_d}{4\pi G N_d m_d^2}\right)^{1/2}. \tag{8.5}$$

Here $c_{sd}^2 = \kappa T_d/m_d$ has been used. From this definition one observes a close analogy with the Debye wavelength for electrostatic interactions — the mass in this case playing the role of electric charge.

In the above intuitive discussion of the Jeans instability I have not considered the possibility of the dust grains being charged. Consequently, the dust thermal speed is used to estimate the timescale for the escape of a particle from the overdense region. However, charged dust grains exhibit collective behaviour, as I discussed at great length in previous chapters. The phase speed of these collective plasma waves can, if sufficiently small so that the dust can couple to the mode, provide the randomizing velocity necessary for a dust grain to escape from the overdense region. In this case I could replace v_{rms} in the above equations with the appropriate phase speed ω/k of the wave. This will be considered in later sections, after the "Jeans swindle" is revisited.

8.2 Revisiting the Jeans swindle

In this section, an attempt is made to be fully consistent, to determine the basic stationary state before small perturbations are studied. The choice of a model that can be worked out is far from trivial, as discussions in the literature attest. Many authors have recently been studying self-gravitation effects in nonuniform gas and plasma layers under various physical conditions [Nakano 1988; Gehman *et al.* 1996]. The basic assumption common to these works is that the nonuniformity of the layer is due to self-gravitation and it is always assumed to be in one direction only. In the lateral directions, parallel to the layer, the medium is considered uniform. Part of the "Jeans swindle" thus sneaks back in.

The model adopted here is a self-gravitating, isothermal plasma cloud in magnetohydrostatic equilibrium, and therefore rather than a multispecies description, which is way too complicated, a single-fluid MHD approach will be used [Čadež *et al.* 1999]. In Cartesian geometry the correct nonlocal treatment can be performed in one dimension only, because there is no additional direction along which the medium could be assumed strictly uniform. The linear perturbations applied to such a system should therefore always propagate in the direction of the density inhomogeneity, unless the perturbations are localized in some sense.

8.2.1   Basic stationary state

The as yet unperturbed massive plasma cloud is assumed isothermal (with constant temperature T_0), of uniform composition but not density, permeated by a nonuniform magnetic field $\mathbf{B}_0$, and with possibly a nonuniform plasma flow $\mathbf{U}_0$ along the field lines. This plasma is treated as a perfect gas, while the spatial extent of the cloud is large enough to provide for a non-vanishing self-gravitation force. The basic state is taken stationary, provided the electrical conductivity is sufficiently large, so that Ohmic dissipation and magnetic diffusivity have a negligible effect on the magnetic field, on time scales relevant to the dynamics of the perturbations introduced later. In Cartesian geometry, the gravitational force and all gradients of physical quantities are taken along the z-axis, whereas the magnetic field and particle flow are oriented in the perpendicular direction, along the x-axis, i.e. $\mathbf{B}_0 = B_0(z)\mathbf{e}_x$ and $\mathbf{U}_0 = U_0(z)\mathbf{e}_x$.

Furthermore, the strength of the magnetic field $B_0(z)$ varies with z in such a way that the ratio of the gas pressure $R_g T_0 \rho_0$ to the magnetic pressure $B_0^2/(2\mu_0)$ is constant. The constancy of the plasma β implies that both the Alfvén speed V_A and the speed of sound c_s are constant. The latter is defined here through $c_s^2 = \gamma R_g T_0$, where $R_g = R/M$ is the gas constant for the cloud, R is the universal gas constant and M is the mean molar mass for the cloud. The constancy of V_A also means that the magnetic field is assumed stronger in the region with a higher plasma density ρ_0 and weaker where the density is lower. Such a distribution of the magnetic field in the cloud is rather realistic and may be found in highly conductive plasmas with frozen-in magnetic fields.

The equations for the basic state described above are therefore the perfect gas law,

$$p_{g0} = R_g \rho_0 T_0, \tag{8.6}$$

the equation of magnetohydrostatic balance,

$$-\nabla p_{g0} - \rho_0 \nabla \phi_0 + \frac{1}{\mu_0}(\nabla \times \mathbf{B}_0) \times \mathbf{B}_0 = 0, \tag{8.7}$$

which reduces to

$$\left(1 + \frac{1}{\beta}\right)\frac{dp_{g0}}{dz} = -\rho_0 \frac{d\phi_0}{dz}, \tag{8.8}$$

and the Poisson equation for the gravitational potential ϕ_0,

$$\frac{d^2\phi_0}{dz^2} = 4\pi G \rho_0. \tag{8.9}$$

Together, (8.6)–(8.9) yield a single equation for ρ_0,

$$\frac{d^2}{dz^2}\ln\rho_0 + \frac{4\pi G\beta\gamma}{(1+\beta)c_s^2}\rho_0 = 0. \tag{8.10}$$

The density profile $\rho_0(z)$ has to satisfy the condition that $\rho_0(z)$ has an extremum at $z = 0$. This follows from the physical reason that the gravitational force vanishes at the center of the cloud and so do the pressure and density gradients if the cloud is isothermal. Hence, the solution of (8.10) is then easily obtained as

$$\rho_0 = \frac{\rho_{00}}{\cosh^2 Z}, \tag{8.11}$$

where $Z = z/H$ and a typical scale length H is defined through

$$H^2 = \frac{(1+\beta)c_s^2}{2\pi\gamma G\beta\rho_{00}}. \tag{8.12}$$

Here ρ_{00} is the density in the center of the cloud and it can be related to the total mass $\mathcal{M}$ per unit surface in the x, y-plane

$$\mathcal{M} = \int_{-\infty}^{+\infty} \rho_0 dz = 2H\rho_{00}. \tag{8.13}$$

As to the magnetic field distribution, it follows immediately from the assumption that β is constant as

$$B_0 = \frac{B_{00}}{\cosh(z/H)}, \tag{8.14}$$

where

$$B_{00} = \left(\frac{2\mu_0 c_s^2}{\beta \gamma} \rho_{00} \right)^{1/2} \tag{8.15}$$

is the field strength at the center of the cloud. Its value is prescribed by β and ρ_{00}.

8.2.2 Linear perturbations

The basic state described in the previous subsection now undergoes linear perturbations whose amplitudes are z-dependent only and governed by the standard set of MHD equations [Baumjohann and Treumann 1996] together with the Poisson equation. Thus I have the following linearized set of equations for the perturbed quantities,

$$\frac{\partial \rho}{\partial t} + \nabla \cdot (\rho_0 \mathbf{v}) + \nabla \cdot (\rho \mathbf{U}_0) = 0,$$

$$\frac{\partial \mathbf{v}}{\partial t} + \mathbf{v} \cdot \nabla \mathbf{U}_0 + \mathbf{U}_0 \cdot \nabla \mathbf{v} + \frac{1}{\rho_0} \nabla p + \frac{\rho}{\rho_0} \nabla \phi_0 + \nabla \phi$$

$$= \frac{1}{\mu_0 \rho_0} (\nabla \times \mathbf{B}_0) \times \mathbf{B} + \frac{1}{\mu_0 \rho_0} (\nabla \times \mathbf{B}) \times \mathbf{B}_0,$$

$$\frac{\partial \mathbf{B}}{\partial t} = \nabla \times (\mathbf{v} \times \mathbf{B}_0) + \nabla \times (\mathbf{U}_0 \times \mathbf{B}),$$

$$\frac{d^2 \phi}{dz^2} = 4\pi G \rho,$$

$$\frac{\partial p}{\partial t} + \mathbf{U}_0 \cdot \nabla p + \mathbf{v} \cdot \nabla p_0 = c_s^2 \left(\frac{\partial \rho}{\partial t} + \mathbf{U}_0 \cdot \nabla \rho + \mathbf{v} \cdot \nabla \rho_0 \right), \quad (8.16)$$

representing the continuity, momentum, magnetic induction, Poisson and energy equations, respectively. The last equation expresses the adiabaticity of the perturbations. In what follows, $\nabla = \mathbf{e}_z \partial/\partial z$ and (8.16) is Fourier transformed in time. In order to write the resulting equations in an uncluttered form, I introduce a normalized magnetic field $\mathbf{b} = \mathbf{B}/B_0$, the total pressure $P = p + V_A^2 \rho_0 b_x$, components for the velocities $\mathbf{v} = (u, v, w)$, a Lagrangian displacement ξ in the z-direction defined through

$$w = -i\omega \xi, \tag{8.17}$$

and a new variable η by

$$\eta = \rho_0 \xi. \tag{8.18}$$

I thus obtain from (8.16) the following set

$$u + \frac{\eta}{\rho_0} \frac{dU_0}{dz} = 0,$$

$$\frac{d\phi_0}{dz}\frac{d\eta}{dz} + \omega^2\eta = \frac{dP}{dz} + \rho_0\frac{d\phi}{dz},$$

$$V_{ms}^2\frac{d\eta}{dz} = \eta\left[\frac{d\ln\rho_0}{dz}\left(\frac{V_A^2}{2} + \frac{\gamma-1}{\gamma}c_s^2\right)\right] - P,$$

$$\frac{d^2\phi}{dz^2} = -4\pi G\frac{d\eta}{dz}, \tag{8.19}$$

where V_{ms} is the magnetosonic velocity. The nonuniform flow speed enters only in the first equation of (8.19) and it only causes induced harmonic motions with speed u in the x-direction. The last equation of (8.19) can be integrated to give

$$\frac{d\phi}{dz} = -4\pi G\eta, \tag{8.20}$$

with the boundary condition

$$\left.\frac{d\phi}{dz}\right|_{\eta=0} = 0. \tag{8.21}$$

From the basic state (8.11) and (8.14) there results

$$\frac{d}{dz}\ln\rho_0 = -\frac{2}{H}\tanh Z,$$

$$\frac{d}{dz}\ln B_0 = -\frac{1}{H}\tanh Z, \tag{8.22}$$

and also then

$$\frac{d\phi_0}{dz} = 4\pi G\rho_{00}H\tanh Z \tag{8.23}$$

from the Poisson equation (8.9). By eliminating P from the two middle equations in (8.19) and using the explicit expressions for the equilibrium quantities I finally obtain a single equation

$$\frac{d^2}{dZ^2}(\eta\cosh Z) - \left(\mathcal{K} - \frac{2}{\cosh^2 Z}\right)(\eta\cosh Z) = 0, \tag{8.24}$$

where I recall that $Z = z/H$ and

$$\mathcal{K} = 1 - \frac{\omega^2 H^2}{V_{ms}^2}. \tag{8.25}$$

Equation (8.24) can easily be solved analytically under some special conditions. For example, simple analytical solutions are obtained in the quasi-uniform region $|Z| \gg 1$, where the coefficients in (8.24) become practically constant, and in the case of $\mathcal{K} = 0$ and $\mathcal{K} = 1$, when the solutions are given in terms of associated Legendre functions.

In what follows, however, it is very instructive to first repeat briefly results that are based on the standard assumption of a uniform and self-gravitating cloud, as used in the literature dealing with the Jeans instability and also, of course, in later attempts to deal with dusty plasmas.

8.2.3   Standard assumption of a uniform basic state

To point out the pitfalls associated with the "Jeans swindle", I repeat the standard derivation commonly used in the literature, when the initial basic state is assumed uniform everywhere. In that case, the Poisson equation is only considered for the perturbations, but not for the determination of the initial state. Thus, I start from the assumption of a uniform initial unperturbed state,

$$\frac{d}{dz}(\phi_0, \rho_0, B_0, U_0) = 0, \tag{8.26}$$

and the set (8.19) reduces to a system with constant coefficients,

$$\frac{dP}{dz} = \omega^2 \eta - \rho_0 \frac{d\phi}{dz},$$

$$V_{ms}^2 \frac{d\eta}{dz} = -P,$$

$$\frac{d^2\phi}{dz^2} = -4\pi G \frac{d\eta}{dz}. \tag{8.27}$$

Because of the uniform equilibrium, the z-dependence of the perturbations can be taken as $\exp(ikz)$, so that (8.27) immediately yields the dispersion law

$$\omega^2 = k^2 V_{ms}^2 - \omega_J^2, \tag{8.28}$$

where the Jeans frequency has been used, defined here through $\omega_J^2 = 4\pi G \rho_0$. Now (8.28) indicates the possibility for gravitational instabilities to set in, when k becomes small enough to render ω^2 negative. In this case the Jeans length λ_J is given through

$$\lambda_J^2 = \frac{\pi V_{ms}^2}{G \rho_0} = \frac{\pi (2 + \beta\gamma)c_s^2}{G \gamma \beta \rho_0}. \tag{8.29}$$

The conclusion regarding the instability is thus based on the assumption that the medium be uniform, at least locally, i.e. over a distance of several wavelengths. On the other hand, we have seen that the basic state cannot intrinsically be uniform on a larger scale, due to the ever-present gravitational forces. According to (8.11) and (8.14), the characteristic length of the nonuniformity H, as given by (8.12), is smaller than λ_J, as given by (8.29). Indeed, the following inequality is obvious,

$$\frac{\lambda_J^2}{H^2} = 2\pi^2 \frac{2 + \beta\gamma}{1 + \beta} > 1. \tag{8.30}$$

This indicates that the required condition for a medium to be locally uniform is not valid for unstable perturbations that satisfy the condition $\lambda > \lambda_J > H$. In the case that $\lambda \ll H$, the local approximation is of course valid but then k is so large that the only correct way of writing the dispersion law (8.28) should be as

$$\omega^2 \simeq k^2 V_{ms}^2, \tag{8.31}$$

i.e. leaving out the gravitational term! Consequently, the standard assumption of a uniform basic state can only be valid locally, in which case the perturbations propagate as stable, fast MHD modes unaffected by gravitation, with phase speed V_{ms}.

8.2.4   Local solutions

Going back to the full model, (8.24) can have local solutions provided the term containing $\cosh^2 Z$ may be neglected. This occurs if the condition

$$|\mathcal{K}| \gg \frac{2}{\cosh^2 Z} \tag{8.32}$$

is satisfied, i.e. when $Z = z/H$ is sufficiently large, depending on $\mathcal{K}$. The solution of (8.24) is then of the form

$$\eta \cosh Z \simeq \exp(\pm ikz). \tag{8.33}$$

The wavenumber k is given by the expression $k^2 = -\mathcal{K}/H^2$, which dispersion law can be rewritten as

$$\omega^2 = k^2 V_{ms}^2 + \frac{V_{ms}^2}{H^2}. \tag{8.34}$$

Hence (8.34) therefore refers to a more general local solution and, contrary to (8.31), it also includes gravitational effects through the length H. The validity of (8.34) requires that $|z| \gg \lambda$ and $|z| \gg H$. According to (8.34), the system is stable to eventual gravitational instabilities at $|z| \gg H$ and perturbations propagate as modified fast magnetosonic waves. The frequency ω has a lower bound at the cut-off frequency V_{ms}/H below which the wave becomes evanescent and can not propagate anymore.

Finally, I can compute the analytical expressions for the displacement $\xi = \eta \rho_0$ and the total pressure perturbation P from (8.19) and (8.33) as

$$\xi = \cosh Z \left(C_1 \exp(ikz) + C_2 \exp(-ikz)\right),$$

$$P = \frac{\rho_{00}}{\cosh Z} \left(\frac{c_s^2 \Gamma}{H} \tanh Z - ikV_{ms}^2\right) C_1 \exp(ikz)$$

$$+ \frac{\rho_{00}}{\cosh Z} \left(\frac{c_s^2 \Gamma}{H} \tanh Z + ikV_{ms}^2\right) C_2 \exp(-ikz), \qquad (8.35)$$

where $C_{1,2}$ are integration constants and $\Gamma = (2 - \gamma)/\gamma$. The solutions (8.35) show that wave amplitudes are z-dependent in the domain $|z| \gg H$, so that the displacement ξ grows with $|z|$ as $\cosh(z/H)$, while the total pressure perturbation decreases with $|z|$ as $1/\cosh(z/H)$.

8.2.5 Global solutions

In the previous section linear perturbations were treated locally through a restricted application of spatial Fourier analysis. The system under investigation could also oscillate as a whole at certain allowed eigenfrequencies, having its energy conserved. Thus, the perturbations have to be localized in the sense that they do not transport energy out of the system, i.e. the energy density of the perturbations has to tend to zero at sufficiently large $|z|/H$. As the local solutions (8.35) for ξ and P are asymptotic expressions of the corresponding global solutions at $|z| \gg H$, this is clearly achieved if the wavenumber k in (8.35) is taken imaginary, $k = i\kappa$. Consequently, the solutions I am now looking for should have the following asymptotic behaviour at $|z|/H \gg 1$,

$$\xi^{(-)} = C_2^{(-)} \exp(\kappa z) \cosh Z,$$

$$P^{(-)} = -\frac{\rho_{00}}{\cosh Z} \left(\frac{c_s^2 \Gamma}{H} + \kappa V_{ms}^2\right) C_2^{(-)} \exp(\kappa z), \qquad (8.36)$$

if $z < 0$ and

$$\xi^{(+)} = C_1^{(+)} \exp(-\kappa z) \cosh Z,$$

$$P^{(+)} = \frac{\rho_{00}}{\cosh Z} \left(\frac{c_s^2 \Gamma}{H} + \kappa V_{ms}^2\right) C_1^{(+)} \exp(-\kappa z), \qquad (8.37)$$

if $z > 0$. Here

$$\kappa = \sqrt{\frac{1}{H^2} - \frac{\omega^2}{V_{ms}^2}} \qquad (8.38)$$

and the frequency ω has to satisfy the inequality $\omega < V_{ms}/H$. These global solutions themselves may be either symmetric or antisymmetric in z due to the symmetry of the basic state, and hence $C_1^{(+)}$ and $C_2^{(-)}$ may differ only in sign. Moreover, one can argue that the global oscillations of the cloud cause displacements of plasma that are antisymmetric with respect to the center of the cloud $z = 0$. In two symmetric points with respect to the center, the fluid particles move simultaneously in opposite directions, either towards or away from the center, but remain at rest at $z = 0$. As to the total pressure perturbation, it is a symmetric function of z, yielding either a local compression or rarefaction at each pair of symmetric points. The final conclusion is now that $C_1^{(+)}$ and $C_2^{(-)}$ should have opposite signs.

The cloud can therefore stably oscillate as a whole with any of the frequencies from the continuous range given by $V_{ms}/H > \omega > 0$. For the two limiting values, $\omega = 0$ and $\omega = V_{ms}/H$, (8.24) has global solutions in a closed form for the whole range of z, given in terms of associated Legendre functions. For any other value of ω, the global solutions of (8.24) can only be obtained numerically.

8.2.6 Caveat

In the one-dimensional model described in this section, a self-gravitating plasma cloud in magnetohydrostatic equilibrium intrinsically has a nonuniform density over a typical length scale H given by (8.12). This density variation cannot be neglected in stability analyses by linear perturbation methods, whenever the wave length of the perturbations becomes comparable to H. Such cases typically occur in derivations of standard Jeans criteria for various self-gravitating plasma configurations in which the plasma density is assumed uniform in the unperturbed state. As a consequence, linear waves yielding instability in a hypothetically uniform plasma cloud are, in fact, stable if the natural density inhomogeneity, resulting from self-gravitation, is taken into account. This further implies that conclusions based on the standard Jeans instability treatments as found in literature, have to be revised accordingly.

This will have repercussions on some of the recent treatments of the Jeans instabilities in dusty plasmas, as discussed in the next sections of this chapter. However, given the vast differences in charge and mass between the ordinary plasma and the charged dust components occurring in dusty plasmas, a multispecies and more-dimensional generalization of the treatment in the present section is called for, but has not obtained yet. To sum up at this stage, the picture is completely confused! Much work needs to be done for complicated plasmas, and it is not clear at all on how to proceed, as long as the initial state cannot be fully determined. Thus, a review will be given in the remaining sections of this chapter of attempts to generalize Jeans modes to multispecies dusty plasmas, the conclusions of which have to be taken with a salutary pinch of salt, and certainly not at face value.

8.3 Self-gravitation of dusty plasmas

8.3.1 Different aspects of gravitation

Various derivations of and modifications to Jeans instabilities in dusty plasmas are found in recent papers [Bliokh and Yaroshenko 1985; Avinash and Shukla 1994; Pandey et al. 1994; Verheest and Shukla 1997; Meuris et al. 1997; Verheest et al. 1997a,1997b; Mace et al. 1998; Verheest et al. 1999a] and even a book [Bliokh et al. 1995]. However, in some of these papers confusion occurred between three different and unrelated aspects of gravitational forces, as they can play a role in solar system plasmas, and presumably also in other astrophysical contexts.

A first aspect has to do with the gravitational attraction between a planet and a particle of its ring system, which in gravito-electrodynamical theories might easily be comparable to the electrodynamical forces between the ring particles. Second, of course, there is the self-gravitational attraction between like grains, and third, there can be influences of self-gravitational forces on perturbations in the system, even though the Newtonian forces might be orders of magnitude smaller than the Coulomb forces, as seen further on.

Before turning to the treatment of various wave modes, however, I need to go somewhat deeper into when self-gravitational attraction could become comparable to the electrostatic repulsion, for two identical particles that are interacting with one another. The condition that the Coulomb and Newtonian forces be equal,

$$\frac{q^2}{4\pi\varepsilon_0 r^2} = \frac{Gm^2}{r^2},\tag{8.39}$$

can be recast into the equality of plasma (ω_p) and Jeans (ω_J) frequencies through

$$\omega_p^2 = \frac{Nq^2}{\varepsilon_0 m} = 4\pi GNm = \omega_J^2.\tag{8.40}$$

To fix the ideas, look at a particle with one electron charge and see that equating the plasma to the Jeans frequency leads to masses of the order

$$m = \frac{e}{\sqrt{4\pi G\varepsilon_0}} = 1.8 \times 10^{-9} \text{ kg} \simeq 10^{18} \, m_p.\tag{8.41}$$

This very well known result shows that in ordinary electron-ion plasmas possible self-gravitational effects are so excruciatingly small that these are never even mentioned! However, for spherical charged dust grains made up of water ice, the radius corresponding to such a mass turns out to be of the order

$$a = \left(\frac{3m}{4\pi \rho_{\text{ice}}} \right)^{1/3} \simeq 80 \ \mu\text{m}. \tag{8.42}$$

For grains with higher charges, say $Q_d = 1000e$, this increases the size by the cube root of 1000 and immediately leads to millimeter-sized radii, rather larger than the micrometer-sized grains upon which attention has focused in solar-system plasmas. However, the existence of dust particles of this size does not seem totally unlikely, given the wide range of cosmic and astrophysical environments in which dusty plasmas are found. So the dusty plasma would be expected to be marginally stable, in the gravitational sense, to all density perturbations except those whose length scales might exceed the appropriate Jeans length, which will turn out to be typically very large — even on astrophysical length scales.

It is important to emphasize here, however, that although for the equilibrium solution of a self-gravitating plasma the gravitational and electrical forces will differ by orders of magnitude for smaller grains, this is not the case for the (linearized) instabilities, as I will show in subsequent sections. Hence the size of the dust grains need not be specified at this stage.

8.3.2 Other factors

When more (plasma) physical detail, such as dust charge states, typical intergrain spacing versus Debye length, magnetic fields, electromagnetic effects, bulk particle flows, and plasma angular momentum is incorporated into the model, one finds that the stability of the dusty plasma (or of a plasma with dust) to collapse is far more complicated than the simple models indicate. It thus becomes necessary to undertake a full investigation into all potentially stabilizing and destabilizing wave modes if we hope to fully understand the stability of cosmic dusty plasmas, and perhaps further our insight into the formation of protoplanets, protostars and protogalaxies.

Another factor to keep in mind is rotation, as it is generally true that astrophysical bodies rotate. Thus, in considering large astrophysical dusty plasma clouds, it is prudent to include also the effects of (differential) rotation. The most fundamental effect that rotation of a dusty plasma has, is to prevent collapse on large length scales. In the (noninertial) frame of the rotating plasma, a centrifugal force $\mathbf{f}_c \propto \mathbf{\Omega}(r) \times (\mathbf{\Omega}(r) \times \mathbf{r})$ is added to the momentum equation, where $\mathbf{\Omega}(r)$ is the angular velocity as a function of the radial distance $r = |\mathbf{r}|$ from the rotation axis. This force opposes the tendency for gravitational collapse and is maximum in the rotational equator plane and zero on the rotation axis, hence those regions most stable to collapse lie in the equatorial plane, leading inevitably to a disk-shaped plasma. Furthermore, this force increases with distance, in contrast to the gravitational force (treating the cloud as a point mass) which decreases with r. Hence, larger distances are more stable to the Jeans instability in a disk-shaped plasma.

In a reasonable first approximation for a slowly, differentially-rotating, magnetized dusty plasma cloud, the rotation and magnetic axes are assumed to coincide, and then rotational effects give an upper length scale, beyond which the plasma (disk) is Jeans stable [Toomre 1964], whereas the simple Jeans criteria provide a lower length scale, beyond which the disk might be Jeans unstable. Hence, the region of plasma which is subject to the Jeans instability is probably limited on both larger and smaller scales, or might even be nonexistent.

To make some headway in an already dubious domain, charge fluctuations of the dust particles will be completely ignored, as well as mass and charge distributions. This is reasonable if the mass distributed dust shows a power law decrease with size, weighted in favour of dust grains at the lower end of the size range. Most of the characteristic frequencies when computed for such grains are then little different from the true values for the whole range [Meuris 1997a], and I will look into those aspects in the next chapter.

8.3.3 Chronology

Before turning to the more mathematical analysis of the various wave modes, let me briefly recall some of the specific results for dusty plasmas in chronological order. A first analysis of this problem has been given by Chhajlani and Parihar [1994] and indicates how charged dust grains in interstellar clouds affect the fragmentation process. For low-frequency modes the effect of the grains is important, decreasing the region of instability and the critical Jeans wavenumber and rendering star formation more difficult, as I will point out below. The model considered, however, is not flawless, because separate self-gravitational potentials have been considered for the ions and the charged dust grains, each obeying their own Poisson equation!

Pandey *et al.* [1994] obtain conditions for stable electrostatic levitation, condensation and dispersion of grains in a plasma background. As the grains start to condense due to self-gravitation, an electrostatic field is set up due to charge separation between grains, electrons and ions. The electrons and ions rush to shield this field, creating density perturbations which if fast enough will tend to smooth out the effects and inhibit the condensation. However, the nonlinear evolution shows a condensation of grains, even when the effect of self-gravitation is annulled by electrostatic repulsion.

More generally, Avinash and Shukla [1994] have studied a purely growing, electrostatic instability, generalizing Jeans instability, which could play a decisive role in levitation/condensation of grains in planetary rings as well as in the formation of galaxies and stars. The description includes a mixture of dust-acoustic and self-gravitational modes, as will be seen, and leads to a rather robust, purely growing instability with a larger growth rate than given by Pandey *et al.* [1994].

Regardless of the validity of the "Jeans swindle", which is accepted in these sections as if completely true, there appear, however, misconceptions (exposed recently *e.g.* in Verheest *et al.* [1997a]) about the physics of this instability and how it should be applied to plasmas in general, and dusty plasmas in particular. Hence, the need for a systematic ordering of the Jeans instabilities and for strict criteria under which these might be important for dusty plasmas.

8.4   Jeans-Buneman modes

8.4.1   Global results

For wave propagation parallel to the external magnetic field we saw that possible self-gravitation only affected the longitudinal modes, with dispersion law (4.72) or

$$
\left(1 - \sum_\alpha \frac{\omega_{p\alpha}^2}{(\omega - kU_\alpha)^2 - k^2 c_{s\alpha}^2}\right)\left(1 + \sum_\alpha \frac{\omega_{J\alpha}^2}{(\omega - kU_\alpha)^2 - k^2 c_{s\alpha}^2}\right)
$$
$$
+ \left(\sum_\alpha \frac{\omega_{p\alpha}\omega_{J\alpha}}{(\omega - kU_\alpha)^2 - k^2 c_{s\alpha}^2}\right)^2 = 0. \tag{8.43}
$$

Without the self-gravitational terms, this has been discussed at great length in Chapter 5. When incorporating all effects now, one still could be tempted to include only the contribution of the dust grains and neglect those of the electrons and ions, seeing that they are so small. While true, it is enlightening to work fully generally, until I want to specialize the results and see what matters.

The dispersion law (8.43) was first derived by Bliokh and Yaroshenko [1985] when addressing spoke formation in the B ring of Saturn. Before discussing (8.43) in more detail, it is seen that if all species are effectively cold ($c_{s\alpha} = 0$) and not streaming ($U_\alpha = 0$), then the last, squared term vanishes exactly and there is total decoupling between electrostatic and pure Jeans modes, regardless of the composition of the plasma [Bliokh *et al.* 1995]. This is because $\omega_{p\alpha}\omega_{J\alpha}$ then sums to zero.

Another way of looking at such a decoupling is by considering (8.43) in the equivalent long-wavelength limit, when $k \to 0$. The high-frequency branch gives the usual electrostatic plasma modes, with dispersion law

$$
\sum_\alpha \frac{\omega_{p\alpha}^2}{(\omega - kU_\alpha)^2 - k^2 c_{s\alpha}^2} \simeq 1 \tag{8.44}
$$

discussed extensively in Chapter 5. On the other hand, when the charge effects do not really matter, the ordinary Jeans instability obeys

$$1 + \sum_\alpha \frac{\omega_{J\alpha}^2}{(\omega - kU_\alpha)^2 - k^2 c_{s\alpha}^2} \simeq 0, \tag{8.45}$$

and is the only global mode surviving in a neutral gas. If all neutral dust or gas is assumed to be of the same kind, then there is no streaming to be considered and the familiar form of the Jeans mode is recovered,

$$\omega^2 \simeq k^2 c_{sd}^2 - \omega_{Jd}^2. \tag{8.46}$$

The Jeans wavenumber is thus

$$k_J = \frac{\omega_{Jd}}{c_{sd}}, \tag{8.47}$$

with the associated neutral Jeans length given by

$$L_J = \frac{2\pi}{k_J} = \frac{2\pi c_{sd}}{\omega_{Jd}}. \tag{8.48}$$

This is the traditional derivation of the critical length.

8.4.2   Cold and massive dust

For a combination of electrostatic and gravitational factors, go back to (8.43) in its full form and specialize the results to a dusty plasma in which the dust grains are so massive that they are treated as a set of non-streaming, cold fluids. In addition, the plasma contains electrons and one species of positive ions. Proceeding step by step, I begin by noting that $\omega_{pi}^2 \omega_{Je}^2 \ll \omega_{pe} \omega_{pi} \omega_{Je} \omega_{Ji} \ll \omega_{pe}^2 \omega_{Ji}^2$, using absolute values for simplicity, and also that $\omega_{Je}^2 \ll \omega_{pe}^2$ and $\omega_{Ji}^2 \ll \omega_{pi}^2$. Then (8.43) is first simplified to

$$\left(1 - \frac{\omega_{pe}^2}{\mathcal{L}_e} - \frac{\omega_{pi}^2}{\mathcal{L}_i} - \frac{1}{\omega^2} \sum_d \omega_{pd}^2\right)\left(1 + \frac{1}{\omega^2}\sum_d \omega_{Jd}^2\right)$$
$$+ \frac{1}{\omega^4}\left(\sum_d \omega_{pd}\omega_{Jd}\right)^2 - \frac{\omega_{pe}^2 \omega_{Ji}^2}{\mathcal{L}_e \mathcal{L}_i} - \left(\frac{\omega_{Je}^2}{\omega^2 \mathcal{L}_e} + \frac{\omega_{Ji}^2}{\omega^2 \mathcal{L}_i}\right)\sum_d \omega_{pd}^2$$
$$+ 2\left(\frac{\omega_{pe}\omega_{Je}}{\omega^2 \mathcal{L}_e} - \frac{\omega_{pi}\omega_{Ji}}{\omega^2 \mathcal{L}_i}\right)\sum_d \omega_{pd}\omega_{Jd} = 0, \tag{8.49}$$

with for the plasma particles

$$\mathcal{L}_\alpha = (\omega - kU_\alpha)^2 - k^2 c_{s\alpha}^2. \tag{8.50}$$

In the next chapter I discuss the possible influences of dust mass distributions, but first assume the dust to be mono-sized here, so that the dispersion law becomes

$$\left(1 - \frac{\omega_{pe}^2}{\mathcal{L}_e} - \frac{\omega_{pi}^2}{\mathcal{L}_i} - \frac{\omega_{pd}^2}{\omega^2}\right)\left(1 + \frac{\omega_{Jd}^2}{\omega^2}\right) + \frac{\omega_{pd}^2\omega_{Jd}^2}{\omega^4} - \frac{\omega_{pe}^2\omega_{Ji}^2}{\mathcal{L}_e\mathcal{L}_i}$$

$$- \left(\frac{\omega_{Je}^2}{\mathcal{L}_e} + \frac{\omega_{Ji}^2}{\mathcal{L}_i}\right)\frac{\omega_{pd}^2}{\omega^2} + 2\left(\frac{\omega_{pe}\omega_{Je}}{\mathcal{L}_e} - \frac{\omega_{pi}\omega_{Ji}}{\mathcal{L}_i}\right)\frac{\omega_{pd}\omega_{Jd}}{\omega^2} = 0. \quad (8.51)$$

Furthermore,

$$\omega_{Je}^2\omega_{pd}^2 \ll \omega_{pe}^2\omega_{Jd}^2,$$
$$\omega_{Ji}^2\omega_{pd}^2 \ll \omega_{pi}^2\omega_{Jd}^2, \quad (8.52)$$

because $\omega_{J\alpha}^2/\omega_{p\alpha}^2$ is proportional to m_α^2/q_α^2, and in all known dusty plasmas the mass-per-charge ratio for the dust species vastly exceeds that of protons and other ions, to say nothing about the electrons,

$$\frac{m_e}{|q_e|} \ll \frac{m_d}{|q_d|}, \qquad \frac{m_i}{q_i} \ll \frac{m_d}{|q_d|}. \quad (8.53)$$

Similarly,

$$\omega_{pe}\omega_{Je}\omega_{pd}\omega_{Jd} \ll \omega_{pe}^2\omega_{Jd}^2,$$
$$\omega_{pi}\omega_{Ji}\omega_{pd}\omega_{Jd} \ll \omega_{pi}^2\omega_{Jd}^2, \quad (8.54)$$

and thus the dispersion law (8.51) is to a high degree of precision approximated by

$$\left(1 - \frac{\omega_{pe}^2}{\mathcal{L}_e} - \frac{\omega_{pi}^2}{\mathcal{L}_i} - \frac{\omega_{pd}^2}{\omega^2}\right)\left(1 + \frac{\omega_{Jd}^2}{\omega^2}\right) + \frac{\omega_{pd}^2\omega_{Jd}^2}{\omega^4} - \frac{\omega_{pe}^2\omega_{Ji}^2}{\mathcal{L}_e\mathcal{L}_i} = 0. \quad (8.55)$$

This will be discussed below in different degrees of approximation.

8.4.3 Boltzmann electrons

First, treat the electrons as an inertialess Boltzmann species and define $A = k^2\lambda_{De}^2/(1 + k^2\lambda_{De}^2)$. The dispersion law can then be simplified to

$$\left[(\omega - kU_i)^2 - k^2c_{si}^2 - A\omega_{pi}^2\right]\left[\omega^2 + \omega_{Jd}^2 - A\omega_{pd}^2\right]$$

$$= A^2\omega_{pi}^2\omega_{pd}^2 - \frac{\omega^2\omega_{Ji}^2}{1 + k^2\lambda_{De}^2}. \quad (8.56)$$

This generalizes the derivation of Pandey and Lakhina [1998] by including both the electrons (albeit through an inertialess Boltzmann description) and by retaining a term in ω_{Ji}^2, which cannot always be neglected without proper justification about the frequencies considered. If I am allowed to do so, the result is

$$\left[(\omega - kU_i)^2 - k^2 c_{si}^2 - A\omega_{pi}^2\right]\left[\omega^2 + \omega_{Jd}^2 - A\omega_{pd}^2\right] = A^2 \omega_{pi}^2 \omega_{pd}^2. \qquad (8.57)$$

For large wavelengths ($k \to 0$) this is the Jeans mode, while for small wavelengths ($k \to \infty$), without self-gravitational effects, the Langmuir-Buneman modes of the previous chapter are recovered. Pandey and Lakhina [1998] found that, in the presence of streaming, the Jeans and the Buneman modes can overlap, and they called the new mode for obvious reasons the Jeans-Buneman mode. Further studies of the Jeans-Buneman mode have included dust distributions [Meuris *et al.* 1997; Pillay *et al.* 1999], discussed in the next chapter. Dust thermal effects were included by Mamun [1998], but lead to similar conclusions. As expected from the general discussion of self-gravitational effects, finite dust temperatures are a stabilizing factor.

8.4.4 Boltzmann electrons and ions

Take now the most simple point of view and suppose that both electrons and (positive) ions are Boltzmann species, the regime for which the dust-acoustic mode occurs without gravitational effects [Rao *et al.* 1990]. Redefine A in terms of the total Debye length λ_D, and find from (8.43) without additional approximations that

$$\omega^2 = A\,\omega_{pd}^2 - \omega_{Jd}^2. \qquad (8.58)$$

Recalling the definition of the dust-acoustic velocity, $c_{da} = \lambda_D \omega_{pd}$, I get for small k from (8.58) that

$$\omega^2 = k^2 c_{da}^2 - \omega_{Jd}^2. \qquad (8.59)$$

This shows how the transition occurs from dust-acoustic to Jeans mode, and essentially reproduces the results of Avinash and Shukla [1994] and Pandey *et al.* [1994].

Intuitively replacing v_{rms} in (8.4) with the dust-acoustic wave speed c_{da} gives the same result, up to numerical factors of order unity. Usually c_{da} exceeds by far the dust thermal speed c_{sd}, because of the much larger electron and ion thermal velocities. Hence the presence of dust-acoustic waves could stabilize the Jeans instability on length scales which would be unstable in the classic Jeans scenario. This is a first and important modification of the usual Jeans criteria.

8.4.5   Oblique electrostatic modes

If the dusty plasma is immersed in a static magnetic field, as in interstellar or galactic environments, other wave modes become possible at the low-frequency end. We have seen that for purely parallel propagation the transverse and longitudinal modes are fully decoupled. Only the latter ones are affected by gravitation, and these have been discussed in the previous subsections. However, for oblique and perpendicular wave propagation I will need to investigate how all the waves are affected.

Rather than cover all possible configurations, I will briefly look at oblique electrostatic modes, before going to perpendicular magnetosonic modes in the next section. Then the oblique dispersion law (4.81) bears a great resemblance to the parallel one (4.72), so that an analogous reduction can be used to obtain

$$\left(1 - \frac{\omega_{pe}^2}{\mathcal{K}_e} - \frac{\omega_{pi}^2}{\mathcal{K}_i} - \frac{\omega_{pd}^2}{\mathcal{K}_d}\right)\left(1 + \frac{\omega_{Jd}^2}{\mathcal{K}_d}\right) + \frac{\omega_{pd}^2\omega_{Jd}^2}{\mathcal{K}_d^2} - \frac{\omega_{pe}^2\omega_{Ji}^2}{\mathcal{K}_e\mathcal{K}_i} = 0. \quad (8.60)$$

This dispersion law has to be treated with some care in the low-frequency limit, and the expressions for $\mathcal{K}_\alpha$ have been given in Chapter 4, in (4.79).

When the common assumptions of Boltzmann electrons and ions and $k^2\lambda_D^2 \ll 1$ are introduced, the dispersion law reduces to

$$\omega^4 - [\Omega_d^2 + k^2(c_{da}^2 + c_{sd}^2) - \omega_{Jd}^2]\omega^2 + [k^2(c_{da}^2 + c_{sd}^2) - \omega_{Jd}^2]\Omega_d^2\cos^2\vartheta = 0. \quad (8.61)$$

For values of k close to the critical values needed to annul $k^2(c_{da}^2 + c_{sd}^2) - \omega_{Jd}^2$, approximate solutions can be obtained as

$$\begin{aligned}
\omega^2 &\simeq \Omega_d^2 + [k^2(c_{da}^2 + c_{sd}^2) - \omega_{Jd}^2]\sin^2\vartheta, \\
\omega^2 &\simeq [k^2(c_{da}^2 + c_{sd}^2) - \omega_{Jd}^2]\cos^2\vartheta.
\end{aligned} \quad (8.62)$$

The latter mode is the first one to become unstable, when

$$k^2(c_{da}^2 + c_{sd}^2) < \omega_{Jd}^2, \quad (8.63)$$

and generalizes (8.59) in two ways. One is rather trivial, because dust thermal velocities have been incorporated, but since the plasma pressures largely exceed the dust pressure, $c_{sd}^2 \ll c_{da}^2$ anyway. However, (8.62) now occurs at all angles of propagation, but slowly disappears as perpendicular propagation is approached. The other mode in (8.62) similarly generalizes (5.94), obtained in Chapter 5 when discussing dust lower-hybrid and related modes. Related investigations were carried out by Salimullah and Shukla [1999].

8.5   Magnetosonic Jeans modes

At propagation strictly perpendicular to the magnetic field there occur
several combinations between Jeans and magnetosonic modes, as we now
discuss [Verheest *et al.* 1997b,1999a; Mace *et al.* 1998]. Opposition to
the gravitational collapse now comes also, and sometimes chiefly, from the
magnetic field pressure. This mechanism could even work in the complete
absence of particle thermal motions. The magnetosonic phase velocity V_{ms}
encountered in Chapter 6 will now play the role previously assigned to $v_{\rm rms}$
or c_{da}.

For the investigation of magnetosonic waves propagating across the external magnetic field, go back to Chapter 4 and put $\vartheta = 90°$ in (4.77),
which yields for the nonzero elements of the symmetric dispersion tensor

$$D_{xx} = 1 - \sum_\alpha \frac{\omega_{p\alpha}^2}{\omega^2 - k^2 c_{s\alpha}^2 - \Omega_\alpha^2},$$

$$D_{xy} = \sum_\alpha \frac{\omega_{p\alpha}^2 \Omega_\alpha}{\omega^2 - k^2 c_{s\alpha}^2 - \Omega_\alpha^2},$$

$$D_{x\psi} = \sum_\alpha \frac{\omega_{p\alpha}\omega_{J\alpha}}{\omega^2 - k^2 c_{s\alpha}^2 - \Omega_\alpha^2},$$

$$D_{yy} = \omega^2 - c^2 k^2 - \sum_\alpha \frac{\omega_{p\alpha}^2(\omega^2 - k^2 c_{s\alpha}^2)}{\omega^2 - k^2 c_{s\alpha}^2 - \Omega_\alpha^2},$$

$$D_{y\psi} = -\sum_\alpha \frac{\omega_{p\alpha}\omega_{J\alpha}\Omega_\alpha}{\omega^2 - k^2 c_{s\alpha}^2 - \Omega_\alpha^2},$$

$$D_{zz} = \omega^2 - c^2 k^2 - \sum_\alpha \omega_{p\alpha}^2,$$

$$D_{\psi\psi} = -1 - \sum_\alpha \frac{\omega_{J\alpha}^2}{\omega^2 - k^2 c_{s\alpha}^2 - \Omega_\alpha^2}. \tag{8.64}$$

The dispersion law (4.76) factorizes into a part $D_{zz} = 0$ or (4.82), describing the high-frequency ordinary mode for which neither thermal nor
self-gravitational effects matter in the fluid description, and another for the
generalized X mode. In what follows, I will be concerned with the latter,
but remark that in unmagnetized plasmas a further factorization occurs.
The dispersion is then reduced to twice that of the O mode, linearly polarized in two independent directions orthogonal to the wave vector, in
addition to the Jeans-Buneman modes discussed in the previous section.

8.5.1   Inertialess electrons and ions

Given the enormous differences in charge-to-mass ratios between ordinary
plasma particles and charged dust grains, I introduce the often used ap-

proximation that the lighter species are inertialess. This was first discussed in Chapter 5 when discussing the dispersion law for ion-acoustic modes. Taking the limit that $m_e \to 0$, such that $m_e c_{se}^2$ remains finite, is tantamount to supposing that the modes are low frequency in the sense that $\omega \ll |\Omega_e|$, and that in addition electron Larmor radius effects can be neglected $(k^2 \rho_e^2 = k^2 c_{se}^2 / \Omega_e^2 \ll 1)$. Analogous conditions are imposed on the ions. For the dust grains I will allow two possibilities, charged and neutral dust, to provide the necessary inertial effects.

Within the limits thus indicated and using charge neutrality in equilibrium, the elements (8.64) of the dispersion tensor are simplified to [Verheest et al. 1999a]

$$D_{xx} = 1 - \frac{\omega_{pd}^2}{\omega^2 - k^2 c_{sd}^2 - \Omega_d^2},$$

$$D_{xy} = \frac{\omega_{pd}^2(\omega^2 - k^2 c_{sd}^2)}{\Omega_d(\omega^2 - k^2 c_{sd}^2 - \Omega_d^2)},$$

$$D_{x\psi} = \frac{\omega_{pd}\omega_{Jd}}{\omega^2 - k^2 c_{sd}^2 - \Omega_d^2},$$

$$D_{yy} = \omega^2 - c^2 k^2(1 + \beta) - \frac{\omega_{pd}^2(\omega^2 - k^2 c_{sd}^2)}{\omega^2 - k^2 c_{sd}^2 - \Omega_d^2},$$

$$D_{y\psi} = -\frac{\omega_{pd}\omega_{Jd}\Omega_d}{\omega^2 - k^2 c_{sd}^2 - \Omega_d^2},$$

$$D_{\psi\psi} = -1 - \frac{\omega_{Jd}^2}{\omega^2 - k^2 c_{sd}^2 - \Omega_d^2} - \frac{\omega_{Jn}^2}{\omega^2 - k^2 c_{sn}^2}. \tag{8.65}$$

In these expressions, I have defined a plasma beta as

$$\beta = \frac{\mu_0(N_i m_i c_{si}^2 + N_e m_e c_{se}^2)}{B_0^2}. \tag{8.66}$$

Using (8.65) to work out the X mode part of the dispersion law I find in general a cubic equation in ω^2. Before discussing that in detail, I will first look at the simpler case where all the dust is charged.

8.5.2 Alfvén-Jeans modes in the absence of neutral gas

When all dust is charged and there is strictly no neutral gas present $(N_n = 0$ and $\omega_{Jn} = 0)$, the dispersion law reduces to a bi-quadratic in ω^2,

$$\omega^4 - A\omega^2 + C = 0, \tag{8.67}$$

with coefficients

$$A = c^2k^2(1+\beta) + k^2c_{sd}^2 + \frac{(\omega_{pd}^2 + \Omega_d^2)^2}{\Omega_d^2} - \omega_{Jd}^2, \tag{8.68}$$

$$C = (\omega_{pd}^2 + \Omega_d^2)\left(c^2k^2(1+\beta) - \frac{\omega_{pd}^2}{\Omega_d^2}(\omega_{Jd}^2 - k^2c_{sd}^2)\right)$$
$$- c^2k^2(1+\beta)(\omega_{Jd}^2 - k^2c_{sd}^2). \tag{8.69}$$

The solutions of (8.67) are given by

$$\omega^2 = \frac{A}{2} \pm \frac{1}{2}\sqrt{A^2 - 4C}. \tag{8.70}$$

As the discriminant $\Delta = A^2 - 4C$ is always positive,

$$\Delta = \left(c^2k^2(1+\beta) - k^2c_{sd}^2 + \omega_{Jd}^2 - \Omega_d^2 + \frac{\omega_{pd}^4}{\Omega_d^2}\right)^2 + \frac{4\omega_{pd}^2(\omega_{pd}^2 + \Omega_d^2)^2}{\Omega_d^2}$$

$$= \left(c^2k^2(1+\beta) - k^2c_{sd}^2 + \omega_{Jd}^2 + \frac{(\omega_{pd}^2 + \Omega_d^2)^2}{\Omega_d^2}\right)^2$$
$$- 4(\omega_{pd}^2 + \Omega_d^2)\left(c^2k^2(1+\beta) - k^2c_{sd}^2 + \omega_{Jd}^2\right), \tag{8.71}$$

the roots for ω^2 are real. The second form of the discriminant will be useful for the computation of approximate solutions to the dispersion law.

Instabilities are only possible whenever $\omega^2 < 0$. The existence of two negative roots for ω^2 would require $A < 0$ and $C > 0$, which cannot occur, as $A \leq 0$ implies $C < 0$, but not the other way around, although $C \geq 0$ means that $A > 0$. The only possibility left to find unstable modes is to have $C < 0$, so that the product of the roots for ω^2 is negative. At extreme values of k I have that

$$C = -\frac{(\omega_{pd}^2 + \Omega_d^2)\omega_{pd}^2\omega_{Jd}^2}{\Omega_d^2} < 0 \qquad (k = 0), \tag{8.72}$$

while

$$C \sim c^2k^4(1+\beta)c_{sd}^2 > 0 \qquad (k \to \infty). \tag{8.73}$$

Thus, there is always an unstable range $0 \leq k < k_{crit}$, and an approximate form for the latter will be given in (8.78).

I now discuss the modes given by (8.70) in more detail and introduce the dust magnetosonic velocity V_{ms}, defined in terms of the dust Alfvén velocity V_{Ad} and a dust-acoustic velocity c_{da} as in (6.46), and here $c_{da}^2 = \beta V_{Ad}^2$. Since for most astrophysical plasmas the charge-per-mass ratio of the dust is very much smaller than for ordinary ions, it is eminently plausible to assume that $V_{Ad} \ll c$ or, in different terms, that $\Omega_d^2 \ll \omega_{pd}^2$, and in addition that $c_{sd} \ll c$. In (8.70) the mode with the plus sign leads to

$$\omega^2 \simeq \frac{c^2}{V_{Ad}^2}\left(\omega_{pd}^2 + k^2 V_{ms}^2\right), \tag{8.74}$$

as far as the leading terms are concerned. Such high frequencies are not compatible with the basic assumptions, since

$$\frac{c^2 \omega_{pd}^2}{V_{Ad}^2} = \frac{\omega_{pd}^4}{\Omega_d^2} = \left(\frac{\omega_{pi}^2}{\Omega_i} + \frac{\omega_{pe}^2}{\Omega_e}\right)^2. \tag{8.75}$$

The mode with the minus sign, however, yields to the same degree of precision that

$$\omega^2 \simeq k^2 c_{sd}^2 + \frac{k^2 V_{ms}^2 \omega_{pd}^2}{\omega_{pd}^2 + k^2 V_{ms}^2 + V_{Ad}^2 \omega_{Jd}^2/c^2} - \omega_{Jd}^2. \tag{8.76}$$

Provided ω_{Jd} does not greatly exceed ω_{pd} in absolute value, which is highly unlikely, the term $\omega_{Jd}^2 V_{Ad}^2/c^2$ in the denominator can safely be neglected. Moreover, the small wavenumber limit ($k^2 V_{ms}^2 \ll \omega_{pd}^2$) is taken, as is more or less imposed by the neglect of electron and ion Larmor radius effects. In this approximation the dispersion law becomes very simple, namely

$$\omega^2 \simeq k^2(V_{ms}^2 + c_{sd}^2) - \omega_{Jd}^2. \tag{8.77}$$

Letting $\omega_{Jd} \to 0$, one recovers a generalization of the fast-magnetosonic mode discussed by Rao [1993b; 1995]. On the other hand, due to the self-gravitational effects, it goes over for very small wavenumbers into a modified form of the Jeans instability. Similar results were obtained by Mamun [1998]. The critical Alfvén-Jeans wavenumber for the changeover is given by

$$k_{AJ} = \frac{\omega_{Jd}}{\sqrt{V_{ms}^2 + c_{sd}^2}}, \tag{8.78}$$

which has to be contrasted with the ordinary neutral Jeans wavenumber [Jeans 1929; Chandrasekhar 1961] involving only the dust thermal velocity. For the ratio of the two wavenumbers, or rather of their inverses, the critical lengths, it is found from the various definitions that

$$\frac{\lambda_{AJ}}{\lambda_{dJ}} = \frac{k_{dJ}}{k_{AJ}} = \sqrt{\frac{P_i + P_e + P_d + B_0^2/\mu_0}{P_d}}. \tag{8.79}$$

It is reasonable to assume that the pure plasma pressures $P_e + P_i$, together with the magnetic field pressure B_0^2/μ_0 will greatly dominate the dust pressure P_d, so that λ_{AJ} will exceed λ_{dJ} by orders of magnitude. These results tally with the form of the dispersion law given by Chandrasekhar [1961] in an MHD formulation at strictly perpendicular propagation. However, a much more explicit form is obtained for the velocities used in the instability criterion, here $V_{Ad}^2 + c_{da}^2 + c_{sd}^2$.

8.5.3   Charged and neutral dust

In the strict absence of charged dust, there is a complete decoupling between the extraordinary mode effects which occur in a simple plasma without massive dust grains, and the usual neutral gas Jeans mode, thus no teaching us anything new. However, some estimates for the dust charges in interstellar plasmas seem to point to average fractional charges, the interpretation being that some grains are charged while others are not. In view of the very dilute character of interstellar plasmas, this is a very reasonable assumption. Hence the need to study mixtures of charged and neutral dust, both present with really non negligible densities. In that case the dispersion law is a cubic in ω^2,

$$\omega^6 - A'\omega^4 + C'\omega^2 + E' = 0, \tag{8.80}$$

with coefficients

$$A' = \frac{c^2 k^2 V_{ms}^2}{V_{Ad}^2} + k^2 c_{sd}^2 + k^2 c_{sn}^2 + \frac{(\omega_{pd}^2 + \Omega_d^2)^2}{\Omega_d^2} - \omega_{Jd}^2 - \omega_{Jn}^2, \tag{8.81}$$

$$
\begin{aligned}
C' = &\left(\frac{(\omega_{pd}^2 + \Omega_d^2)\omega_{pd}^2}{\Omega_d^2} + \frac{c^2 k^2 V_{ms}^2}{V_{Ad}^2} \right) \left(k^2 c_{sd}^2 + k^2 c_{sn}^2 - \omega_{Jd}^2 - \omega_{Jn}^2 \right) \\
&+ (\omega_{pd}^2 + \Omega_d^2) \left(\frac{c^2 k^2 V_{ms}^2}{V_{Ad}^2} + k^2 c_{sn}^2 - \omega_{Jn}^2 \right) \\
&- \omega_{Jd}^2 k^2 c_{sn}^2 - \omega_{Jn}^2 k^2 c_{sd}^2 + k^4 c_{sd}^2 c_{sn}^2,
\end{aligned} \tag{8.82}
$$

$$
\begin{aligned}
E' = &\left(\frac{(\omega_{pd}^2 + \Omega_d^2)\omega_{pd}^2}{\Omega_d^2} + \frac{c^2 k^2 V_{ms}^2}{V_{Ad}^2} \right) \left(\omega_{Jd}^2 k^2 c_{sn}^2 + \omega_{Jn}^2 k^2 c_{sd}^2 - k^4 c_{sd}^2 c_{sn}^2 \right) \\
&+ (\omega_{pd}^2 + \Omega_d^2) \frac{c^2 k^2 V_{ms}^2}{V_{Ad}^2} (\omega_{Jn}^2 - k^2 c_{sn}^2).
\end{aligned} \tag{8.83}
$$

In the limit $k = 0$ the dispersion law (8.80) reduces to

$$\omega^2 \left(\omega^2 - \frac{(\omega_{pd}^2 + \Omega_d^2)^2}{\Omega_d^2} \right) \left(\omega^2 + \omega_{Jd}^2 + \omega_{Jn}^2 \right) \simeq 0. \tag{8.84}$$

Here also there is a high-frequency solution, given for $k \neq 0$ by (8.74), not considered for reasons detailed already in the previous subsection.

There are now two coupled low-frequency modes which I need to discuss. To the same degree of precision as used in the previous section in the absence of neutral gas (i.e. $\Omega_d^2 \ll \omega_{pd}^2$ and $\omega_{Jn}^2 \sim \omega_{Jd}^2 \ll \omega_{pd}^4/\Omega_d^2$) and discarding from the outset the high-frequency solution (so that $\omega^2 \ll \omega_{pd}^4/\Omega_d^2 + c^2 k^2(1 + \beta)$), I find that (8.80) reduces to

$$\omega^4 + \left(\omega_{Jd}^2 - k^2 c_{sd}^2 + \omega_{Jn}^2 - k^2 c_{sn}^2 - \frac{k^2 V_{ms}^2 \omega_{pd}^2}{\omega_{pd}^2 + k^2 V_{ms}^2}\right) \omega^2$$
$$+ k^4 c_{sd}^2 c_{sn}^2 - \omega_{Jd}^2 k^2 c_{sn}^2 - \omega_{Jn}^2 k^2 c_{sd}^2$$
$$- (\omega_{Jn}^2 - k^2 c_{sn}^2) \frac{k^2 V_{ms}^2 \omega_{pd}^2}{\omega_{pd}^2 + k^2 V_{ms}^2} = 0. \tag{8.85}$$

In the small wavenumber limit ($k^2 V_{ms}^2 \ll \omega_{pd}^2$) this can be written as

$$\left[\omega^2 + \omega_{Jd}^2 - k^2(c_{sd}^2 + V_{ms}^2)\right]\left[\omega^2 + \omega_{Jn}^2 - k^2 c_{sn}^2\right] = \omega_{Jd}^2 \omega_{Jn}^2, \tag{8.86}$$

nicely showing how the Alfvén-Jeans mode in the charged dust fluid and the pure Jeans mode in the neutral dust fluid are coupled together. Again, the discriminant of the new bi-quadratic equation (8.86) is always positive,

$$\Delta' = \left[\omega_{Jd}^2 - \omega_{Jn}^2 - k^2(c_{sd}^2 + V_{ms}^2) + k^2 c_{sn}^2\right]^2 + 4\omega_{Jd}^2 \omega_{Jn}^2 > 0. \tag{8.87}$$

The reasoning is then completely analogous, and for an unstable mode to exist, the constant term in (8.86) needs to be negative, leading to

$$k^2 < k_{AJnd}^2 = \frac{\omega_{Jd}^2}{c_{sd}^2 + V_{ms}^2} + \frac{\omega_{Jn}^2}{c_{sn}^2}$$
$$= (k_{AJ}^2)_{\text{charged dust}} + (k_{nJ}^2)_{\text{neutral gas}} \tag{8.88}$$

If both the charged and the neutral dust originate from the same species, I specify $\sigma = N_d/(N_d + N_n)$ as the fraction of the dust density referring to the charged component. It seems eminently reasonable to assume both dust species to share a common temperature, so that $c_{sd} = c_{sn} = c_s$. In terms of the total dust Jeans frequency, given through $\omega_J^2 = 4\pi G(N_d + N_n)m_d$, and in the approximation $c_s \ll V_{ms}$, I rewrite (8.88) as

$$k_{AJnd}^2 \simeq \left(\frac{1 - \sigma}{c_s^2} + \frac{\sigma}{V_{ms}^2}\right) \omega_J^2. \tag{8.89}$$

Unless the dust is very predominantly charged, which can be defined here by $\sigma \simeq 1 - c_s^2/V_{ms}^2$, it is obvious that the first term between brackets in

(8.89) will dominate, and k_{AJnd} will usually be much larger than when all the dust would be charged. Interestingly enough, this brings the Jeans wavenumber much closer to the ordinary neutral Jeans wavenumber than could have been anticipated *a priori*.

8.5.4 Interstellar dust

In order to get a feeling for the sizes of the different length scales involved, I will give some simple numerical estimates for the different cases involving (8.78) and (8.88), based on average parameters for interstellar dust [Whitten 1994; Evans 1996]. Assume that the interstellar dust (charged and/or neutral) has a total density of $N_d + N_n = 10^{-1}$ m^{-3}, an average mass for water-ice, micron-sized grains of $m_d = 4 \times 10^{-15}$ kg and carries when charged $Z_d = 10^4$ electron charges per grain to fix the ideas. The latter value is purely speculative, as there are no reliable estimates here. The interstellar magnetic field is of the order of $B_0 = 3 \times 10^{-10}$ T, and the dust temperature is $T_d \simeq T_n = 30$ K, to be compared to an electron temperature of $T_e = 10^4$ K. This results in a total dust Jeans frequency of $\sqrt{\omega_{Jd}^2 + \omega_{Jn}^2} = 5.8 \times 10^{-13}$ s^{-1} and a dust thermal velocity of $c_{sd} \simeq c_{sn} = 3.2 \times 10^{-4}$ m/s. For the sake of comparison, the gyrofrequencies turn out to be $\Omega_i = 2.9 \times 10^{-2}$ s^{-1} for the ions and $\Omega_d = 1.2 \times 10^{-10}$ s^{-1} for the dust.

To start with, if there were only neutral dust, this would give a critical wavenumber $k_{nJ} = 1.8 \times 10^{-9}$ m^{-1}, based on the dust thermal velocity. The corresponding Jeans length then is $\lambda_{nJ} = 2\pi/k_{nJ} = 3.3 \times 10^6$ km, a small fraction of an astronomical unit.

On the other hand, if all the dust were charged, the dust Alfvén velocity is $V_{Ad} = 13.4$ m/s and the dust-acoustic velocity $c_{da} = 1.8$ m/s, giving a dust magnetosonic velocity of $V_{ms} = 13.5$ m/s. This clearly shows the preponderance of the magnetic effects in lowering the critical wavenumber, given by (8.78), to $k_{AJ} = 4.5 \times 10^{-14}$ m^{-1}. The corresponding length scales hence are $\lambda_{AJ} = 1.4 \times 10^{11}$ km, vastly larger than the dimensions of the solar system. The conclusions here clearly point to the magnetic field as an important inhibitor of gravitational collapse, provided that all the dust grains are effectively charged indeed. This is not an entirely unreasonable assumption, as dust grains placed in a radiative and plasma environment inevitably get charged [Spitzer 1941]. However, whether the timescales would allow for that and whether all the dust would become charged is an open question.

For the intermediate case suppose, that half of the dust is charged, to fix the ideas, the other half remaining neutral ($\sigma = 0.5$). The changes are that now $V_{ms} = 19$ m/s, and following from (8.89) that $k_{AJnd} = 1.3 \times 10^{-9}$ m^{-1}. This reduces the critical lengths by orders of magnitude, to $\lambda_{AJnd} =$

4.9×10^6 km, again a small fraction of an astronomical unit but slightly larger than obtained for completely neutral gas at the same temperature.

Of course, the rather large unreliability of the present interstellar data about (charged) dust cautions against taking the above numbers too literally, but nevertheless the above numbers give already some indications.

8.5.5 Summing up

Before going to nonlinear developments, let us sum up what has been learned in the previous sections, where the "Jeans swindle" has been used. Essentially, three possible mechanisms to counter the gravitational collapse of dusty plasmas were discussed: (i) the classic Jeans scenario with sound waves in a neutral gas, (ii) low-frequency dust-acoustic waves in charged dusty plasmas, and (iii) dust magnetosonic modes in mixtures of charged and neutral dust in the presence of other plasma constituents. By definition, the dust-acoustic wave phase speed exceeds the dust thermal speed and, provided these waves are naturally excited to a significant level over a wide-enough range of wavelengths, the plasma will be stabilized on length scales larger than indicated by the usual neutral Jeans length.

More generally, a careful analysis and comparison of both excited wavenumbers and of wave speeds will be necessary to determine the length scales on which gravitational collapse will occur. The magnetosonic mechanism could oppose collapse even in a truly cold ($T_d = 0$) dusty plasma, relying on the dust magnetosonic velocity as the largest among the low-frequency phase velocities. Nevertheless, the influence of neutral gas quickly brings the critical lengths back to usual values, and the combined influence is not as different from standard analysis as could be supposed initially.

However, the fundamental discussion remains about the validity and real use of the basic model assumption of a homogeneous equilibrium, when considering wave perturbations that turn out to give rather large critical lengths before the gravitational instability could set in. If the wavelengths are small enough for the waves to be stable, then the Jeans frequencies only serve to lower the effective frequencies and no great harm is done. I think that efforts should now go into trying to develop proper stationary states for complex plasmas, where self-gravitation modifies the equilibria, although it promises to be far from straightforward. Only then can the (in)stability of these configurations be studied with full confidence.

Before going on to nonlinear developments, mention should be made of few results with purport to deal with inhomogeneous dusty plasmas in the presence of self-gravitational forces [Shukla and Rahman 1996; Mamun et al. 1999]. These are not really satisfactory, because the density gradients are assumed to be weak, and the equilibrium is not determined self-consistently from the proper Poisson's equations. There is thus an improvement over homogeneous models, but still within the framework of the "Jeans swindle", and in the end better will be needed.

8.6 Nonlinear modes

Very little has been done on the nonlinear evolution of self-gravitational modes [Verheest and Shukla 1997]. In order to make some progress, I assume that the grain sizes are such that self-gravitation does not really affect the linear behaviour of the dust-acoustic modes, but comes into play at the nonlinear level. This is motivated by the fact that the charge-to-mass ratios of even heavy dust grains are usually too large to make the Coulomb and gravitational forces between grains fully comparable. The gravitational effects will thus only enter at higher orders.

For electrostatic modes parallel to the external magnetic field in the z direction the basic equations are the continuity equations (5.32), the equations of motion,

$$\frac{\partial u_\alpha}{\partial t} + u_\alpha \frac{\partial u_\alpha}{\partial z} = -\frac{1}{n_\alpha m_\alpha}\frac{\partial p_\alpha}{\partial z} - \frac{q_\alpha}{m_\alpha}\frac{\partial \phi}{\partial z} - \frac{\partial \psi}{\partial z} \qquad (8.90)$$

and the two Poisson's equations (5.35) and (4.56). For the pressures a polytropic law of the form $p_\alpha = C_\alpha n_\alpha^{\gamma_\alpha}$ is adopted. Use is made of the standard perturbation technique involving the stretching of coordinates and time (5.53) and the expansions of the dependent variables (5.54). As indicated, gravitational effects will only enter at higher orders, and thus to linear order all quantities will be as developed in Chapter 5, when dealing with nonlinear dust-acoustic modes.

To the next significant order a nonlinear equation is derived for the lowest-order electrostatic potential ϕ, which has the structure of the derivative of a KdV equation with an extra source term,

$$\frac{\partial}{\partial \xi}\left\{ A(v)\frac{\partial \phi}{\partial \tau} + B(v)\phi \frac{\partial \phi}{\partial \xi} + \frac{1}{2}\frac{\partial^3 \phi}{\partial \xi^3}\right\} + C\phi = 0. \qquad (8.91)$$

The subscript on the leading term in the expansion of ϕ has been omitted, $A(v)$ is given by (5.61) with all $U_\alpha = 0$, and $B(v)$ is rewritten from (5.59) as

$$B(v) = \sum_\alpha \frac{\omega_{p\alpha}^2 q_\alpha [3v^2 + (\gamma_\alpha - 2)c_{s\alpha}^2]}{2m_\alpha(v^2 - c_{s\alpha}^2)^3}. \qquad (8.92)$$

The new coefficient is

$$C(v) = \frac{1}{2}\left(\sum_\alpha \frac{\omega_{p\alpha}\omega_{J\alpha}}{v^2 - c_{s\alpha}^2}\right)^2. \qquad (8.93)$$

An equation, identical in structure but with quite different coefficients, has recently been derived by Ida *et al.* [1996], when properly describing the higher frequency branch of perpendicular magnetosonic waves in a bi-ion plasma.

I now again specialize these general results to the standard model of a dusty plasma, with $c_{sd} \ll v \sim c_{da} \ll c_{si} \ll c_{se}$ and treat electrons and ions as isothermal ($\gamma_e = \gamma_i = 1$), inertialess Boltzmann species. On the other hand, the dust grains are treated as almost cold. The coefficients are then simplified to

$$A(v) = \frac{\omega_{pd}^2}{c_{da}^3},$$

$$B(v) = \frac{e}{2\kappa T_i \lambda_{Di}^2} - \frac{e}{2\kappa T_e \lambda_{De}^2} + \frac{3q_d}{2m_d c_{da}^2 \lambda_D^2},$$ (8.94)

$$C(v) = \frac{\omega_{Jd}^2}{2c_{da}^2 \lambda_D^2}.$$

It is easy to prove that for negative dust the coefficient $B(v)$ can be rewritten as

$$B = \frac{\varepsilon_0}{2N_i e \lambda_{Di}^4} \left[1 - \frac{N_i \lambda_{Di}^4}{N_e \lambda_{De}^4} - \frac{3N_i}{N_d Z_d} \left(1 + \frac{\lambda_{Di}^2}{\lambda_{De}^2} \right)^2 \right]$$

$$< -\frac{\varepsilon_0}{N_i e \lambda_{Di}^4} - \frac{\varepsilon_0}{2N_e e \lambda_{De}^4}$$ (8.95)

and hence is strictly negative. A possible nondimensional form of (8.91) is obtained when lengths are measured in units of λ_D, time by ω_{pd}^{-1}, the electrostatic and gravitational potentials by $\kappa T_{\text{eff}}/e$ and velocities by c_{da}. The extended KdV equation (8.91) is then

$$\frac{\partial}{\partial \xi} \left\{ \frac{\partial \varphi}{\partial \tau} - B^* \varphi \frac{\partial \varphi}{\partial \xi} + \frac{1}{2} \frac{\partial^3 \varphi}{\partial \xi^3} \right\} + \frac{\omega_{Jd}^2}{2\omega_{pd}^2} \varphi = 0,$$ (8.96)

with the nondimensional constant $B^* > 0$. For a further discussion of the mathematical properties of this equation I refer to Ida et al. [1996], but would like to remark that the usual boundary conditions, which demand that $\varphi \to 0$ when $|\xi| \to \infty$, single out solutions for which

$$\int_{-\infty}^{+\infty} \varphi \, d\xi = 0,$$ (8.97)

in other words, soliton or double layer solutions in the classical sense are not possible at all. We have also not succeeded in finding a sufficient number of conservation laws, so that most certainly (8.91) is not integrable. In view of the fact that the Jeans frequency is small compared to the dust plasma frequency, for most types of dust, one could also attempt to solve (8.91) in a perturbative way, by focusing on a solution of the ordinary KdV equation (embedded in our equation when the extra term drops out) and computing corrections in an iterative way.

Put in a different way, the self-gravitational force acting on the dust grains is responsible for an additional, nonlocal term $\int (\omega_{Jd}^2/2\omega_{pd}^2)\varphi d\xi$ in the usual KdV equation which governs the dynamics of nonlinear dust-acoustic waves. This extra term hinders the formation of localized coherent structures. However, (8.91) could be used to investigate the long-term behaviour of finite amplitude dust-acoustic disturbances, when the gravitational force, although weaker than the electrostatic force, cannot be completely discounted.

CHAPTER 9

MASS AND SIZE DISTRIBUTIONS

9.1 Dust mass distributions

As we pointed out from the very beginning in the introduction (Chapter 1), or when dealing with space observations (Chapter 3) and with discussions about the general multispecies framework (Chapter 4), dust grains in space come in all sizes, in an almost continuous spectrum ranging from macromolecules to rock fragments and asteroids. This is much less the case for the laboratory experiments, where the range of dust particles used is controlled within tight limits. Several chapters were then devoted to many different waves in dusty plasmas, ending in the previous chapter with the incorporation for heavier grains of possible self-gravitation. These results have all been derived by treating the dust grains as mono-sized, point-like heavy (negative) ions, as a reasonable first approximation to the true behaviour of dusty plasmas. In the preceding sentence, the qualification 'reasonable' is used not so much on real physical grounds, but relates to the mathematical difficulties that arise when a more realistic modelling is attempted.

Surprisingly enough maybe, this naive approach has given us very valuable insight, mainly because of the vast differences in typical length and time scales associated with the dust, compared to the other elements of the plasma. This allowed a rather neat separation between dust and plasma effects. However, while many papers have tried to account in some way for charge fluctuations, as reviewed in Chapter 7, comparatively little has been done about other facets of dust grains, like their distributions in mass and size. The occasion and need has now come to take a closer look at these, to obtain estimates of how good average grains really are at describing wave phenomena in dusty plasmas, and of what the important modifications might be. Needless to add, accounting for a variety of sizes and masses by any form of tractable distribution is a daunting task, that has yielded only very preliminary and of necessity incomplete results so far. Few attempts at a proper kinetic theory for such complex systems have been made, and the ones that have become available have not yet achieved universally accepted status, far from it. In addition, there remains the stumbling-block of how to deal with the basic interactions between dust grains before a proper kinetic theory can be worked out.

Nevertheless, to recite again our mantra, rather than wait for perfection, scientific curiosity marches on, and in the absence of fully self-consistent kinetic theories, I will adopt a poor man's perspective and consider charged dust as a limited number of discrete species, or else continuously distributed in a limited size range, by using a power-law decrease in the number of particles of larger sizes. Because not much is encountered in the literature, changes in dispersion laws will only be indicated for some of the better-known dusty plasma modes. Self-gravitation will be incorporated where appropriate, so that plasmas with different charged and neutral dust grains can also be discussed. In tune with the available literature, I start with parallel modes in dusty plasmas with various grain mass distributions.

9.2 Parallel modes in a fluid description

9.2.1 Dust-acoustic modes

Many authors have devoted quite a lot of attention to streaming effects on dusty plasma waves, including self-gravitational influences, which as discussed in Chapters 5 and 8. Hence I will omit here all Buneman-type instabilities, and go back for parallel electrostatic modes to the dispersion law (8.49). Besides the obvious inequalities used to simplify that dispersion law, one can prove for a mixture of different dust grains [Meuris et al. 1997, Meuris 1997a] that

$$\omega_{Je}^2 \sum_d \omega_{pd}^2 \ll \omega_{pe}^2 \sum_d \omega_{Jd}^2,$$

$$\omega_{Ji}^2 \sum_d \omega_{pd}^2 \ll \omega_{pi}^2 \sum_d \omega_{Jd}^2, \tag{9.1}$$

because in all known dusty plasmas the mass-per-charge ratio for the dust species is much larger than for ions and certainly for electrons. As seen in the previous chapter, this is certainly the case for a mono-sized dust distribution, and, as I now show, a mass distribution for the different dust grains renders (9.1) even easier to fulfill than in the mono-dust case.

When for all grain sizes $a \ll \lambda_{De}$ is assumed, and the standard charging model is adhered to, I can express mass and charge of a dust particle as

$$m_d(a) = \frac{4}{3}\pi\rho_{\text{mat}}a^3 \propto a^3, \qquad q_d(a) = 4\pi\varepsilon_0 V_0 a \propto a. \tag{9.2}$$

Here ρ_{mat} is the mass density of the grain material and V_0 the electric surface potential at equilibrium. Both characteristics are assumed to be constant and equal for all grains, to simplify the discussion as otherwise the problem becomes well nigh impossible to deal with.

To gain some insight, I first deal with two different dust species, with respective grain sizes a_1 and a_2, and densities N_{d1} and N_{d2}. The average size is logically given by

$$\bar{a} = \frac{N_{d1}a_1 + N_{d2}a_2}{N_{d1} + N_{d2}},\tag{9.3}$$

because this conserves the total dust charge in the system, as size is proportional to charge. I thus can use (9.2) to define

$$
\begin{aligned}
m_{d1} &= m(a_1), & q_{d1} &= q(a_1),\\
m_{d2} &= m(a_2), & q_{d2} &= q(a_2),\\
\bar{m}_d &= m(\bar{a}), & \bar{q}_d &= q(\bar{a}),
\end{aligned}
\tag{9.4}
$$

where the average mass and charge is taken for a dust particle of average size. The plasma and Jeans frequencies are then, on average, defined through

$$\bar{\omega}_{pd}^2 = \frac{(N_{d1} + N_{d2})\bar{q}_d^2}{\varepsilon \bar{m}_d}, \qquad \bar{\omega}_{Jd}^2 = 4\pi G(N_{d1} + N_{d2})\bar{m}_d.\tag{9.5}$$

As observations indicate that there are more grains of smaller size than larger ones, the following quantities

$$\delta = \frac{a_1}{a_2} < 1, \qquad \nu = \frac{N_{d2}}{N_{d1}} \le \delta < 1\tag{9.6}$$

can be introduced. The assumption that $\nu \le \delta$ amounts to the observation that $N_{d1}a_1 > N_{d2}a_2$, in other words the densities fall steeper with radius than linear, as mentioned already earlier. I need to investigate the ratio

$$
\begin{aligned}
R &= \frac{\sum_d \omega_{Jd}^2}{\sum_d \omega_{pd}^2} \bigg/ \frac{\bar{\omega}_{Jd}^2}{\bar{\omega}_{pd}^2}\\
&= \frac{(N_{d1}a_1^3 + N_{d2}a_2^3)(N_{d1} + N_{d2})^4 a_1 a_2}{(N_{d1}a_1 + N_{d2}a_2)^4(N_{d1}a_2 + N_{d2}a_1)}\\
&= \frac{(\delta^3 + \nu)(1 + \nu)^4 \delta}{(\delta + \nu)^4(1 + \delta\nu)}\\
&= 1 + \varepsilon + \varepsilon',
\end{aligned}
\tag{9.7}
$$

with both

$$
\begin{aligned}
\varepsilon &= \frac{\nu(1 - \delta)^2 \left[\delta^2(1 - \delta) + \nu^2(2\delta + 4\delta^2 - \nu)\right]}{(\delta + \nu)^4(1 + \delta\nu)} > 0,\\
\varepsilon' &= \frac{\nu(1 - \delta)^2 \delta \left[1 + \delta + \nu(2 + \delta)(2 + \nu^2)\right]}{(\delta + \nu)^4(1 + \delta\nu)} > 0.
\end{aligned}
\tag{9.8}
$$

As can be seen, R is larger than one under the assumptions made. That means that explicit use of two dust species increase the ratio R, and the assumption (9.1) is even easier to fulfill, as *e.g.*

$$\frac{\omega_{Ji}^2}{\omega_{pi}^2} \ll \frac{\overline{\omega}_{Jd}^2}{\overline{\omega}_{pd}^2} < \frac{\sum_d \omega_{Jd}^2}{\sum_d \omega_{pd}^2}. \tag{9.9}$$

Of course, the same reasoning *a fortiori* holds for three or more dust species, by repeating the argument whenever two dust species are replaced by their averages. Analogous results are obtained for power law dust distributions and will be discussed below.

Along the same paths, one can prove [Meuris 1997a] that

$$\omega_{pe}\omega_{Je} \sum_d \omega_{pd}\omega_{Jd} \ll \omega_{pe}^2 \sum_d \omega_{Jd}^2,$$

$$\omega_{pi}\omega_{Ji} \sum_d \omega_{pd}\omega_{Jd} \ll \omega_{pi}^2 \sum_d \omega_{Jd}^2. \tag{9.10}$$

Taking all this into account, the dispersion law (8.49) is simplified to

$$\left(1 - \frac{\omega_{pe}^2}{\mathcal{L}_e} - \frac{\omega_{pi}^2}{\mathcal{L}_i} - \frac{1}{\omega^2} \sum_d \omega_{pd}^2\right)\left(1 + \frac{1}{\omega^2} \sum_d \omega_{Jd}^2\right)$$

$$+ \frac{1}{\omega^4}\left(\sum_d \omega_{pd}\omega_{Jd}\right)^2 = 0. \tag{9.11}$$

In the usual dust-acoustic regime ($c_{sd} \ll \mathrm{Re}\ \omega/k \ll c_{si}, c_{se}$), and in the same notations used for (8.58), (9.11) yields [Meuris *et al.* 1997]

$$\omega^4 + \left(\omega_{Jdg}^2 - A\ \omega_{pdg}^2\right)\omega^2 - \frac{A}{2}\sum_d \sum_{d'}(\omega_{pd}\omega_{Jd'} - \omega_{pd'}\omega_{Jd})^2 = 0. \tag{9.12}$$

Global dust plasma and Jeans frequencies are defined for all dust species together, by putting

$$\omega_{pdg}^2 = \sum_d \omega_{pd}^2, \qquad \omega_{Jdg}^2 = \sum_d \omega_{Jd}^2. \tag{9.13}$$

As the product of the roots of (9.12) for ω^2 is negative, there is always one positive solution for ω^2, corresponding to a stable mode, as well as a purely imaginary mode ($\omega^2 < 0$). The roots are given by

$$\omega^2 = \frac{\omega_{pdg}^2}{2}\left(A - B \pm \sqrt{(A-B)^2 + 2AC}\right), \tag{9.14}$$

with

$$A = \frac{k^2\lambda_D^2}{1 + k^2\lambda_D^2} < 1,$$

$$B = \frac{\omega_{Jdg}^2}{\omega_{pdg}^2},$$

$$C = \frac{1}{\omega_{pdg}^4} \sum_d \sum_{d'} (\omega_{pd}\omega_{Jd'} - \omega_{pd'}\omega_{Jd})^2. \tag{9.15}$$

Both coefficients B and C would be very small for typical micron-sized dust grains, and it is only for highly charged millimeter-sized dust that $B \simeq 1$. However, even though small, C can only vanish provided that for all dust species $\omega_{pd}/\omega_{Jd} = \omega_{pd'}/\omega_{Jd'}$, in other words all species have the same charge-to-mass ratio q_d/m_d, which would be a very special case indeed. Since

$$2B - C = \frac{2}{\omega_{pdg}^4} \left(\sum_d \omega_{pd}\omega_{Jd} \right)^2 > 0, \tag{9.16}$$

there are simple bounds for the discriminant in (9.14), namely

$$(A - B)^2 \le (A - B)^2 + 2AC < (A + B)^2. \tag{9.17}$$

In addition, the special case when $A = B$ gives rise to a critical wavenumber

$$k_c = \frac{1}{\lambda_D} \sqrt{\frac{\omega_{Jd}^2}{\omega_{pd}^2 - \omega_{Jd}^2}}. \tag{9.18}$$

Similarly, a general dust-acoustic frequency ω_{da} can be defined through

$$\omega_{da}^2 = A\omega_{pdg}^2 = \frac{k^2\lambda_D^2\omega_{pdg}^2}{1 + k^2\lambda_D^2}. \tag{9.19}$$

The description of the modes then goes as follows [Meuris et al. 1997].

- When $A > B$ or $\omega_{da} > \omega_{Jdg}$, the plasma effects dominate, and the permissible range is $k_c < k$. In the solution of (9.14) the square root with the plus sign gives a generalized dust-acoustic wave [Rao et al. 1990], recovered for $C = 0$. The dust distribution tends to increase the effective plasma frequency, but the self-gravitational effects counteract that increase somewhat.

  On the other hand, the minus sign in front of the square root gives rise to a weak unstable mode, the frequency of which vanishes when C and the dust mass distribution go to zero. This new mode is hence a dust distribution instability, but probably not of great importance.

- In the special case that $A = B$ or $k = k_c$, the dust distribution leads to stable and unstable modes, with

$$\omega^2 = \pm \omega_{pdg}\omega_{Jdg}\sqrt{\frac{C}{2}}. \tag{9.20}$$

Without dust distribution these are all zero-frequency modes.

- For the case $A < B$ the self-gravitational effects dominate, and the root with the plus sign gives a new stable mode, which does not exist without dust mass distribution. For the limits on k I have to distinguish between the possibilities that $A < B < 1$, with then $k < k_c$, whereas for $A < 1 \leq B$ there are no restrictions on k, and that, of course, comes about because the dust species have been treated as cold. This new stable mode is highly unlikely for the usual micron-sized grains.

  The mode with the minus sign is a modified Jeans mode, and the dust mass distribution tends to increase its growth rate.

For the usual micron-sized dust grains, even spread in range, both B and C are small, so that very small k values are needed, with correspondingly very large structures, both for the new stable mode as well as for the Jeans instability itself.

Qualitatively similar results are obtained when one considers charged and neutral dust together, each kind of dust being represented by at least one species. Then C is reduced to $C = 2\omega_{Jdn}^2/\omega_{pdc}^2$, where the subscripts c and n refer to charged and neutral dust. If $A > B$, the solution of (9.14) with the plus sign gives the generalized dust-acoustic mode, whereas the minus sign is essentially the Jeans instability in the neutral gas. The latter cannot be counterbalanced, as all dust thermal effects have been omitted. For $A < B$ the plus sign gives a new stable mode that does not exist without neutral dust, and the minus sign is a modified Jeans mode, the growth rate of which is enhanced by the neutral gas.

9.2.2 Continuous mass distribution

Rather than describing the dust as a number of discrete species, attempts have been made to treat charged dust in a size range $[a_{min}, a_{max}]$. As discussed in Chapter 3, the differential density usually goes as a power-law distribution like

$$n_d(a)da = Ka^{-\beta}da, \tag{9.21}$$

such that the equilibrium density of all dust grains is given by

$$N_d = \int_{a_{min}}^{a_{max}} n_d(a)da. \tag{9.22}$$

Such distributions have been observed in heliospheric dusty plasmas, with power law indices $\beta = 4.6$ for the F ring of Saturn [Showalter and Cuzzi 1992], $\beta = 7$ [Gurnett et al. 1983] or $\beta = 6$ [Showalter et al. 1993] for the G ring, and $\beta = 3.4$ in cometary environments [McDonnell et al. 1987]. Further references are given by Meuris [1997a] and have been mentioned in the chapter on dust observations in the solar system.

One can again introduce an average grain size, this time through

$$\bar{a} = \int_{a_{\min}}^{a_{\max}} n_d(a)a\,da \Big/ \int_{a_{\min}}^{a_{\max}} n_d(a)\,da, \tag{9.23}$$

which also conserves the total charge on all the dust grains together, by dint of the linear relation between charge and size.

Just to give a flavour of the reasoning with continuous distributions [Meuris 1997a], define the total plasma frequency through

$$\omega_{pd}^2 = \int_{a_{\min}}^{a_{\max}} \frac{n_d(a)q_d^2(a)}{\varepsilon_0 m_d(a)}\,da, \tag{9.24}$$

and compute the ratio with the averaged plasma frequency squared. For a power-law distribution this is

$$\begin{aligned}
R &= \omega_{pd}^2 \Big/ \frac{N_{\text{tot}}\bar{q}_d^2}{\varepsilon_0 \overline{m}_d} \\
&= \frac{(\beta-1)^2}{\beta(\beta-2)} \cdot \frac{(a_{\max}^{-\beta} - a_{\min}^{-\beta})(a_{\max}^{-\beta+2} - a_{\min}^{-\beta+2})}{(a_{\max}^{-\beta+1} - a_{\min}^{-\beta+1})^2} \\
&= \frac{(\beta-1)^2}{\beta(\beta-2)} \cdot \frac{(c^{-\beta} - 1)(c^{-\beta+2} - 1)}{(c^{-\beta+1} - 1)^2},
\end{aligned} \tag{9.25}$$

where the total dust number density is given by

$$N_{\text{tot}} = \int_{a_{\min}}^{a_{\max}} n_d(a)\,da, \tag{9.26}$$

and $c = a_{\max}/a_{\min}$ governs the spread in dust sizes. It is also supposed that β does not equal one of the critical values occurring in the expression for R, otherwise slight changes have to be made. It turns out that $R > 1$ when $\beta > 1$ and $c > 1$. From the last expression for R, it is also obvious that for reasonably large β the ratio tends to 1, and the influence of the dust distribution and of c itself is of little importance. The results are then heavily weighted towards the smaller sized grains, of which there more. There is an observational problem here, since most experiments to measure dust impacts on space missions have a cutoff for lower sizes that cannot reliably be detected [Havnes et al. 1990]. Great care should thus be exercised when using values for a_{min} in any of the subsequent treatment of part of the dispersion law by continuous integrals over the dust size spectrum.

I now follow the reasoning of Bliokh and Yaroshenko [1996] and look at Langmuir oscillations without streaming, described by (5.1) or, alternatively, by the electrostatic part of (8.43),

$$\frac{\omega_{pe}^2}{\omega^2 - k^2 c_{se}^2} + \frac{\omega_{pi}^2}{\omega^2 - k^2 c_{si}^2} + \sum_d \frac{\omega_{pd}^2}{\omega^2 - k^2 c_{sd}^2} = 1, \qquad (9.27)$$

in which thermal effects have been kept but self-gravitation not. The idea is to replace the sum over several dust species by an integral, so that

$$\frac{\omega_{pe}^2}{\omega^2 - k^2 c_{se}^2} + \frac{\omega_{pi}^2}{\omega^2 - k^2 c_{si}^2} + \int_{a_{min}}^{a_{max}} \frac{n_d(a) q_d^2(a)}{\varepsilon_0 \omega^2 m_d(a) - k^2 \varepsilon_0 \kappa T_d} da = 1, \quad (9.28)$$

where a common (low) dust temperature has been taken.

Before going on, it must be admitted that this procedure is questionable, because a fluid picture somehow assumes that there are enough particles of a given kind, so that these can be more or less treated together as one fluid. In a multispecies description several of these fluids are encountered together in the same physical space. Saying now that grains are distributed in a size range probably implies that there are not enough of each kind to allow for a fluid picture to hold. If that is so, a kinetic theory would be necessary, but that is the subject of a subsequent section.

Nevertheless, let us take (9.28) at face value, so that with the assumed dependencies of mass and charge on size, given in (9.2), the integral is in obvious notation of the form

$$\int_{a_{min}}^{a_{max}} \frac{n_d(a) a^2}{a^3 - b} da \rightarrow \int_{z_{min}}^{z_{max}} \frac{f(z)}{z - b} dz, \qquad (9.29)$$

the latter expression having been obtained by using $z = a^3$ and all superfluous coefficients have been omitted. Note that the differential density $n_d(a)$ leading to $f(z)$ is not specified, so that the reasoning is not restricted to power-law dependencies.

The analogy with the Landau integral in velocity space [Stix 1962] is now immediate, and hence there is a analogous, new kind of damping mechanism if the pole in $z = b$ is accessible, $z_{min} \leq b \leq z_{max}$, leading to

$$\int_{z_{min}}^{z_{max}} \frac{f(z)}{z - b} dz \simeq \mathcal{P} \int_{z_{min}}^{z_{max}} \frac{f(z)}{z - b} dz - i\pi f(b). \qquad (9.30)$$

Even though this closely parallels kinetic theory and Landau damping, the physical mechanism is quite different. At this stage it has to be remarked that, although Landau damping is strictly speaking outside the reach of fluid theory, it can nevertheless be recovered in a formal way through the Dawson modes for an infinite number of beams [Dawson 1960].

There are hence three kind of damping mechanisms for the various Langmuir waves, namely the ordinary (velocity space) Landau damping, the charge fluctuation damping due to variable dust charges [Melandsø et al. 1993a; Varma et al. 1993] and a dust distribution damping [Bliokh and Yaroshenko 1996]. The latter is not to be confused with the Jeans-like dust distribution modes discussed in the previous subsection, as here self-gravitation has not even been considered. In a different vein, Brattli et al. [1997] considered dust size distribution from a kinetic point of view, and found that the dust Landau damping dominates at short wavelengths, whereas for larger wavelengths the dust charge variations become more important.

Analogous results have been obtained for electromagnetic modes through the gyrofrequencies [Tripathi and Sharma 1996b], to be discussed in the next subsection, and for streaming instabilities [Yaroshenko 1997]. Following the latter reasoning in a slightly more general form, the starting point is the dispersion law (5.1),

$$\frac{\omega_{pi}^2}{\omega^2 - k^2 c_{si}^2} + \sum_d \frac{\omega_{pd}^2}{(\omega - kU_d)^2} = 1 + \frac{1}{k^2 \lambda_{De}^2}, \qquad (9.31)$$

where the electrons are Boltzmann distributed, and the different dust species are streaming with respect to the plasma. Again assuming that the sum over the dust can be replaced by an integral over a continuous distribution, leads to

$$\frac{\omega_{pi}^2}{\omega^2 - k^2 c_{si}^2} + \int_{a_{min}}^{a_{max}} \frac{n_d(a)q_d^2(a)}{\varepsilon_0 m_d(a)[\omega - kU_d(a)]^2} da = 1 + \frac{1}{k^2 \lambda_{De}^2}. \qquad (9.32)$$

Now the question is to find some acceptable dependence of $U_d(a)$ on a. As Yaroshenko [1996] applies this to a dusty plasma in a narrow planetary ring, the streaming of the dust with respect to the co-rotating plasma in the planet's magnetosphere is nearly Keplerian. Balancing the centrifugal, gravitational and Lorentz forces at the orbit with radius R of a dust particle, indicates that the frequencies are connected through

$$\Omega^2 = \Omega_K^2 + \Omega_d(\Omega - \Omega_p). \qquad (9.33)$$

Here Ω is the angular frequency of the dust grain, Ω_K the Keplerian frequency given through $\Omega_K^2 = GM_p/R^3$ and Ω_p the rotation frequency of the planet itself. The gyrofrequency Ω_d depends on the size of the dust as $1/a^2$. For nearly Keplerian motion

$$\Omega \simeq \Omega_K \left(1 + \frac{\Omega_d(\Omega_K - \Omega_p)}{2\Omega_K}\right) \qquad (9.34)$$

yields for the assumed size dependencies that

$$U_d(a) = \Omega R \simeq V_K \left(1 + \frac{C}{a^2}\right), \tag{9.35}$$

where V_K is the Keplerian velocity in an orbit with radius R, and C takes care of all the other constants. The integral in (9.32) is now of the form

$$\int_{a_{min}}^{a_{max}} \frac{n_d(a)a^3}{(a^2 - b)^2} da \rightarrow \int_{z_{min}}^{z_{max}} \frac{f(z)}{(z - b)^2} dz, \tag{9.36}$$

by putting $z = a^2$. An integration by parts immediately gives a Landau type integral.

9.2.3   Circularly polarized electromagnetic modes

To study some of the low-frequency electromagnetic modes [Shukla 1992] in the presence of more dust species, I go back to the relevant dispersion law (6.24) in Chapter 6, omit all streaming effects, and find

$$\omega^2 \pm \omega \frac{V_{Ap}^2}{c^2} \sum_d \frac{\omega_{pd}^2}{\Omega_d} - k^2 V_{Ap}^2 \simeq \omega^2 \mp (1 - \delta)\Omega_i \omega - k^2 V_{Ap}^2 = 0. \tag{9.37}$$

Here $\delta = N_e/N_i$ reflects the imbalance in densities and V_{Ap} is the Alfvén velocity in the plasma without charged dust. For the RHCP mode the solutions of (9.37) are

$$\omega = \tfrac{1}{2}\left\{(1 - \delta)\Omega_i \pm \sqrt{(1 - \delta)^2\Omega_i^2 + 4k^2V_{Ap}^2}\right\}. \tag{9.38}$$

For the further discussion assume that all the dust together is negatively charged (implying $\delta < 1$) and distinguish two extreme wavelength regimes.

At small wavenumbers $4k^2V_{Ap}^2 \ll (1 - \delta)^2\Omega_i^2$ the roots (9.38) are approximately

$$\omega_1 = (1 - \delta)\Omega_i + \frac{k^2V_{Ap}^2}{(1 - \delta)\Omega_i}, \qquad \omega_2 = -\frac{k^2V_{Ap}^2}{(1 - \delta)\Omega_i}. \tag{9.39}$$

The validity of our approximations in deriving (9.37) and (9.38) requires for the forward propagating mode that $|\Omega_d| \ll \omega_1 \sim (1 - \delta)\Omega_i \ll \Omega_i$, in other words most of the mass density should be in the dust ($N_im_i \ll \sum_d N_dm_d$), but at low levels of electron depletion ($\delta \lesssim 1$). For the backward propagating mode the conditions are equally stringent, since the ordering of the various small terms turns out to be

$$|\Omega_d| \ll \frac{k^2V_{Ap}^2}{(1 - \delta)\Omega_i} \ll (1 - \delta)\Omega_i. \tag{9.40}$$

On the other hand, the small wavelength regime is easier to deal with, implying $(1 - \delta)^2\Omega_i^2 \ll 4k^2V_{Ap}^2$, so that the roots are then given by

$$\omega \simeq \tfrac{1}{2}(1 - \delta)\Omega_i \pm kV_{Ap}. \tag{9.41}$$

These results do not really reveal the contributions of the different dust species, and hence this treatment could equally well have been given in Chapter 6 already!

Inspired by the treatment of Tripathi and Sharma [1996b], I rewrite the dispersion law (6.23) for parallel electromagnetic modes as

$$\omega^2 = c^2k^2 + \frac{\omega\omega_{pe}^2}{\omega \pm \Omega_e} + \frac{\omega\omega_{pi}^2}{\omega \pm \Omega_i} + \sum_d \frac{\omega\omega_{pd}^2}{\omega \pm \Omega_d}, \tag{9.42}$$

and replace the sum over the dust components by an integral over a range of dust sizes,

$$\omega^2 = c^2k^2 + \frac{\omega\omega_{pe}^2}{\omega \pm \Omega_e} + \frac{\omega\omega_{pi}^2}{\omega \pm \Omega_i} + \int_{a_{min}}^{a_{max}} \frac{\omega n_d(a)q_d^2(a)}{\varepsilon_0[\omega m_d(a) \pm q_d(a)B_0]}da. \tag{9.43}$$

With the assumed dependencies on size, the integral becomes

$$\int_{a_{min}}^{a_{max}} \frac{n_d(a)a}{a^2 \pm b}da \rightarrow \int_{z_{min}}^{z_{max}} \frac{f(z)}{z \pm b}dz, \tag{9.44}$$

immediately of the Landau type with $z = a^2$.

Finally, other electromagnetic modes have not yet been investigated in sufficient detail to be incorporated here, except for some preliminary results at oblique propagation by Tripathi and Sharma [1996b].

9.2.4 Nonlinear dust-acoustic modes

Also here there is little to report. Of course, since in Chapter 5 a fully multispecies treatment has been discussed, it is straightforward to adapt the Sagdeev pseudo-potential results [Verheest and Hellberg 1999] to different dust species. Up to fourth order in the electrostatic potential we had obtained a mixed KdV-mKdV equation (5.71) of the form

$$\frac{1}{2}\left(\frac{d\varphi}{d\xi}\right)^2 + \frac{1}{2}A(V)\varphi^2 + \frac{1}{3}B(V)\varphi^3 + \frac{1}{4}C(V)\varphi^4 = 0. \tag{9.45}$$

Here $\xi = x - Vt$ is the coordinate in a reference system co-moving with the nonlinear structure, and the coefficients are, for Boltzmann distributed electrons and ions, given by

$$A(V) = \sum_d \frac{\omega_{pd}^2}{V^2} \frac{1}{\lambda_D^2} \simeq -\frac{2\mu^2}{\lambda_D^2},$$

$$B(V) = \frac{3}{2V^4} \sum_d \frac{\omega_{pd}^2 q_d}{m_d} - \frac{e}{2m_e c_{se}^2 \lambda_{De}^2} + \frac{e}{2m_i c_{si}^2 \lambda_{Di}^2},$$

$$C(V) = \frac{5}{2V^6} \sum_d \frac{\omega_{pd}^2 q_d^2}{m_d^2} - \frac{e^2}{6m_e^2 c_{se}^4 \lambda_{De}^2} - \frac{e^2}{6m_i^2 c_{si}^4 \lambda_{Di}^2}. \tag{9.46}$$

In the expression for A I have used the common knowledge that in a consistent description only slightly supersonic solitons are possible, defined in this context as relative to the generalized dust-acoustic velocity $c_{da} = \lambda_D \omega_{pdg}$, involving the global dust plasma frequency and plasma Debye length, so that $V = c_{da}(1 + \mu^2)$ and $\mu \sim \mathcal{O}(\varphi)$. For the solutions I can refer you back to (5.73) in Chapter 5.

9.3   Kinetic equations for dust distributions

Only a few, partially self-consistent kinetic descriptions are available in the literature, pioneered by the Tromsø group [Aslaksen and Havnes 1994; Aslaksen 1995]. In these papers, the dust grains are supposed to have the same mass, but with different charges, and so results are qualified as partially self-consistent. The charges themselves are treated as if they were ionization levels, and the charge becomes an additional variable in phase space. Generalizations of the standard kinetic equations have been obtained, by using a Krook-like collision integral to describe the interactions between the various charge levels, but without real charge fluctuations as introduced in Chapter 7. The formalism is impressive but very mathematical, and maybe for that reason has not received the following up that it deserved. Aspects of it had to be studied numerically, where it turned out that the effects of the charge dispersive terms are smaller than expected.

In this chapter, I have reviewed the consequences of dust mass and size distributions for some of the well-known modes propagating in dusty plasmas. For lack of better alternatives in these extremely complicated problems, I have used or discussed very simple models with either a discrete number of different dust species, or with a continuous, but bounded range of sizes. Clearly much remains to be done, especially in kinetic theory, although the complexities are intimidating, as in many areas of dusty plasma theory beyond the present frontier.

CHAPTER 10

OTHER MODES

10.1 Inhomogeneous plasmas and mode coupling

In the previous chapters I have tried to give a systematic progression in the description of dusty plasma waves, starting with the simpler electrostatic and electromagnetic modes, and afterwards including specific complications in dusty plasmas, like fluctuating grain charges, self-gravitational effects and mass and size distributions. Evidently, these are by far not the only wave phenomena that can occur in dusty plasmas. For various reasons, of which lack of space is one, I can only devote an all too short chapter to some of the other modes, which require extensive modifications to the basic model used up to now. Then there is the current state of the literature, where the modes discussed below have not received as much attention as other, apparently more popular wave types. At a different level, my own expertise and research have only recently evolved towards waves in inhomogeneous plasmas.

What I propose to briefly review here are first of all two specific phenomena in inhomogeneous plasmas, the formation of coherent structures like vortices, and also surface waves. The discussion about vortices will revolve around extensions of the familiar Kelvin-Helmholtz and Rayleigh-Taylor instabilities. Somewhat related are then self-similar expansions of dusty plasmas. Finally, I want to devote some attention to lattice waves in plasma crystals, because these rely on radically different physical mechanisms to excite and sustain the waves, and have led to challenging new ideas and formalisms.

There are other topics that can unfortunately not be covered here, although this should in no way be construed as implying that these phenomena would be less important or interesting. Scattering of different waves off dusty plasmas in the Solar system is influenced by the charge imbalance between plasma electrons and ions, even in the regime of relatively small grain charges. Since the scattering cross section can be smaller for negative grains than for positive ones at the same absolute charge, negative grains experience a lower incident radiation pressure. This could lead to separation of the two charge components under the Sun's radiation and hence to cometary tail filamentation, to cite but one example. There could also be enhanced radar backscattering from noctilucent clouds, as was discussed in Chapter 3.

Potentially important is the possibility that different modes interact through mode coupling and lead to parametric and modulational instabilities, especially between the higher-frequency plasma modes and the typical low-frequency response of the charged dust. This would open new avenues for the transition to strong turbulence, but obviously needs more investigation.

10.2   Rayleigh-Taylor, Kelvin-Helmholtz instabilities and vortices

10.2.1   Rayleigh-Taylor instabilities

One of the simplest situations in which inhomogeneous plasmas naturally occur is under the influence of gravity, leading to internal gravity waves in a stratified fluid. This is covered in many plasma physics textbooks [see e.g. Choudhuri 1998; Krishan 1999] and I will recall only the essential features. When a parcel of the fluid is displaced vertically in such a way as to maintain adiabatic pressure balance, it undergoes vertical oscillations at the characteristic Brunt-Väisälä frequency ω_{BV} defined through

$$\omega_{BV}^2 = g \left[\frac{1}{\gamma} \frac{d \ln p_0}{dz} - \frac{d \ln \rho_0}{dz} \right]. \tag{10.1}$$

Here p_0 and ρ_0 are the equilibrium pressure and mass density at height z and γ is the adiabatic exponent.

As the definition of the Brunt-Väisälä frequency indicates, unstable oscillations are possible whenever the density scale height $H_\rho = \rho_0 (d\rho_0/dz)^{-1}$ is smaller than the pressure scale height $\gamma H_p = p_0 (dp_0/dz)^{-1}$, in other words, when the entropy gradient dS/dz is negative, as can be seen from the alternative formulation

$$\omega_{BV}^2 = \frac{g}{\gamma} \frac{d}{dz} \ln \left(\frac{p_0}{\rho_0^\gamma} \right). \tag{10.2}$$

In particular, this is true when the density increases with height, as in the case of a heavier fluid on top of a lighter one, or when the pressure decreases too fast, like when a colder fluid sits on top of a hotter one. This is the classic, neutral fluid Rayleigh-Taylor instability.

For magnetized fluids and plasmas, the magnetic field effects also have to be included in the picture. A horizontal magnetic field can support a heavier fluid on top of a lighter one against gravity, since the sinking of a parcel of the heavier fluid bends the magnetic field lines, for which energy is needed. Supposing that both fluids are uniform, and separated by a sharp interface at $z = 0$, the oscillations there are governed by [Krishan 1999]

$$\omega_{RT}^2 = \frac{1}{\rho_1 + \rho_2} \left[gk(\rho_1 - \rho_2) + \frac{2k_\perp^2 B_0^2}{\mu_0} \right], \tag{10.3}$$

where ρ_2 is the mass density of the upper and ρ_1 of the lower fluid, $k_\perp$ and $k_\parallel$ are the horizontal and vertical wavenumbers, such that $k^2 = k_\perp^2 + k_\parallel^2$, and B_0 is as usual the static magnetic field, here oriented in the horizontal direction. Instability is possible provided

$$\rho_1 + \frac{2k_\perp^2 B_0^2}{gk\mu_0} < \rho_2, \tag{10.4}$$

showing that the heavier fluid mass density has to compensate for the equivalent magnetic field density before the system can become unstable, for perturbations having at least a horizontal component.

The previous conclusions treat both plasmas in their respective half spaces as single charged fluids through an MHD description. Physical refinements and mathematical complications can be included when the plasma species are described separately. Especially the much heavier grains could play a decisive role, quite different from the lighter plasma particles.

A first treatment of the Rayleigh-Taylor instability in dusty plasmas indicates that the negatively charged dust considerably reduces the range of unstable wavenumbers, while the opposite is true for positive dust grains [D'Angelo 1993]. Self-consistent charge fluctuations coupled to dust dynamics influence the Rayleigh-Taylor instability [Jana et al. 1995], leading to a rapid decrease of the unstable regime. On the other hand, charge fluctuations can drive drift waves unstable, under conditions that might prevail in the tail of comet 21P/Giacobini-Zinner.

Varma and Shukla [1995a] then show that the dust grains directly respond to gravity, not really hindered by magnetic fields, whereas electrons and ions, which are magnetized, can do so only through the Rayleigh-Taylor instability. Static dust grains reduce not only the increment of this instability, but also the frequency of convective cells. For the nonlinear development similar conclusions hold, also when dust inertia is properly taken into account. In a companion paper, Varma and Shukla [1995b] study a novel flute-like instability that is different from the usual Rayleigh-Taylor instability. It comes about because in a nonuniform plasma, held in equilibrium by a magnetic field against gravity, there is a polarization electric field in the direction of gravity for negatively charged dust grains. The familiar $\mathbf{E} \times \mathbf{B}$ drift then induces this new instability, in combination with the positive current and compressible dust dynamics.

10.2.2 Kelvin-Helmholtz instabilities

A further complication to the Rayleigh-Taylor instabilities is obtained by including horizontal flows between the upper and lower fluids. The excess kinetic energy in the flows could drive systems unstable that otherwise might be Rayleigh-Taylor stable. The new, flow-induced instability

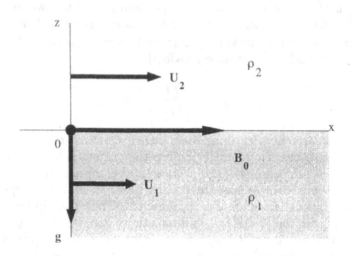

Figure 10.1. Plasma configuration for the Kelvin-Helmholtz instability

is known as the Kelvin-Helmholtz instability, with dispersion law [Krishan 1999]

$$
\left(\omega_{KH}^2 - k_\perp \frac{\rho_1 U_1 + \rho_2 U_2}{\rho_1 + \rho_2} \right)^2
$$
$$
= \frac{gk(\rho_1 - \rho_2)}{\rho_1 + \rho_2} + \frac{2k_\perp^2 B_0^2}{\mu_0(\rho_1 + \rho_2)} - \frac{k_\perp^2 \rho_1 \rho_2 (U_1 - U_2)^2}{(\rho_1 + \rho_2)^2}. \tag{10.5}
$$

The instability criterion is now that

$$
\rho_1 + \frac{2k_\perp^2 B_0^2}{gk\mu_0} < \rho_2 + \frac{k_\perp^2 \rho_1 \rho_2 (U_1 - U_2)^2}{gk(\rho_1 + \rho_2)} \tag{10.6}
$$

and this shows how the magnetic field tends to stabilize the mode, whereas the normalized relative kinetic energy between the two fluids is destabilizing. Even when both fluids have the same mass density, the Kelvin-Helmholtz instability is possible provided

$$
4V_A^2 < (U_1 - U_2)^2, \tag{10.7}
$$

with V_A representing the then common Alfvén velocity. On the other hand, the Kelvin-Helmholtz instability is also possible in neutral fluids, if

$$
\rho_1 < \rho_2 + \frac{k_\perp^2 \rho_1 \rho_2 (U_1 - U_2)^2}{gk(\rho_1 + \rho_2)} \tag{10.8}
$$

is fulfilled. In contrast to the purely Rayleigh-Taylor instability, included as a special case in the Kelvin-Helmholtz instability, the latter has a real part to the frequency, corresponding to a global Doppler-shift,

$$\omega_R = k_\perp \frac{\rho_1 U_1 + \rho_2 U_2}{\rho_1 + \rho_2}. \tag{10.9}$$

Turning now to some treatments of the Kelvin-Helmholtz instability in dusty plasmas, the simplest case occurs when adjacent plasma layers are in relative motion, but the dust is stationary [D'Angelo and Song 1990]. For negative dust grains, the critical ion shear increases without upper bound, whereas for positively charged grains it is reduced, so that excitation of the instability becomes easier. This could be important for wavelike motions in cometary tails, as it would seem that the sign of the charged dust grains is a function of the distance to the cometary nucleus. For comet 1P/Halley the grain potential varies from −8 V at a distance of 15,000 km, changes sign at about 30,000 km, and continues to increase with distance up to a value of about +7 V in the undisturbed solar wind [Ellis and Neff 1990]. Thus the stability conditions of various regions of cometary plasmas can be assessed.

Sheared positive or negative dust rather than ion flows are discussed by Rawat and Rao [1993], by also taking the dust dynamics fully into account, contrary to what was done earlier by Bharuthram and Shukla [1992b]. The relative velocity between adjacent layers of dust should be at least equal to the dust-acoustic rather than to the ion-acoustic velocity, and the instability would hence be easier to excite. This could arise in comets where dust grains flow through the solar wind plasma.

Finally here, low frequency drift instabilities in various configurations are discussed by Rosenberg and Krall [1996], in magnetized plasmas with negatively charged grains where there is locally an electron density gradient which is opposite to the dust density gradient. Two different equilibria are considered, by comparing the dust Larmor radius and the dust density scale length, and taking the former very small or very large compared to the latter. Possible applications include the spoke regions of the B ring of Saturn in the vicinity of synchronous orbit. For typical parameters the dust Larmor radius is larger than the density scale length, and the instability might grow in a couple of minutes, rendering the ring edges in this regions unstable to the perturbations considered.

10.2.3   Vortex formation

The nonlinear stationary state of the Kelvin-Helmholtz instability in dusty plasmas can be represented as coherent dipolar vortices. The numerical study of these vortices shows that they tend to occur in those regions of parameter space where the linear Kelvin-Helmholtz instability is not excited [Bharuthram and Shukla 1992b].

This brings me naturally to other occurrences of vortices in inhomogeneous plasmas. Linear and nonlinear properties of low-frequency motion in inhomogeneous, magnetized, dusty plasmas are investigated by Shukla et al. [1991] and Bharuthram and Shukla [1992b]. A new dust-drift wave is shown to exist, similar to impurity-drift modes. The nonlinear mode coupling equations indicate the possibility of solitary vortex structures, as might occur in astrophysical and cometary plasmas, in view of the claims that very low frequency drift-like waves have been observed near comet Halley [Grard et al. 1986; Klimov et al. 1986]. The Kelvin-Helmholtz instability leads to circular dipolar vortices, that can trap and transport particles over large distances.

Purely damped convective cell modes acquire a real frequency in the presence of static charged dust grains [Shukla and Varma 1993]. This frequency is induced by the plasma density gradient and corresponds to charge density waves, the dynamics of which is governed by a generalized Navier-Stokes equation. The dust inhomogeneity provides the possibility of a two-dimensional dipolar vortex. In a further paper, Shukla et al. [1993] show that Alfvén waves can be coupled to finite-frequency convective cell modes, leading again to dipolar vortices, with frequencies proportional to the gradient of the dust.

Shukla and Rao [1993] derive nonlinear equations for short-wavelength electrostatic fluctuations in the presence of density and magnetic-field aligned dust flow gradients, thereby extending the work of Rawat and Rao [1993]. Similarly, Lakshmi et al. [1993] investigate nonlinear potential structures, keeping the grain charges constant. The model follows that of Shukla et al. [1991], with the inclusion now of magnetic curvature effects. In addition to solitary dipole vortex solutions, modified convective cell vortex structures are shown to be possible in the complementary region of parameter space. Similar conclusions were reached by Mamun et al. [1996b] and Mirza et al. [1998], showing that the current convective cells arise because there is a phase lag between the electron velocity and the wave potential perturbations, on account of the electron velocity gradient.

In the above papers, dust charge fluctuations were not included, although several authors surmised that it would lead to some form of non-Landau damping, as is the case for homogeneous plasma modes studied in the previous chapters. Koide and Sato [1998] then included charge fluctuations explicitly, and show that this indeed leads to a damping of the structures, unless the attachment frequencies are sufficiently small, so that the net effect is to reduce the speed of the vortices through an increase in the vorticity itself.

Dust-acoustic shocks can occur as a consequence of the balance between nonlinear wave breaking and energy dissipation due to grain charge fluctuations [Melandsø and Shukla 1995]. When the dust-acoustic frequency is much smaller than the charging frequency, the shock wave propagation is described by a Burgers equation. In the complementary regime numerical studies show that also there shock waves are possible.

10.3 Surface waves

Many of the theoretical wave modes have been developed for infinite and homogeneous plasmas, or in media with slowly varying characteristics, but space plasmas are bounded and separated from surrounding regions by boundary layers. Typical examples are comet tails downstream from the nucleus, or the plasma torus emanating from Io in the Jovian magnetosphere, or when highly structured flux tubes occur due to the presence of an external magnetic field. Surprisingly enough, surface waves on these boundaries of a dusty plasma have not been studied in great detail. By surface waves we mean waves that travel along the interface or perpendicular to the density gradient, and are exponentially decaying along this gradient.

Bharuthram and Shukla [1993] deal with low-frequency surface waves in a plasma with warm electrons and ions, in addition to cold dust, and occupying a half space with a sharp vacuum-plasma interface. This is a valid approximation when the wavelengths are larger than the transition zone. A novel surface dust-acoustic mode is derived, in which the electron and ion pressures provide the tension, and the massive dust the inertia to maintain the wave. The dispersion law is, in our notations,

$$\omega^2 = \frac{2k^2 c_{da}^2}{1 + 2k^2\lambda_D^2 \pm \sqrt{1 + 4k^4\lambda_D^4}}. \tag{10.10}$$

In the long wavelength limit, the dispersion of the surface dust-acoustic mode becomes identical with that of the volume dust-acoustic mode [Rao et al. 1990]. Intended applications could be interstellar clouds, if a sharp boundary exists between a vacuum and an expanding plasma.

Several papers address low-frequency Alfvén surface waves in a structured magnetized dusty plasma, characterized either by sharp interfaces between dusty plasma and vacuum or dielectric regions [Ostrikov and Yu 1998; Ostrikov et al. 1999], or by rather narrow but smooth transitions between a dusty plasma and a vacuum [Cramer and Vladimirov 1996a; Cramer et al. 1998]. In their analysis, Ostrikov and Yu [1998] also include the possibility of dust charge variations. The presence of charged dust increases the skin depth for ion-acoustic surface modes, due to electron depletion. At low pressures, the skin depths and the sheath widths become comparable, so that a nonlocal kinetic treatment would become necessary, but that is not given yet.

The Alfvén surface waves in a magnetized dusty plasma studied by Cramer and Vladimirov [1996a] can be of importance when studying space plasmas with rather sharp interfaces, or when highly structured flux tubes occur due to the presence of an external magnetic field. The most important effect of negatively charged dust occurs at low parallel wavenumber k_z. Waves propagating in the surface at an acute angle with the magnetic field (viewed from the vacuum region) show a cutoff frequency as $k_z \rightarrow 0$, which does not exist in a dust-free plasma. Conversely, for propagation at obtuse angles, the surface wave phase velocity $V \rightarrow 0$ as $k_z \rightarrow 0$, contrasting with the dust-free case where $V > V_A$.

Kotsarenko *et al.* [1998a] then extend the treatment of Bharuthram and Shukla [1993] to include also an external magnetic field, both in a plane and in a cylindrical geometry. Electrons and ions shield the variable potential due to the low frequency motion of the dust grains, in a range determined by the dust cyclotron and plasma frequencies. One type of waves propagate along the magnetic field, if it is parallel to the boundary delineating the plasma region, with velocities smaller than the volume dust-acoustic velocity. When the magnetic field is perpendicular to the surface, then surface waves with higher phase velocities are possible. As an application of these ideas, wavelike structures in the comet plasma tails have been considered, where wavelengths of several thousand km have been observed. There is good agreement with theory, for typical parameters.

Similarly, Ostrikov *et al.* [1999] showed that charged dust particles carrying a considerable proportion of the negative charge of a structured magnetized plasma can lead to low-frequency electromagnetic surface waves that do not exist otherwise. These waves are Alfvén-like and propagate across the static magnetic field with frequencies below the ion gyrofrequency but well above the admittedly extremely low dust gyrofrequency. Since highly structured magnetized plasmas are common in space, this is a topic that should certainly be pursued in much more detail.

10.4 Self-similar expansions

Several papers purport to address self-similar expansion of a dusty plasma in vacuum, starting with the one by Lonngren [1990], based upon the model proposed by Rao *et al.* [1990]. Such expansions are essentially governed by the mass of the dust and the pressure of the plasma particles. Later, Luo and Yu [1992] gave a kinetic description, showing that the velocity distribution of the dust quickly becomes highly asymmetric and narrow, leading to a vanishing dust density. Moreover, if dust grains of different masses and charges were present in an expanding plasma, they would rapidly be separated into beams of similar charge-to-mass ratios. By looking at the scaling effects, one can see that the plasma expansion is hindered by the massive dust grain inertia.

While previous papers address one-dimensional or planar expansion, Yu and Bharuthram [1994] work in a cylindrical geometry, to essentially reach similar conclusions. Later, Yu and Luo [1995] include charge variations, to find that these can significantly alter the expansion process. Also, the cooling effect arising from pressure variations is an important feature. In two related papers [Bharuthram and Rao 1995; Rao and Bharuthram 1995], not only temperature but also static magnetic field effects are investigated. Isothermal rather than adiabatic pressures lead to expansions over larger distances. As expected, the magnetic field constrains the expansion perpendicular to the field.

Although no longer self-similar, El-Zein *et al.* [1997] basically address the analogous nonlinear expansion of a three-component dusty plasma, this time by numerical means. The main differences with the previous self-similar treatments, as far as the conclusions are concerned, is that only part of the dust grains are accelerated. The electrons and ions that initially leave the dusty plasma before the dust particles have a chance to move sets up an electric field, and this field then tries to accelerate some of the dust grains. As the free electrons will neutralize some of the plasma ions, the field that is available for the dust acceleration is reduced, compared with the self-similar solutions, which assumed that all electrons had accreted onto the dust grains. Hence the bulk of the dust expands slowly, whereas a small fraction seems to be accelerated together with the electrons and ions.

Recently Pillay *et al.* [1997] have extended various treatments discussed in this section by reworking the problem in cylindrical and spherical geometries, taking charge fluctuations into account. As the governing equations quickly become very involved, numerical simulations were used to investigate how the expansion behaves. In the initial stages of the cylindrical expansion, the pressure forces dominate and the dust density decreases very rapidly, until all densities become constant. This is to be contrasted with the spherical expansion, where at similar stages the dust density vanishes completely and the electron and ion densities become equal.

10.5 Lattice waves

Recently, after having explored the possibilities of familiar wave types in dusty plasmas, attention has turned to waves in strongly coupled plasmas. The most prominent of these are the dust-lattice waves, that rely on a different physical mechanism. Rather than fully collective gas-type influences, the charged grains form a one- or two-dimensional lattice and interact with their neighbours. Following the original work by Melandsø [1996], I will discuss a one-dimensional chain of identical dust grains, with lattice constant d. The electrons and ions are typically assumed to be Boltzmann distributed, as for many other wave modes at the lower frequencies characteristic of the dust. With these simplifications Poisson's equation becomes

$$\nabla^2 \varphi = k_D^2 \varphi - \frac{Q_d}{\varepsilon_0} \sum_i \delta(\mathbf{x} - \mathbf{x}_i), \tag{10.11}$$

where k_D is the inverse of the global plasma Debye length λ_D, and the instantaneous coordinates of the dust grains are $\mathbf{x}_i$. The solution of this equation goes via Green's function theory and yields for point particles the total electrostatic potential as

$$V = \frac{Q_d^2}{8\pi\varepsilon_0} \sum_i \sum_j{}' G(|\mathbf{x}_i - \mathbf{x}_j|). \tag{10.12}$$

The accent on the summations indicate that the values $i = j$ are excluded.

The preceding expression will be simplified in two ways, in order to sketch the method. First, only nearest-neighbour interactions are retained in a linear chain along the x axis, so that V reduces to

$$V = \frac{Q_d^2}{4\pi\varepsilon_0} \sum_i G(|x_{i+1} - x_i|), \tag{10.13}$$

and secondly, the grains are supposed to be strongly coupled, in the sense that oscillations η_i around the equilibrium positions X_i are small. With $x_i = X_i + \xi_i$ and the equilibrium intergrain separation d, the potential V is Taylor expanded to

$$V = V(d) + \frac{A(d)}{2} \sum_i (\xi_{i+1} - \xi_i)^2, \tag{10.14}$$

where $A(d)$ is given by [Melandsø 1996]

$$A(d) = \frac{Q_d^2}{4\pi\varepsilon_0 d^3} \exp(-k_D d)(1 + k_D d + k_D^2 d^2/2). \tag{10.15}$$

The linear terms in the separations vanish for a periodic lattice. In the same way one could include interactions between more distant neighbours or expand to higher orders. Using now (10.14) in the equations of motion for the grains gives

$$\frac{d^2 \xi_i}{dt^2} = \frac{\beta(d)}{m_d}(\xi_{i+1} - 2\xi_i + \xi_{i-1}), \tag{10.16}$$

which yields for plane-wave solutions $\xi_i \propto \exp i(kX_i - \omega t)$ the dust-lattice dispersion law

$$\omega = 2\sqrt{\frac{A(d)}{m_d}} \sin \frac{kd}{2}. \tag{10.17}$$

Hence the dust-lattice wave is dispersive, except for very long wavelengths, compared to the intergrain distance ($kd \ll 1$).

Several authors then have extended these results in various ways. It would lead me too far to mention all of them, so only some can be quoted. Amplitude modulation was studied by Amin *et al.* [1998] in a one dimensional horizontal chain of a dusty plasma crystal, and the dust grains are treated individually. In the long wavelength approximation, where the dispersion is small, the continuous limit yields a Korteweg-de Vries equation, as found already by Melandsø [1996]. When separating the fast time and space scales, associated with the dust-lattice wave, from the slow scales due to amplitude modulation, a nonlinear Schrödinger equation is obtained. Both envelope holes and shocks are possible, and these nonlinear structures are stable against perturbations and collisions.

It is not surprising that efforts have also gone into developing a overall description of strongly coupled dusty plasmas that could account for both dust-acoustic and dust-lattice waves in the appropriate limits [Otani *et al.* 1999]. However, this has not yet led to unanimous agreement, to say the least, and this quandary closely mimics the difficulties mentioned in Chapter 4 concerning the proper modelling of interactions between charged grains. An interesting but wide open field!

CHAPTER 11

CONCLUSIONS AND OUTLOOK

11.1 Comparison with observations

At the end of this book, it is worth recapitulating from the discussion of different modes and instabilities some of the results which invite comparisons with observations, albeit mostly in qualitative form rather than quantitative.

11.1.1 Spokes in Saturn's B ring

The spokes in the B ring of Saturn gave a tremendous impetus to the field of dusty plasma physics. It is thus not surprising that many and competing explanations have been advanced for these intriguing phenomena, either in terms of combinations of gravitational and electromagnetic forces on individual charged grains, or else through the generation of wave modes.

First of all, single particle dynamics of charged dust grains has been invoked, rather than collective interactions and waves. Consider a macroscopic ring particle in the neighbourhood of which a dense plasma cloud develops, caused by meteor impact on the ring particles, or by other mechanisms like lightning discharges. The dense plasma electrostatically levitates small dust grains that consequently drift relative to the plasma. The latter corotates with the planet, while the dust has angular velocities between the Keplerian and corotation velocities. The ensuing charge separation and electric field generate $\mathbf{E} \times \mathbf{B}$ drifts in the plasma, driving the dust grains away from synchronous orbit [Goertz 1989; Northrop 1992].

Other explanations have involved density waves in a multistream model to account not only for spoke formation but also for the very many narrow rings and gaps in the B ring [Bliokh and Yaroshenko 1985; Bliokh et al. 1995]. Although such waves of dust charge density can probably not be strong enough to fully account for the spokes as observed, the idea of dust-driven plasma waves provides an intriguing mechanism for ion heating at the expense of the gravitational energy of the orbiting dust grains.

Nonlinear developments of oblique dust-acoustic modes are described by KdV equations, and have been applied by Kotsarenko et al. [1998b] to the creation of condensates in cometary tails and to Saturn's rings, where it provides an alternative explanation for spoke formation. Whereas in the model of Bliokh and Yaroshenko [1985] the spokes are interpreted through the excitation of space charge waves, Kotsarenko et al. [1998b]

239

attribute the spokes to cylindrical solitons in an ion-dust plasma, most of the electrons having been accreted onto the dust grains. Interestingly enough, both models give spatial sizes of the order of 1000 km, close to the observed sizes. Of course, plasma data are not tightly constrained by observations yet, and thus a little stretching or interpretation is always possible. However, the models differ in the lifetimes of the spoke structures, for which Bliokh and Yaroshenko [1985] obtain some 1000 s or fractions of an hour, whereas Kotsarenko et al. [1998b] predict 10^5 s or a few days. Hence the debate is still open to what constitutes the most appropriate way to describe the spoke formation. In a similar vein, condensates in cometary tails are simulated with a spatial extent of the right order, also around 1000 km.

11.1.2 Planetary rings

Besides spokes in the B ring of Saturn, braids, kinks and clumps have been observed in a few Voyager images of the F ring. These strange features have up to this time defied a generally accepted explanation. Avinash and Sen [1994] explain this phenomenon based on the balance of the pinch pressure due to the dust ring current and the electrostatic pressure. Purely gravitational theories rely upon shepherding satellites and the resonances their motions induce in the rings. However, possible kinks and clumps formed in this way are washed out too quickly for braiding to occur. A combined gravito-electrodynamical approach does not work either, as it requires high dust charges that can only occur for isolated grains. In addition, to complicate the picture, Showalter [1998] indicates that centimeter-sized meteoroid impacts in the F ring are too short lived to relate to the longer lived clumps.

Another feature of planetary rings is that repelling electrostatic forces between charged dust grains would prevent collapse to very thin sheets, so that planetary rings retain a finite thickness transverse to the equatorial plane, up to hundred km. This could be important for Jupiter's and some of Uranus' rings, and even for others too, if the uncertainties in the relevant parameters are taken into account [Melandsø and Havnes 1991; Melandsø et al. 1993b]. The thickness of the rings is determined by a balance between the gravity component toward the central plane and the expanding electrostatic force of the dust grains. If the ambient plasma conditions change, natural oscillations occur in the thickness profiles, which become increasingly complicated for denser rings. Resonances between oscillations in transverse thickness and Keplerian frequencies can produce gaps and prominent features in the ring structures at Jupiter and Saturn [Melandsø and Havnes 1991]. In the plasma that connects Jupiter with its satellite Io, the detection of the electrostatic ion cyclotron instability would signal the presence of negative dust, as without it the mode would be stable [Chow and Rosenberg 1995].

Charged dust influences two-stream instabilities between electrons and ions, or can itself form a drifting beam, as in planetary rings. In all cases the presence of dust grains enhances the growth of the instabilities, as well as the velocity ranges over which the instability can occur. Stationary dust, as in the magnetosphere of Neptune, also modifies the propagation properties [Bharuthram *et al.* 1992]. Rosenberg and Krall [1995] conclude that the edges of the ringlets in the F ring of Saturn are stable against the excitation of the high-frequency drift mode, whereas the low-frequency two-stream instability could be of importance in the E ring, depending on local dust conditions.

Addressing a totally different problem, Buti [1997] has applied a general DNLS equation to explore the effect of really massive and heavily charged dust grains on chaotic Alfvén waves. Interestingly, because of the presence of charged dust, chaos is not only reduced but disappears, even at very low dust grain densities due to inertial stabilization by the heavy dust. The computations were done for typical parameters in the rings of Saturn and in cometary cases.

11.1.3 Comets

Coming to the influence of charged dust in cometary physics, missions to comets 21P/Giacobini-Zinner and 1P/Halley have shown that the smaller grains are not distributed in a symmetric fashion with respect to the orbital plane of the comet, as expected if only radiation pressure is taken into account. The preferential gathering to one side could be due to a downward push on negative grains, induced because the magnetic field and velocity of the solar wind are not aligned.

Streaming between the solar wind and cometary dust grains has other ramifications. At the high dust densities which may be found in cometary comae, the dust grains can be important charge carriers, leading to electron depletion. The resulting instability drastically enhances the coupling between the solar wind and the dust, which favours small dust grains to be swept along with the solar wind plasma. Turbulent drag could form dust striae in cometary tails, leading to narrower ones in a high-velocity solar wind, whereas one would expect wider tails in lower velocity solar winds, where gravity, radiation pressure and expansion dominate. It is estimated that the dust-solar wind interaction would be higher for comets at larger heliocentric distances, outside the inner part of the solar system [Havnes 1988].

Double layers can accelerate particles, as in the Earth's auroral regions where double layers are believed to be responsible for the observed energetic electrons and ions [Raadu 1989]. Furthermore, models for solar flares have been advanced in which the triggering mechanism is a double layer related phenomenon. Hence the question whether the intense dust jets of comet

1P/Halley could be explained this way.

Mass loading of the solar wind in cometary atmospheres is accompanied by electromagnetic turbulence, the plasma surrounding the comet being made up of solar wind ions, electrons and energetic heavier ions, such as water, of cometary origin. Dust grains may also be present, close to the comet in the Sunward direction, or downstream of the comet in the dust tail. Even if the dust charge density is small, it can have a large effect on hydromagnetic shear and compressional Alfvén waves propagating at frequencies well below the ion gyrofrequency [Pilipp et al. 1987; Cramer and Vladimirov 1997], not only in cometary environments but also in interstellar molecular clouds. More important changes occur if a higher proportion of the negative charges resides on the dust [Cramer and Vladimirov 1996b].

The effect of charged dust on the nonresonant firehose instabilities [Verheest and Meuris 1998; Cramer et al. 1999] indicates that dust at low densities is unlikely to be of any noticeable importance in cometary physics. Of course, this conclusion holds on the basis of the scarce data available from the few cometary missions flown up to now and might well have to be revised for different comets in the future, when more and accurate data become available. However, electromagnetic instabilities where the Doppler-shifted wave frequency resonates with the water group gyrofrequency are rather significantly altered by charged dust [Cramer et al. 1999]. Variable dust charges also influence electromagnetic waves, the instabilities of which could generate spatial structures over a wide range of hundreds to tens of million kilometers in dusty cometary tails [Reddy et al. 1996].

The Kelvin-Helmholtz instability in dusty plasmas has been used to explain wavelike motions in cometary tails, as the sign of the charged dust is a function of the distance to the cometary nucleus. For comet 1P/Halley the grain potential is negative at a distance of 15,000 km, and becomes positive at about 30,000 km and further outwards [Ellis and Neff 1990]. In this way the stability conditions of various regions of cometary plasmas can be assessed.

Finally, nonlinear mode coupling indicates the possibility of solitary vortex structures, as might occur in astrophysical and cometary plasmas, in view of the claims that very low frequency drift-like waves have been observed near comet Halley [Grard et al. 1986; Klimov et al. 1986].

11.1.4   Interstellar dust

Havnes [1980] examined the conditions for the onset of the instability in a plasma model in which two charged dust distributions stream relative to each other. Small grains are brought to rest in very short distances, while larger grains are practically unaffected, and this two-stream instability may be important for grain destruction in high velocity clouds. Collisions between the two grain populations may be energetic enough to destroy a large fraction of the total grain content. Shocks travelling through interstellar gas cause grain separation, subsequent destruction and return of the material to the cloud in gas form, explaining why some elements are observed to be underabundant in low-velocity clouds but have near normal abundances in clouds of high velocity.

Dust is also supposed to play a crucial role in the formation of stars, solar or planetary systems [Alfvén 1981], and several attempts have been made to adapt Jeans' instability criterion to the presence of charged dust. Chhajlani and Parihar [1994] indicate how charged dust in interstellar clouds affect the fragmentation process. For low-frequency modes the effect of the grains is important, decreasing the region of instability and the critical Jeans wavenumber and rendering star formation more difficult. Similar conclusions were reached by other authors [Avinash and Shukla 1994; Meuris *et al.* 1997]. When all the dust grains are charged, the magnetic field acts as an important inhibitor of gravitational collapse [Verheest *et al.* 1997b,1999a]. Of course, the rather large unreliability of the present interstellar data about (charged) dust cautions against taking these conclusions too literally, and moreover a mixture of charged and neutral gas behaves very much as if all the material were neutral.

11.2 Summing up

The growth of dusty plasma physics was stimulated by observations of spokes and braids in planetary rings, and wavelike features in comet tails. These are difficult to explain in a satisfactorily manner without the presence of charged dust. Nevertheless, reality commands us to note that at present, barring the proverbial exception, there is no quantitative agreement between the observed phenomena and the theoretical explanations, nor are there many theoretical papers which specifically purport to give such detailed explanations.

So we are still in the position, as I pointed out some years ago [Verheest 1996], that the theoretical studies concerning waves and instabilities in dusty (space) plasmas are far ahead of what observations corroborate at present. New observations will come from the two solar system missions specifically concerned with planetary or cometary exploration. Cassini-Huygens left the Earth already and is on its way to Saturn, whereas Rosetta is due to be launched in 2003, but will not reach its target, comet 46P/Wirtanen, before 2011, if all goes to plan. Because fears for the integrity of the spacecraft rule out a too close encounter with planetary rings, other methods need to be devised to probe these from farther afield. One of these is the possible detection of Mach cones [Havnes *et al.* 1995] formed by the wake of larger boulders in the rings, when lighter dust grains are deflected. It is hoped that the Cassini mission will actually discover such wakes and measure their opening angles from a safe distance.

While this dearth of observational data leaves the theoreticians free to explore many avenues, one would like to see a closer relation to space data and a consensus between different theoretical views, if only to design on board experiments which might be able to discriminate between different opinions. Vigorous progress in laboratory dusty plasma research will fill gaps in our understanding before space observations can do that. In particular, the work on plasma dust crystal structures looks very promising in furthering our knowledge about dust charging. It could be worth to speculate whether plasma dust crystal structures occur in space and where, except in exotic but unscrutable surroundings like neutron stars.

11.3 Outlook

The real conclusion of this book simply is that there are more open problems than solved ones! I could only review the field as it has developed up to now, and that this monograph has turned out to be rather theoretical is a mere reflection of the present state of the literature.

Among the questions that need addressing are the gap between theory and observations, pointed out on several occasions in this monograph and by different people. Closing this gap is of paramount importance, as observations are the ultimate check to calibrate theoretical advances that otherwise risk becoming not more than fertile leaps of imagination. Moreover, we want to understand what Nature is telling us.

Further areas awaiting intenser scrutiny are the role of charged dust in the formation of planetary systems around other stars than our Sun, especially now that solid astrophysical evidence is emerging that we are not alone in this respect. Intimately related are various aspects of galaxy formation, for which present-day models are not adequate.

Of course, there remain many challenging problems in connection with the fundamental description at its most basic level. These are maybe not of immediate importance for astrophysics, but need to be resolved if we want to achieve a deeper understanding of the fascinating world of dusty plasmas.

BIBLIOGRAPHY

Ablowitz, M.J. and P.A. Clarkson (1991) *Solitons, nonlinear evolution equations and inverse scattering.* Cambridge University Press, Cambridge.

Alfvén, H. (1981) *Cosmic plasma.* Reidel, Dordrecht, pp. 118–121.

Allen, J.E. (1992) Probe theory – The orbital motion approach, *Physica Scripta*, **45**, 497–503.

Amin, M.R. (1996) Low-frequency electrostatic waves in a hot magnetized dusty plasma, *Phys. Rev. E*, **54**, R2422-R2235.

Amin, M.R., G.E. Morfill and P.K. Shukla (1998) Amplitude modulation of dust-lattice waves in a plasma crystal, *Phys. Plasmas*, **5**, 2578–2581.

Anderson, R.R. and W.S. Kurth (1989) Ultra-low frequency waves at comets, in: *Plasma waves and instabilities at comets and in magnetospheres.* Ed. B.T. Tsurutani and H. Oya, Geophys. Monograph Series 53, Washington, pp. 81–117.

Aslaksen, T.K. (1995) Effect of charge dispersion in dusty plasmas, *J. Plasma Phys.*, **54**, 373–391.

Aslaksen, T.K. and O. Havnes (1994) Kinetic theory for a distribution of ionized dust particles, *J. Plasma Phys.*, **51**, 271–290.

Avinash, K. and A. Sen (1994) A model for the fine structure of Saturn's rings, *Phys. Lett. A*, **194**, 241–245.

Avinash, K. and P.K. Shukla (1994) A purely growing instability in a gravitating dusty plasma, *Phys. Lett. A*, **189**, 470–472.

Baboolal, S., R. Bharuthram and M.A. Hellberg (1988) Arbitrary-amplitude rarefactive ion-acoustic double layers in warm multi-fluid plasmas, *J. Plasma Phys.*, **40**, 163–178.

Balsiger, H., K. Altwegg, F. Bühler, J. Geiss, A.G. Ghielmetti, B.E. Goldstein, R. Goldstein, W.T. Huntress, W.H. Ip, A.J. Lazarus, A. Meier, M. Neugebauer, U. Rettenmund, H. Rosenbauer, R. Schwenn, R.D. Sharp, E.G. Shelley, E. Ungstrup and D.T. Young (1986) Ion composition and dynamics at comet Halley, *Nature*, **321**, 330–334.

Barkan, A., N. D'Angelo and R.L. Merlino (1994) Charging of dust grains in a plasma, *Phys. Rev. Lett.*, **73**, 3093–3096.

Barkan, A., R.L. Merlino and N. D'Angelo (1995a) Laboratory observation of the dust-acoustic wave mode, *Phys. Plasmas*, **2**, 3563–3565.

Barkan, A., N. D'Angelo and R.L. Merlino (1995b) Laboratory experiments on electrostatic ion cyclotron waves in dusty plasmas, *Planet. Space Sci.*, **43**, 905–908.

Barkan, A., N. D'Angelo and R.L. Merlino (1996) Experiments on ion-acoustic waves in dusty plasmas, *Planet. Space Sci.*, **44**, 239–242.

Baumjohann, W. and R.A. Treumann (1996) *Basic space plasma physics.* Imperial College Press, London.

Baumjohann, W. and R.A. Treumann (1997) *Advanced space plasma physics.* Imperial College Press, London.

Baynham, A.C. and A.D. Boardman (1971) *Plasma effects in semiconductors.* Taylor and Francis, London.

Bergstralh, J.T., E.D. Miner and M.S. Matthews (1991) *Uranus.* University of Arizona Press, Tucson.

Bernhardt, P.A., G. Ganguli, M.C. Kelley and W.E. Swartz (1995) Enhanced radar backscatter from space shuttle exhaust in the ionosphere, *J. Geophys. Res.* **100**, 23811–238.

Bharuthram, R. (1997) Low frequency electrostatic instabilities in a dusty plasma driven by crossfield beams, *Planet. Space Sci.*, **45**, 379–383.

Bharuthram, R. and T. Pather (1996) The kinetic dust-acoustic instability in a magnetized dusty plasma, *Planet. Space Sci.*, **44**, 137–146.

Bharuthram, R. and N.N. Rao (1995) Self-similar expansion of a warm dusty plasma — I. Unmagnetized case, *Planet. Space Sci.*, **43**, 1079–1085.

Bharuthram, R. and M. Rosenberg (1998) A note on the generation of fluctuations by Space Shuttle exhaust in the ionosphere, *Planet. Space Sci.*, **46**, 425–427.

Bharuthram, R. and P.K. Shukla (1992a) Large amplitude double layers in dusty plasmas, *Planet. Space Sci.*, **40**, 465–471.

Bharuthram, R. and P.K. Shukla (1992b) Vortices in non-uniform dusty plasmas, *Planet. Space Sci.*, **40**, 647–654.

Bharuthram, R. and P.K. Shukla (1992c) Large amplitude ion-acoustic solitons in a dusty plasma, *Planet. Space Sci.*, **40**, 973–977.

Bharuthram, R. and P.K. Shukla (1993) Low-frequency surface waves on a warm dusty plasma, *Planet. Space Sci.*, **41**, 17–19.

Bharuthram, R., H. Saleem and P.K. Shukla (1992) Two-stream instabilities in unmagnetized dusty plasmas, *Physica Scripta*, **45**, 512–514.

Bhatt, J.R. (1997) Langmuir waves in a dusty plasma with variable grain charge, *Phys. Rev. E*, **55**, 1166–1169.

Bhatt, J.R. and B.P. Pandey (1994) Self-consistent charge dynamics and collective modes in a dusty plasma, *Phys. Rev. E*, **50**, 3980–3983.

Biermann, L., B. Brosowski and H.U. Schmidt (1967) The interaction of the solar wind with a comet, *Solar Phys.*, **1**, 254–284.

Bliokh, P.V. and V.V. Yaroshenko (1985) Electrostatic waves in Saturn's rings, *Sov. Astron.*, **29**, 330–336.

Bliokh, P.V. and V.V. Yaroshenko (1996) Damping of low-frequency electrostatic waves in dusty plasmas, *Plasma Phys. Rep.*, **22**, 411–416.

Bliokh, P., V. Sinitsin and V. Yaroshenko (1995) *Dusty and self-gravitational plasmas in space.* Kluwer, Dordrecht.

Boehnhardt, H. and H. Fechtig (1987) Electrostatic charging and fragmentation of dust near P/Giacobini–Zinner and P/Halley, *Astron. Astrophys.*, **187**, 824–828.

Booker, H.G. (1984) *Cold plasma waves.* Nijhoff, Dordrecht, pp. 2–12.

Boss A.P. (1987) Protostellar formation in rotating interstellar clouds. VI. Nonuniform initial conditions, *Astrophys. J.*, **319**, 149–161.

Bouchoule, A. (1999) *Dusty plasmas. Physics, chemistry and technological impacts in plasma processing.* Wiley, Chichester.

Brattli, A., O. Havnes and F. Melandsø (1997) The effect of a dust-size distribution on dust acoustic waves, *J. Plasma Phys.*, **58**, 691–704.

Brinca, A.L., L. Borda de Água and D. Winske (1993) On the stability of nongyrotropic ion populations: A first (analytic and simulation) assessment, *J. Geophys. Res.*, **98**, 7549–7560.

Buneman, O. (1958) Instability, turbulence, and conductivity in current-carrying plasmas, *Phys. Rev. Lett.*, **1**, 8–9.

Burlaga, L.F., L.W. Klein, R.P. Lepping and K.W. Behannon (1984) Large-scale interplanetary magnetic fields: Voyager 1 and 2 observations between 1 AU and 9.5 AU, *J. Geophys. Res.*, **89**, 10659–10668.

Burns, J.A., M.R. Showalter, D.P. Hamilton, P.D. Nicholson, I. De Pater, M.E. Ockert-Bell and P.C. Thomas (1999) The formation of Jupiter's faint rings, *Science*, **284**, 1146–1150.

Buti, B. (1997) Control of chaos in dusty plasmas, *Phys. Lett. A*, **235**, 241–247.

Čadež, V.M. (1990) Applicability problem of Jeans criterion to a stationary self-gravitating cloud, *Astron. Astrophys.*, **235**, 242–244.

Čadež, V.M., F. Verheest and G. Jacobs (1999) Stability of self-gravitating plasma clouds: The "Jeans swindle" revisited, *Astrophys. Space Sci.*, submitted.

Chandrasekhar, S. (1961) *Hydrodynamic and hydromagnetic stability.* Clarendon Press, Oxford.

Chhajlani, R.K. and A.K. Parihar (1994) Magnetogravitational instability of self-gravitating dusty plasma, *Astrophys. J.*, **422**, 746–750.

Choi, J. and M.J. Kushner (1994) A particle-in-cell simulation of dust charging and shielding in low pressure glow discharges, *IEEE Trans. Plasma Sci.*, **22**, 151–158.

Choudhuri, A.R. (1998) *The physics of fluids and plasmas. An introduction for astrophysicists.* Cambridge University Press, Cambridge (UK).

Chow, V.W. and M. Rosenberg (1995) Electrostatic ion cyclotron instability in dusty plasmas, *Planet. Space Sci.*, **43**, 613–618.

Chow, V.W. and M. Rosenberg (1996) A note on the electrostatic ion cyclotron instability in dusty plasmas: comparison with experiment, *Planet. Space Sci.*, **44**, 465–467.

Chow, V.W., D.A. Mendis and M. Rosenberg (1993) Role of grain size and particle velocity distribution in secondary electron emission in space plasmas, *J. Geophys. Res.*, **98**, 19065–19076.

Chu, J.H. and L. I (1994) Coulomb lattice in a weakly ionized colloidal plasma, *Physica A*, **205**, 183–190.

Chu, J.H., J.B. Du and L. I (1994) Coulomb solids and low-frequency fluctuations in RF dusty plasmas, *J. Phys. D: Appl. Phys.*, **27**, 296–300.

Clark, R.N. (1980) Ganymede, Europa, Callisto and Saturn's rings: Compositional analysis from reflectance spectroscopy, *Icarus*, **44**, 388–409.

Clark, R.N. and T.B. McCord (1980) The rings of Saturn: New near-infrared reflectance measurements and a 0.326–4.08 μm summary, *Icarus*, **43**, 161–168.

Collins, S.A., A.F. Cook, J.N. Cuzzi, G.E. Danielson, G.E. Hunt, T.V. Johnson, D. Morrison, T. Owen, J.B. Pollack, B.A. Smith and R.J. Terrile (1980) First Voyager view of the rings of Saturn, *Nature*, **288**, 439–442.

Colwell, J.E., M. Horányi and E. Grün (1998) Jupiter's exogenic dust ring, *J. Geophys. Res.*, **103**, 20023–20030.

Coppa, G.G.M., V. Riccardo and G. Lapenta (1996) Kinetic theory of charged particles of variable shape, *Phys. Plasmas*, **3**, 2229–2238.

Cramer, N.F. and S.V. Vladimirov (1996a) Alfvén surface waves in a magnetized dusty plasma, *Phys. Plasmas*, **3**, 4740–4747.

Cramer, N.F. and S.V. Vladimirov (1996b) The Alfvén resonance in a magnetized dusty plasma, *Physica Scripta*, **53**, 586–590.

Cramer, N.F. and S.V. Vladimirov (1997) Alfvén waves in dusty interstellar clouds, *Publ. Astron. Soc. Austr.*, **14**, 170–178.

Cramer, N.F., F. Verheest and S.V. Vladimirov (1999) Instabilities of Alfvén and magnetosonic waves in dusty cometary plasmas with an ion ring beam, *Phys. Plasmas*, **6**, 36–43.

Cravens, T.E. (1997) *Physics of solar system plasmas.* Cambridge University Press, Cambridge (UK).

Cruikshank, D.P. (1996) *Neptune and Triton.* University of Arizona Press, Tucson.

Cui, C. and J. Goree (1994) Fluctuations of the charge on a dust grain in a plasma, *IEEE Trans. Plasma Sci.*, **22**, 151–158.

D'Angelo, N. (1967) Recombination instability, *Phys. Fluids*, **10**, 719–723.

D'Angelo, N. (1990) Low-frequency electrostatic waves in dusty plasmas, *Planet. Space Sci.*, **38**, 1143–1146.

D'Angelo, N. (1993) The Rayleigh-Taylor instability in dusty plasmas, *Planet. Space Sci.*, **41**, 469–474.

D'Angelo, N. (1994) Ion-acoustic waves in dusty plasmas, *Planet. Space Sci.*, **42**, 507–511.

D'Angelo, N. (1995) Coulomb solids and low-frequency fluctuations in RF dusty plasmas, *J. Phys. D: Appl. Phys.*, **28**, 1009–1010.

D'Angelo, N. (1997) Ionization instability in dusty plasmas, *Phys. Plasmas*, **4**, 3422–3426.

D'Angelo, N. (1998) Current-driven electrostatic dust-cyclotron instability in a collisional plasma, *Planet. Space Sci.*, **46**, 1671–1676.

D'Angelo, N. and B. Song (1990) The Kelvin-Helmholtz instability in dusty plasmas, *Planet. Space Sci.*, **38**, 1577–1579.

D'Angelo, N., S. Von Goeler and T. Ohe (1966) Propagation and damping of ion waves in a plasma with negative ions, *Phys. Fluids*, **9**, 1605–1606.

Das, A.C., A.K. Misra and K.S. Goswami (1996) Streaming instability in the grain charge fluctuations, *Phys. Plasmas*, **3**, 457–460.

Das, G.C., J. Sarma and C. Uberoi (1997) Explosion of soliton in multicomponent plasma, *Phys. Plasmas*, **4**, 2095–2100.

Dawson, J.M. (1960) Plasma oscillations of a large number of electron beams, *Phys. Rev.*, **118**, 381–389.

De Angelis, U. (1992) The physics of dusty plasmas, *Physica Scripta*, **45**, 465–474.

De Angelis, U. and A. Forlani (1998) Grain charge in dusty plasmas, *Phys. Plasmas*, **5**, 3068–3069.

De Angelis, U., V. Formisano and M. Giordano (1988) Ion plasma waves in dusty plasmas: Halley's comet, *J. Plasma Phys.*, **40**, 399–406.

Deconinck, B., P. Meuris and F. Verheest (1993a) Oblique nonlinear Alfvén waves in strongly magnetized beam plasmas. Part 1. Nonlinear vector evolution equation, *J. Plasma Phys.*, **50**, 445–455.

Deconinck, B., P. Meuris and F. Verheest (1993b) Oblique nonlinear Alfvén waves in strongly magnetized beam plasmas. Part 2. Soliton solutions and integrability, *J. Plasma Phys.*, **50**, 457–476.

De Juli, M.C. and R.S. Schneider (1998) The dielectric tensor for dusty magnetized plasmas with variable charge on dust particles, *J. Plasma Phys.*, **60**, 243–263.

Divine, N. (1993) Five populations of interplanetary meteoroids, *J. Geophys. Res.*, **98**, 17029–17048.

Draine, B.T. and E.E. Salpeter (1979) On the physics of dust grains in hot gas, *Astrophys. J.*, **231**, 77–94.

Ellis, T.A. and J.S. Neff (1991) Numerical simulation of the emission and motion of neutral and charged dust from P/Halley, *Icarus*, **91**, 280–296.

El-Zein, Y., S. Yi and K.E. Lonngren (1997) Expansion of a dusty plasma into a vacuum: Effects of charge nonneutrality, *Planet. Space Sci.*, **45**, 251–254.

Ferrari, C. and A. Brahic (1994) Azimuthal brightness asymmetries in planetary rings. I: Neptune arcs and narrow rings, *Icarus*, **111**, 193–210.

Forlani, A., U. De Angelis and V.N. Tsytovich (1992) Waves in dusty plasmas, *Physica Scripta*, **45**, 509–511.

Galeev, A.A. (1991) Plasma processes in the outer coma, in: *Comets in the post-Halley era.* Eds. R.L. Newburn, Jr., M. Neugebauer and J. Rahe, Kluwer Academic Publishers, Dordrecht, pp. 1145–1169.

Gary, S.P., C.W. Smith, M.A. Lee, M.L. Goldstein and D.W. Forslund (1984) Electromagnetic ion beam instabilities, *Phys. Fluids*, **27**, 1852–1862. (Erratum (1985) *Phys. Fluids*, **28**, 438).

Gehman, C.S., F.C. Adams and R. Watkins (1996) Linear gravitational instability of filamentary and sheetlike molecular clouds with magnetic fields, *Astrophys. J.*, **472**, 673–683.

Gehrels, T. (1976) *Jupiter.* University of Arizona Press, Tucson.

Gehrels, T. and M.S. Matthews (1984) *Saturn.* University of Arizona Press, Tucson.

Goertz, C.K. (1989) Dusty plasmas in the solar system, *Rev. Geophys.*, **27**, 271–292.

Goertz, C.K. and W.H. Ip (1984) Limitations of electrostatic charging of dust particles in a plasma, *Geophys. Res. Lett.*, **11**, 349–352.

Goertz, C.K. and G. Morfill (1983) A model for the formation of spokes in Saturn's ring, *Icarus*, **53**, 219–229.

Goertz, C.K., L. Shan and O. Havnes (1988) Electrostatic forces in planetary rings, *Geophys. Res. Lett.*, **15**, 84–87.

Goldstein, B.E., M. Neugebauer, J.L. Phillips, S. Bame, J.T. Gosling, D. McComas, Y.-M. Wang, N.R. Sheeley and S.T. Suess (1996) Ulysses plasma parameters: Latitudinal, radial, and temporal variations, *Astron. Astrophys.*, **316**, 296–303.

Grard, R., A. Pedersen, J.G. Trotignon, C. Beghin, M. Mogilevsky, Y. Mikhaïlov, O. Molchanov and V. Formisano (1986) Observations of waves and plasma in the environment of comet Halley, *Nature*, **321**, 290–291.

Grard, R., H. Laakso, A. Pedersen, J.G. Trotignon and Y. Mikhaïlov (1989) Observations of the plasma environment of comet Halley during the Vega flybys, *Ann. Geophys.*, **7**, 141–149.

Greenberg, R. and A. Brahic (1984) *Planetary rings*. University of Arizona Press, Tucson.

Grün, E., N. Pailer, H. Fechtig and J. Kissel (1980) Orbital and physical characteristics of micrometeoroids in the inner solar system as observed by *Helios 1*, *Planet. Space Sci.*, **28**, 333–349.

Grün, E., G.E. Morfill, R.J. Terrile, T.V. Johnson and G. Schwehm (1983) The evolution of spokes in Saturn's B Ring, *Icarus*, **54**, 227–252.

Grün, E., H.A. Zook, H. Fechtig and R.H. Giese (1985) Collisional balance of the meteoritic complex, *Icarus*, **62**, 244–272.

Grün, E., H.A. Zook, M. Baguhl, A. Balogh, S.J. Bame, H. Fechtig, R. Forsyth, M.S. Hanner, M. Horanyi, J. Kissel, B.A. Lindblad, D. Linkert, G. Linkert, I. Mann, J.A.M. McDonnell, G.E. Morfill, J.L. Phillips, C. Polanskey, G. Schwehm, N. Siddique, P. Staubach, J. Svestka and A. Taylor (1993) Discovery of Jovian dust streams and interstellar grains by the Ulysses spacecraft, *Nature*, **362**, 428–430.

Grün, E., P. Staubach, M. Baguhl, D.P. Hamilton, H.A. Zook, S. Dermott, B.A. Gustafson, H. Fechtig, J. Kissel, D. Linkert, G. Linkert, R. Srama, M.S. Hanner, C. Polanskey, M. Horányi, B.A. Lindblad, I. Mann, J.A.M. McDonnell, G.E. Morfill and G. Schwehm (1997) South-North and radial traverses through the interplanetary dust cloud, *Icarus*, **129**, 270–288.

Grün, E., H. Krüger, A.L. Graps, D.P. Hamilton, A. Heck, G. Linkert, H.A. Zook, S. Dermott, H. Fechtig, B.A. Gustafson, M.S. Hanner, M. Horányi, J. Kissel, B.A. Lindblad, D. Linkert, I. Mann, J.A.M. McDonnell, G.E. Morfill, C. Polanskey, G. Schwehm and R. Srama (1998) Galileo observes electromagnetically coupled dust in the Jovian magnetosphere, *J. Geophys. Res.*, **103**, 20011–20022.

Gurnett, D.A., E. Grün, D. Gallagher, W.S. Kurth and F.L. Scarf (1983) Micron-sized particles detected near Saturn by the Voyager plasma wave instrument, *Icarus*, **53**, 236–254.

Gurnett, D.A., W.S. Kurth, F.L. Scarf, J.A. Burns, J.N. Cuzzi and E. Grün (1987) Micron-sized particle impacts detected near Uranus by the Voyager 2 plasma wave instrument, *J. Geophys. Res.*, **92**, 14959–14968.

Gurnett, D.A., W.S. Kurth, L.J. Granroth, S.C. Allendorf and R.L. Poynter (1991) Micron-sized particles detected near Neptune by the Voyager 2 plasma wave instrument, *J. Geophys. Res.*, **96**, 19177–19186.

Hartquist, T.W., W. Pilipp and O. Havnes (1997) Dusty plasmas in interstellar clouds and star forming regions, *Astrophys. Space Sci.*, **246**, 243–289.

Havnes, O. (1980) On the motion and destruction of grains in interstellar clouds, *Astron. Astrophys.*, **90**, 106–112.

Havnes, O. (1984) Charges on dust particles, *Adv. Space Res.*, **4**, (9)75–(9)83.

Havnes, O. (1988) A streaming instability interaction between the solar wind and cometary dust, *Astron. Astrophys.*, **193**, 309–312.

Havnes, O., C.K. Goertz, G.E. Morfill, E. Grün and W. Ip (1987) Dust charges, cloud potential, and instabilities in a dust cloud embedded in a plasma, *J. Geophys. Res.*,

92, 2281–2287.

Havnes, O., T.K. Aanesen and F. Melandsø (1990) On dust charges and plasma potentials in a dusty plasma with dust size distribution, *J. Geophys. Res.*, **95**, 6581–6585.

Havnes, O., T. Aslaksen, T.W. Hartquist, F. Li, F. Melandsø, G.E. Morfill and T. Nitter (1995) Probing the properties of planetary ring dust by the observation of Mach cones, *J. Geophys. Res.*, **100**, 1731–1734.

Havnes, O., J. Trøim, T. Blix, W. Mortensen, L.I. Næsheim, E. Thrane and T. Tønnesen (1996) First detection of charged dust particles in the Earth's mesosphere, *J. Geophys. Res.*, **101**, 10839–10847.

Hazelton, R.C. and E.J. Yadlowsky (1994) Measurement of dust grain charging in a laboratory plasma, *IEEE Trans. Plasma Sci.*, **22**, 91–96.

Hill, J.R. and D.A. Mendis (1981) Electrostatic disruption of a charged conducting spheroid, *Can. J. Physics*, **59**, 897–901.

Homann, A., A. Melzer, S. Peters and A. Piel (1997) Determination of the dust screening length by laser-excited lattice waves, *Phys. Rev. E*, **56**, 7138–7141.

Homann, A., A. Melzer and A. Piel (1999) Measuring the charge on single particles by laser-excited resonances in plasma crystals, *Phys. Rev. E*, **59**, R3835–R3838.

Horányi, M. (1996) Charged dust dynamics in the solar system, *Annu. Rev. Astron. Astrophys.*, **34**, 383–418.

Horányi, M., S. Robertson and B. Walch (1995) Electrostatic charging properties of simulated lunar dust, *Geophys. Res. Lett.*, **22**, 2079–2082.

Horányi, M., S. Robertson and B. Walch (1998) *Physics of dusty plasmas.* AIP Conference Proceedings **446**, Woodbury.

Houpis, H.L.F. and E.C. Whipple Jr. (1987) Electrostatic charge on a dust size distribution in a plasma, *J. Geophys. Res.*, **92**, 12057–12068.

Huddleston, D.E., A.J. Coates, A.D. Johnstone and F.M. Neubauer (1993) Mass loading and velocity diffusion models for heavy pickup ions at comet Grigg-Skjellerup, *J. Geophys. Res.*, **98**, 20995–21002.

Humes, D.H. (1980) Results of Pioneer 10 and 11 meteoroid experiments: Interplanetary and near-Saturn, *J. Geophys. Res.*, **85**, 5841–5852.

I, L., W.-T. Juan, C.-H. Chiang and J.H. Chu (1996) Microscopic particle motions in strongly coupled dusty plasmas, *Science*, **272**, 1626–1628.

Ikezi, H. (1986) Coulomb solid of small particles in plasmas, *Phys. Fluids*, **29**, 1764–1766.

Ishihara, O. (1998) Instability due to the dust-particulate–phonon interaction, *Phys. Rev. E*, **58**, 3733–3738.

Ivlev, A.V., D. Samsonov, J. Goree, G. Morfill and V.E. Fortov (1999) Acoustic modes in a collisional dusty plasma, *Phys. Plasmas*, **6**, 741–750.

James, C.R. and F. Vermeulen (1968) A microparticle plasma, *Canad. J. Phys.*, **46**, 855–863.

Jana, M.R., A. Sen and P.K. Kaw (1993) Collective effects due to charge-fluctuation dynamics in a dusty plasma, *Phys. Rev. E*, **48**, 3930–3933.

Jana, M.R., A. Sen and P.K. Kaw (1995) Influence of grain charge fluctuation dynamics on collective modes in a magnetized dusty plasma, *Physica Scripta*, **51**, 385–389.

Jeans, J.H. (1929) *Astronomy and cosmogony.* Cambridge University Press.

Johnstone, A.D. (1991) *Cometary plasma processes.* Geophys. Monogr. Ser., vol. 61, American Geophysical Union, Washington, D.C.

Juhász, A. and M. Horányi (1999) Magnetospheric screening of cosmic dust, *J. Geophys. Res.*, **104**, 12577–12583.

Jurac, S., A. Baragiola, R.E. Johnson and E.C. Sittler Jr. (1995) Charging of ice grains by low-energy plasmas: Application to Saturn's E-ring, *J. Geophys. Res.*, **100**, 14821–14831.

Kakutani, T., H. Ono, T. Taniuti and C.-C. Wei (1968) Reductive perturbation analysis in nonlinear wave propagation II. Application to hydromagnetic waves in cold plasma, *J. Phys. Soc. Japan*, **24**, 1159–1166.

Kallenrode, M.-B. (1998) *Space physics. An introduction to plasmas and particles in the heliosphere and magnetospheres.* Springer-Verlag, Berlin.

Kaw, P.K. (1992) Non-Abelian screening and colour oscillations in a quark gluon plasma, *Plasma Phys. Contr. Fusion*, **34**, 1795–1802.

Kaw, P.K. and A. Sen (1998) Low frequency modes in strongly coupled dusty plasmas, *Phys. Plasmas*, **5**, 3552–3559.

Kaw, P.K. and R. Singh (1997) Collisional instabilities in a dusty plasma with recombination and ion-drift effects, *Phys. Rev. Lett.*, **79**, 423–426.

Khrapak, S.A., A.P. Nefedov, O.F. Petrov and O.S. Vaulina (1999) Dynamical properties of random charge fluctuations in a dusty plasma with different charging mechanisms, *Phys. Rev. E*, **59**, 6017–6022.

Kimura, H. and I. Mann (1998) The electric charging of interstellar dust in the solar system and consequences for its dynamics, *Astrophys. J.*, **499**, 454–462.

Kimura, H., H. Ishimoto and T. Mukai (1997) A study on solar dust ring formation based on fractal models, *Astron. Astrophys.*, **326**, 263–270.

Klimov, S., S. Savin, Ya. Aleksevich, G. Avanesova, V. Balebanov, M. Balikhin, A. Galeev, B. Gribov, M. Nozdrachev, V. Smirnov, A. Sokolov, O. Vaisberg, P. Oberc, Z. Krawczyk, S. Grzedzielski, J. Juchniewicz, K. Nowak, D. Orlowski, B. Parfianovich, D. Woźniak, Z. Zbyszynski, Ya. Voita and P. Triska (1986) Extremely-low-frequency plasma waves in the environment of comet Halley, *Nature*, **321**, 292–293.

Koide, M. and M. Sato (1998) Influence of dust-charge fluctuation on coherent structures in magnetized, inhomogeneous dusty plasmas, *J. Phys. Soc. Japan*, **67**, 1968–1972.

Kotsarenko, A.N., N.Ya. Kotsarenko and S.A. Silich (1998a) Low frequency surface waves in the magnetized dusty plasma, *Planet. Space Sci.*, **46**, 399–403.

Kotsarenko, N.Ya., S.V. Koshevaya, G.A. Stewart and D. Maravilla (1998b) Electrostatic spatially limited solitons in a magnetized dusty plasma, *Planet. Space Sci.*, **46**, 429–433.

Krishan, V. (1999) *Astrophysical plasmas and fluids.* Kluwer, Dordrecht.

Krivov, A., I. Mann and H. Kimura (1998a) The circumsolar dust complex and solar magnetic field, *Earth Planets Space*, **50**, 551–554.

Krivov, A., H. Kimura and I. Mann (1998b) Dynamics of dust near the Sun, *Icarus*, **134**, 311–327.

Laframboise, J.G. and L.J. Sonmor (1993) Current collection by probes and electrodes in space magnetoplasmas: A review, *J. Geophys. Res.*, **98**, 337–357.

Lakhina, G.S. and F. Verheest (1988) Alfvén wave instabilities and ring current during solar wind-comet interaction, *Astrophys. Space Sci.*, **143**, 329–338.

Lakshmi, S.V. and R. Bharuthram (1994) Arbitrary amplitude rarefactive dust-acoustic solitons, *Planet. Space Sci.*, **42**, 875–881.

Lakshmi, S.V., R. Bharuthram and M.Y. Yu (1993) Nonlinear potential structures in a dusty plasma, *Astrophys. Space Sci.*, **209**, 71–78.

Lakshmi, S.V., R. Bharuthram, N.N. Rao and P.K. Shukla (1997) Kinetic theory of nonlinear dust-acoustic waves in a dusty plasma, *Planet. Space Sci.*, **45**, 355–360.

Li, F., O. Havnes and F. Melandsø (1994) Longitudinal waves in a dusty plasma, *Planet. Space Sci.*, **42**, 401–407.

Lonngren, K.E. (1990) Expansion of a dusty plasma into a vacuum, *Planet. Space Sci.*, **38**, 1457–1459.

Luo, H. and M.Y. Yu (1992) Kinetic theory of self-similar expansion of a dusty plasma, *Phys. Fluids B*, **4**, 1122–1125.

Lyons, W.A., T.E. Nelson, E.R. Williams, J.A. Cramer and T.R. Turner (1998) Enhanced positive cloud-to-ground lightning in thunderstorms ingesting smoke from fires, *Science*, **282**, 77–80.

Ma, J.X. and J. Liu (1997) Dust-acoustic soliton in a dusty plasma, *Phys. Plasmas*, **4**, 253–255.

Ma, J.X. and M.Y. Yu (1994a) Self-consistent theory of ion acoustic waves in a dusty plasma, *Phys. Plasmas*, **1**, 3520–3522.

Ma, J.X. and M.Y. Yu (1994b) Langmuir wave instability in a dusty plasma, *Phys. Rev. E*, **50**, R2431–R2434.

Mace, R.L. and M.A. Hellberg (1993) Dust-acoustic double layers: Ion inertial effects, *Planet. Space Sci.*, **41**, 235–244.

Mace, R.L., F. Verheest and M.A. Hellberg (1998) Jeans stability of dusty space plasmas, *Phys. Lett. A*, **237**, 146–151.

Malik, H.K. and R. Bharuthram (1998) Small amplitude solitons in a magnetized dusty plasma with two-ion species, *Phys. Plasmas*, **5**, 3560–3564.

Malik, H.K., K.D. Tripathi and S.K. Sharma (1998) Dust acoustic solitons in a magnetized dusty plasma, *J. Plasma Phys.*, **60**, 265–273.

Mamun, A.A. (1998) Effects of dust temperature and fast ions on gravitational instability in a self-gravitating magnetized dusty plasma, *Phys. Plasmas*, **5**, 3542–3546.

Mamun, A.A., R.A. Cairns and P.K. Shukla (1996a) Solitary potentials in dusty plasmas, *Phys. Plasmas*, **3**, 702–704.

Mamun, A.A., R.A. Cairns and P.K. Shukla (1996b) Effects of vortex-like and non-thermal ion distributions on non-linear dust-acoustic waves, *Phys. Plasmas*, **3**, 2610–2614.

Mamun, A.A., M. Salahuddin and M. Salimullah (1999) Ultra-low-frequency electrostatic waves in a self-gravitating magnetized and inhomogeneous dusty plasma, *Planet. Space Sci.*, **47**, 79–83.

Mann, I. (1995) The out of ecliptic distribution of interplanetary dust and its relation to other bodies in the solar system, *Earth, Moon, and Planets*, **68**, 419–426.

Mann, I. (1996) Interstellar grains in the solar system: Requirements for an analysis, *Space Sci. Rev.*, **78**, 259–264.

McDonnell, J.A.M., R. Beard, S.F. Green and G.H. Schwehm (1992) Cometary coma particulate modelling for the ROSETTA mission aphelion rendezvous, *Ann. Geophys.*, **10**, 150–156.

Melandsø, F. (1996) Lattice waves in dust plasma crystals, *Phys. Plasmas*, **3**, 3890–3901.

Melandsø, F. and O. Havnes (1991) Oscillations and resonances in electrostatically supported dust rings, *J. Geophys. Res.*, **96**, 5837–5845.

Melandsø, F. and P.K. Shukla (1995) Theory of dust-acoustic shocks, *Planet. Space Sci.*, **43**, 635–648.

Melandsø, F., T. Aslaksen and O. Havnes (1993a) A new damping effect for the dust-acoustic wave, *Planet. Space Sci.*, **41**, 321–325.

Melandsø, F., T.K. Aslaksen and O. Havnes (1993b) A kinetic model for dust acoustic waves applied to planetary rings, *J. Geophys. Res.*, **98**, 13315–13323.

Melzer, A., T. Trottenberg and A. Piel (1994) Experimental determination of the charge on dust particles forming Coulomb lattices, *Phys. Lett. A*, **191**, 301–308.

Mendis, D.A. (1988) A postencounter view of comets, *Annu. Rev. Astron. Astrophys.*, **26**, 11–49.

Mendis, D.A. and M. Rosenberg (1994) Cosmic dusty plasma, *Annu. Rev. Astron. Astrophys.*, **32**, 419–463.

Mendis, D.A., H.L.F. Houpis and J.R. Hill (1982) The gravito-electrodynamics of charged dust in planetary magnetospheres, *J. Geophys. Res.*, **87**, 3449–3455.

Merlino, R.L., A. Barkan, C. Thompson and N. D'Angelo (1998) Laboratory studies of waves and instabilities in dusty plasmas, *Phys. Plasmas*, **5**, 1607–1614.

Meuris, P. (1997a) The influence of a dust size distribution on the dust-acoustic mode, *Planet. Space Sci.*, **45**, 1171–1174.

Meuris, P. (1997b) *Wave phenomena in dusty space plasmas.* PhD Thesis, Universiteit Gent, pp. 1–213.

Meuris, P. (1998) Structural aspects of solitons in dusty plasmas, *Physica Scripta*, **T75**, 186–188.

Meuris, P. and F. Verheest (1996a) Electrostatic instabilities due to charge fluctuations in dusty plasmas, in: *The physics of dusty plasmas.* Eds. P.K. Shukla, D.A. Mendis and V.W. Chow, World Scientific, Singapore, pp. 159–170.

Meuris, P. and F. Verheest (1996b) Korteweg-de Vries equation for magnetosonic modes in dusty plasmas, *Phys. Lett. A,* **219**, 299–302.

Meuris, P., N. Meyer-Vernet and J. Lemaire (1996) The detection of dust grains by a wire dipole antenna: The radio dust analyzer, *J. Geophys. Res.,* **101**, 24471–24477.

Meuris, P., F. Verheest and G.S. Lakhina (1997) Influence of dust mass distributions on generalized Jeans-Buneman instabilities in dusty plasmas, *Planet. Space Sci.,* **45**, 449–454.

Meyer-Vernet, N. (1982) "Flip-flop" of electric potential of dust grains in space, *Astron. Astrophys.,* **105**, 98–106.

Meyer-Vernet, N. (1984) Some constraints on particles in Saturn's Spokes., *Icarus,* **57**, 422–431.

Meyer-Vernet, N., M.G. Aubier and B.M. Pedersen (1986) Voyager-2 at Uranus: Grain impacts in the ring plane, *Geophys. Res. Lett.,* **13**, 617–620.

Meyer-Vernet, N., A. Lecacheux and B.M. Pedersen (1996) Constraints on Saturn's E ring from the Voyager 1 radio astronomy instrument, *Icarus,* **123**, 113–128.

Meyer-Vernet, N., A. Lecacheux and B.M. Pedersen (1998) Constraints on Saturn's G ring from the Voyager 2 radio astronomy instrument, *Icarus,* **132**, 311–320.

Mio, K., T. Ogino, K. Minami and S. Takeda (1976) Modified nonlinear Schrödinger equation for Alfvén waves propagating along the magnetic field in cold plasmas, *J. Phys. Soc. Japan,* **41**, 265–271.

Mirza, A.M., R.T. Faria, Jr. P.K. Shukla and G. Murtaza (1998) Vortex formation in sheared flow driven fluctuations in nonuniform magnetized dusty gases, *Phys. Rev. E,* **57**, 1047–1052.

Mofiz, U.A., M. Islam and Z. Ahmed (1993) Nonlinear propagation of ion-acoustic waves and low-frequency electrostatic modes in a dusty plasma, *J. Plasma Phys.,* **50**, 37–44.

Motschmann, U. and K-H. Glassmeier (1993) Nongyrotropic distribution of pickup ions at comet P/Grigg-Skjellerup: A possible source of wave activity, *J. Geophys. Res.,* **98**, 20977–20983.

Mott-Smith, H.M. and I. Langmuir (1926) The theory of collectors in gaseous discharges, *Phys. Rev.,* **28**, 727–763.

Mukai, T. (1981) On the charge distribution of interplanetary grains, *Astron. Astrophys.,* **99**, 1–6.

Mukai, T., W. Miyake, T. Terasawa, M. Kitayama and K. Hirao (1986) Plasma observation by Suisei of solar-wind interaction with comet Halley, *Nature,* **321**, 299–303.

Nakamura, Y., T. Yokota and P.K. Shukla (2000) *Frontiers in dusty plasmas.* Elsevier Science, Amsterdam.

Nakano, T. (1988) Gravitational instability of magnetized gaseous disks, *Publ. Astron. Soc. Japan,* **40**, 593–604.

Ness, N.F. (1994) Intrinsic magnetic fields of the planets: Mercury to Neptune, *Phil. Trans. R. Soc. Lond. A,* **349**, 249–260.

Newburn Jr., R.L., M. Neugebauer and J. Rahe (1991) *Comets in the post-Halley era.* Kluwer Academic Publishers, Dordrecht.

Nicholson, D.R. (1983) *Introduction to plasma theory.* Wiley, New York.

Northrop, T.G. (1992) Dusty plasmas, *Physica Scripta,* **45**, 475–490.

Northrop, T.G., D.A. Mendis and L. Schaffer (1989) Gyrophase drifts and the orbital evolution of dust at Jupiter's gossamer ring, *Icarus,* **79**, 101–115.

Ockert, M.E., J.N. Cuzzi, C.C. Porco and T.V. Johnson (1987) Uranian ring photometry: Results from Voyager 2, *J. Geophys. Res.,* **92**, 14696–14978.

Ockert-Bell, M.E., J.A. Burns, I.J. Daubar, P.C. Thomas, J. Veverka, M.J.S. Belton and K.P. Klaasen (1999) The structure of Jupiter's ring system as revealed by the Galileo imaging experiment, *Icarus*, **138**, 188–213.

Öpik, E.J. (1956) Interplanetary dust and terrestrial accretion of meteoritic matter, *Irish Astron.*, **4**, 84–135.

Ostrikov, K.N. and M.Y. Yu (1998) Ion-acoustic surface waves on a dielectric-dusty plasma interface, *IEEE Trans. Plasma Sci.*, **26**, 100–103.

Ostrikov, K.N., S.V. Vladimirov and M.Y. Yu (1999) Low-frequency surface waves in a structured magnetized dusty plasma, *J. Geophys. Res.*, **104**, 593–596.

Otani, N.F., A. Bhattacharjee and X. Wang (1999) A unified model of acoustic and lattice waves in a one-dimensional strongly coupled dusty plasma, *Phys. Plasmas*, **6**, 409–412.

Pandey, B.P. and G.S. Lakhina (1998) Jeans-Buneman instability in a dusty plasma, *Pramana*, **50**, 191–204.

Pandey, B.P., K. Avinash and C.B. Dwivedi (1994) Jeans instability of a dusty plasma, *Phys. Rev. E*, **49**, 5599–5606.

Parker, L.W. and B.L. Murphy (1967) Potential buildup on an electron-emitting ionospheric satellite, *J. Geophys. Res.*, **72**, 1631–1636.

Parks, G.K. (1991) *Physics of space plasmas: An introduction.* University of Washington, Seattle, Washington.

Pedersen, B.M., N. Meyer-Vernet, M.G. Aubier and P. Zarka (1991) Dust distribution around Neptune: Grain impacts near the ring plane measured by the Voyager planetary radio astronomy experiment, *J. Geophys. Res.*, **96**, 19187–19196.

Pieper, J.B. and J. Goree (1996) Dispersion of plasma dust acoustic waves in the strong-coupling regime, *Phys. Res. Lett.*, **77**, 3137–3140.

Pilipp, W., T.W. Hartquist, O. Havnes and G.E. Morfill (1987) The effects of dust on the propagation and dissipation of Alfvén waves in interstellar clouds, *Astrophys. J.*, **314**, 341–351.

Pillay, S.R., S.V. Singh, R. Bharuthram and M.Y. Yu (1997) Self-similar expansion of dusty plasmas, *J. Plasma Phys.*, **58**, 467–474.

Pillay, S.R., R. Bharuthram and F. Verheest (1999) The Jeans-Buneman instability in the presence of an ion beam in a dusty plasma and the influence of dust-size distributions, *Physica Scripta*, in press.

Popel, S.I. and M.Y. Yu (1995) Ion acoustic solitons in impurity-containing plasmas, *Contrib. Plasma Phys.*, **35**, 103–108.

Raadu, M.A. (1989) The physics of double layers and their role in astrophysics, *Phys. Reports*, **178**, 25–97.

Rao, N.N. (1993a) Hydromagnetic waves and shocks in magnetized dusty plasmas, *Planet. Space Sci.*, **41**, 21–26.

Rao, N.N. (1993b) Low-frequency waves in magnetized dusty plasmas, *J. Plasma Phys.*, **49**, 375–393.

Rao, N.N. (1993c) Dust-magnetoacoustic waves in magnetized dusty plasmas, *Phys. Scripta*, **48**, 363–366.

Rao, N.N. (1995) Magnetoacoustic modes in a magnetized dusty plasma, *J. Plasma Phys.*, **53**, 317–334.

Rao, N.N. (1998) Linear and nonlinear dust-acoustic waves in non-ideal dusty plasmas, *J. Plasma Phys.*, **59**, 561–574.

Rao, N.N. (1999) Dust-Coulomb waves in dense dusty plasmas, *Phys. Plasma*, **6**, 4414–4417.

Rao, N.N. and R. Bharuthram (1995) Self-similar expansion of a warm dusty plasma — II. Magnetized case, *Planet. Space Sci.*, **43**, 1087–1093.

Rao, N.N. and P.K. Shukla (1994) Nonlinear dust-acoustic waves with dust charge fluctuations, *Planet. Space Sci.*, **42**, 221–225.

Rao, N.N., P.K. Shukla and M.Y. Yu (1990) Dust-acoustic waves in dusty plasmas, *Planet. Space Sci.*, **38**, 543–546.

Rawat, S.P.S. and N.N. Rao (1993) Kelvin-Helmholtz instability driven by sheared dust flow, *Planet. Space Sci.*, **41**, 137–140.

Reddy, R.V., G.S. Lakhina, F. Verheest and P. Meuris (1996) Alfvén modes in dusty cometary and planetary plasmas, *Planet. Space Sci.*, **44**, 129–135.

Richardson, J.D. (1995) An extended plasma model for Saturn, *Geophys. Res. Lett.*, **22**, 1177–1180.

Richter, K., W. Curdt and H.U. Keller (1991) Velocity of individual large dust particles ejected from Comet P/Halley, *Astron. Astrophys.*, **250**, 548–555.

Rosenberg, M. (1993) Ion- and dust-acoustic instabilities in dusty plasmas, *Planet. Space Sci.*, **41**, 229–233.

Rosenberg, M. and V.W. Chow (1999) Collisional effects on the electrostatic dust cyclotron instability, *J. Plasma Phys.*, **61**, 51–63.

Rosenberg, M. and G. Kalman (1997) Dust acoustic waves in strongly coupled dusty plasmas, *Phys. Rev. E*, **56**, 7166–7173.

Rosenberg, M. and N.A. Krall (1994) High frequency drift instabilities in a dusty plasma, *Planet. Space Sci.*, **42**, 889–894.

Rosenberg, M. and N.A. Krall (1995) Modified two-stream instabilities in dusty space plasmas, *Planet. Space Sci.*, **43**, 619–624.

Rosenberg, M. and N.A. Krall (1996) Low frequency drift instabilities in a dusty plasma, *Phys. Plasmas*, **3**, 644–649.

Rubinstein, J. and J.G. Laframboise (1982) Theory of a spherical probe in a collisionless magnetoplasma, *Phys. Fluids*, **25**, 1174–1182.

Salimullah, M. (1996) Low-frequency dust-lower-hybrid modes in a dusty plasma, *Phys. Lett. A*, **215**, 296–298.

Salimullah, M. and A. Sen (1992) Low frequency response of a dusty plasma, *Phys. Lett. A*, **163**, 82–86.

Salimullah, M. and P.K. Shukla (1999) On the stability of self-gravitating magnetized dusty plasmas, *Phys. Plasmas*, **6**, 686–691.

Salimullah, M., M.H.A. Hassan and A. Sen (1992) Low-frequency electrostatic modes in a magnetized dusty plasma, *Phys. Rev. A*, **45**, 5929–5934.

Sharma, O.P. and V.L. Patel (1986) Low-frequency electromagnetic waves driven by gyrotropic gyrating ion beams, *J. Geophys. Res.*, **91**, 1529–1534.

Shkarofsky, I.P., T.W. Johnston and M.P. Bachynski (1966) *The particle kinetics of plasmas.* Addison-Wesley, Reading, Mass.

Showalter, M.R. (1995) Arcs and clumps in the Uranian λ ring, *Science*, **267**, 490–493.

Showalter, M.R. (1998) Detection of centimeter-sized meteoroid impact events in Saturn's F ring, *Science*, **282**, 1099–1102.

Showalter, M.R. and J.N. Cuzzi (1993) Seeing ghosts: Photometry of Saturn's G-ring, *Icarus*, **103**, 124–143.

Showalter, M.R., J.A. Burns, J.N. Cuzzi and J.B. Pollack (1985) Discovery of Jupiter's 'gossamer' ring, *Nature*, **316**, 526–528.

Showalter, M.R., J.A. Burns, J.N. Cuzzi and J.B. Pollack (1987) Jupiter's ring system: New results on structure and particle properties, *Icarus*, **69**, 458–498.

Showalter, M.R., J.N. Cuzzi and S.M. Larson (1991) Structure and particle properties of Saturn's E ring, *Icarus*, **94**, 451–473.

Showalter, M.R., J.B. Pollack, M.E. Ockert, L.R. Doyle and J.B. Dalton (1992) A photometric study of Saturn's F-ring, *Icarus*, **100**, 394–411.

Shukla, P.K. (1992) Low-frequency modes in dusty plasmas, *Physica Scripta*, **45**, 504–507.

Shukla, P.K. and H.U. Rahman (1996) Low-frequency electromagnetic waves in nonuniform gravitating dusty magnetoplasmas, *Planet. Space Sci.*, **44**, 469–472.

Shukla, P.K. and N.N. Rao (1993) Vortex structures in magnetized plasmas with sheared dust flow, *Planet. Space Sci.*, **41**, 401–403.

Shukla, P.K. and V.P. Silin (1992) Dust ion-acoustic wave, *Physica Scripta*, **45**, 508.

Shukla, P.K. and R.K. Varma (1993) Convective cells in nonuniform dusty plasmas, *Phys. Fluids B*, **5**, 236–237.

Shukla, P.K., M.Y. Yu and R. Bharuthram (1991) Linear and nonlinear dust drift waves, *J. Geophys. Res.*, **96**, 21343–21346.

Shukla, P.K., R.K. Varma, V. Krishan and J.F. McKenzie (1993) Alfvén vortices in nonuniform dusty magnetoplasmas, *Phys. Rev. E*, **47**, 750–752.

Shukla, P.K., D.A. Mendis and V.W. Chow (1996) *The physics of dusty plasmas*. World Scientific, Singapore.

Shukla, P.K., D.A. Mendis and T. Desai (1997a) *Advances in dusty plasmas*. World Scientific, Singapore.

Shukla, P.K., G.T. Birk and G. Morfill (1997b) Dust-acoustic waves in partially ionized dusty plasmas, *Physica Scripta*, **56**, 299–301.

Singh, N., B.I. Vashi and L.C. Leung (1994) Three-dimensional numerical simulation of current collection by a probe in a magnetized plasma, *Geophys. Res. Lett.*, **21**, 833–836.

Singh, P.D., A.A. De Almeida and W.F. Hübner (1992) Dust release rates and dust-to-gas mass ratios of eight comets, *Astron. J.*, **104**, 848–858.

Singh, P.D., W.F. Hübner, D.D. Costa, J.C. Landaberry and J.A. de Freitas Pacheco (1997) Gas and dust release rates and color of dust in comets P/Halley (1986 III), P/Giacobini-Zinner (1985 XIII), and P/Hartley-Good (1985 XVII), *Planet. Space Sci.*, **45**, 455–467.

Singh, S.V. and N.N. Rao (1997) Adiabatic dust-acoustic solitons, *Phys. Lett. A*, **235**, 164–168.

Sittler Jr., E.C., K.W. Ogilvie and J.D. Scudder (1983) Survey of low-energy plasma electrons in Saturn's magnetosphere: Voyagers 1 and 2, *J. Geophys. Res.*, **88**, 8847–8870.

Smith, B.A., L. Soderblom, R. Beebe, J. Boyce, G. Briggs, A. Bunker, S.A. Collins, C.J. Hansen, T.V. Johnson, J.L. Mitchell, R.J. Terrile, M. Carr, A.F. Cook II, J. Cuzzi, J.B. Pollack, G.E. Danielson, A. Ingersoll, M.E. Davies, G.E. Hunt, H. Masursky, E. Shoemaker, D. Morrison, T. Owen, C. Sagan, J. Veverka, R. Strom and V.E. Suomi (1981) Encounter with Saturn: Voyager 1 imaging science results, *Science*, **212**, 163–191.

Smith, B.A., L. Soderblom, R. Batson, P. Bridges, J. Inge, H. Masursky, E. Shoemaker, R. Beebe, J. Boyce, G. Briggs, A. Bunker, S.A. Collins, C.J. Hansen, T.V. Johnson, J.L. Mitchell, R.J. Terrile, A.F. Cook II, J. Cuzzi, J.B. Pollack, G.E. Danielson, A. Ingersoll, M.E. Davies, G.E. Hunt, D. Morrison, T. Owen, C. Sagan, J. Veverka, R. Strom and V.E. Suomi (1982) A new look at the Saturn system: The Voyager 2 images, *Science*, **215**, 504–537.

Sodha, M.S. and S. Guha (1971) Physics of colloidal plasmas, in: *Advances in plasma physics*. Eds. A. Simon and W.B. Thompson, Interscience, New York, vol. 4, pp. 219–309.

Sonmor, L.J. and J.G. Laframboise (1991) Exact current to a spherical electrode in a collisionless, large-Debye-length magnetoplasma, *Phys. Fluids B*, **3**, 2472–2490.

Sorrell, W.H. (1997) Interstellar grains as amino acid factories and the origin of life, *Astrophys. Space Sci.*, **253**, 27–41.

Spitzer, L. (1941) The dynamics of the interstellar medium. I. Local equilibrium, *Astrophys. J.*, **93**, 369–379.

Spitzer, L. (1978), *Physical Processes in Interstellar Medium*. Wiley, New York.

Stix, T.H. (1962) *The theory of plasma waves*. McGraw-Hill, New York.

Stix, T.H. (1992) *Waves in plasmas*. American Institute of Physics, New York.

Tagare, S.G. (1997) Dust-acoustic solitary waves and double layers in dusty plasmas consisting of cold dust particles and two-temperature isothermal ions, *Phys. Plasmas*, **4**, 3167–3172.

Thomas, H., G.E. Morfill, V. Demmel, J. Goree, B. Feuerbacher and D. Möhlmann (1994) Plasma crystal: Coulomb crystallization in a dusty plasma, *Phys. Rev. Lett.*, **73**, 652–655.

Thomsen, M.F., C.K. Goertz, T.G. Northrop and J.R. Hill (1982) On the nature of particles in Saturn's spokes, *Geophys. Res. Lett.*, **9**, 423–426.

Thompson, C., A. Barkan, N. D'Angelo and R.L. Merlino (1997) Dust acoustic waves in a direct current glow discharge, *Phys. Plasmas*, **4**, 2331–2335.

Thorne, R.M. and B.T. Tsurutani (1987) Resonant interactions between cometary ions and low frequency electromagnetic waves, *Planet. Space Sci.*, **35**, 1501–1511.

Toida, M. and Y. Ohsawa (1994) KdV equations for high- and low-frequency magnetosonic waves in a multi-ion plasma, *J. Phys. Soc. Japan*, **63**, 573–582.

Toomre, A. (1964) On the gravitational stability of a disk of stars, *Astrophys. J.*, **139**, 1217–1238.

Tripathi, K.D. and S.K. Sharma (1996a) Self-consistent charge dynamics in magnetized dusty plasmas: Low-frequency electrostatic modes, *Phys. Rev. E*, **53**, 1035–1041.

Tripathi, K.D. and S.K. Sharma (1996b) Dispersion properties of low-frequency waves in magnetized dusty plasmas with dust size distribution, *Phys. Plasmas*, **3**, 4380–4385.

Tsintikidis, D., D. Gurnett, L.J. Granroth, S.C. Allendorf and W.S. Kurth (1994) A revised analysis of micron-sized particles detected near Saturn by the Voyager 2 plasma wave instrument, *J. Geophys. Res.*, **99**, 2261–2270.

Tsintikidis, D., W.S. Kurth, D.A. Gurnett and D.D. Barbosa (1995) Study of dust in the vicinity of Dione using the Voyager 1 plasma wave instrument, *J. Geophys. Res.*, **100**, 1811–1822.

Tsurutani, B.T. (1991) Comets: A laboratory for plasma waves and instabilities, in: *Cometary plasma processes*, Geophys. Monogr. Ser., vol. 61. Ed. A.D. Johnstone, American Geophysical Union, Washington, D.C., pp. 189–209.

Tsurutani, B.T. and E.J. Smith (1986) Strong hydromagnetic turbulence associated with Comet Giacobini-Zinner, *Geophys. Res. Lett.*, **13**, 259–262.

Tsurutani, B.T., R.M. Thorne, E.J. Smith, J.T. Gosling and H. Matsumoto (1987) Steepened magnetosonic waves at comet Giacobini-Zinner, *J. Geophys. Res.*, **92**, 11074–11082.

Tsytovich, V.N. and U. De Angelis (1999) Kinetic theory of dusty plasmas: I. General approach, *Phys. Plasmas*, **6**, 1093–1106.

Tsytovich, V.N. and O. Havnes (1993) Charging processes, dispersion properties and anomalous transport in dusty plasmas, *Comm. Plasma Phys. Contr. Fusion*, **15**, 267–280.

Tsytovich, V.N., U. De Angelis, R. Bingham and D. Resendes (1999) Long-range correlations in dusty plasmas, *Phys. Plasmas*, **4**, 3882–3894.

Vaisberg, O.L., V.N. Smirnov, L.S. Gorn, M.V. Iovlev, M.A. Balikchin, S.I. Klimov, S.P. Savin, V.D. Shapiro and V.I. Shevchenko (1986) Dust coma structure of comet Halley from SP-1 detector measurements, *Nature*, **321**, 274–276.

Varma, R.K. and P.K. Shukla (1995a) Linear and nonlinear Rayleigh-Taylor modes in nonuniform dusty magnetoplasmas, *Physica Scripta*, **51**, 522–525.

Varma, R.K. and P.K. Shukla (1995b) A new dust-dynamics-induced interchange instability in dusty plasmas, *Phys. Lett. A*, **196**, 342–345.

Varma, R.K., P.K. Shukla and V. Krishan (1993) Electrostatic oscillations in the presence of grain-charge perturbations in dusty plasmas, *Phys. Rev. E*, **47**, 3612–3616.

Verheest, F. (1967) General dispersion relations for linear waves in multicomponent plasmas, *Physica*, **34**, 17–35.

Verheest, F. (1988) Ion-acoustic solitons in multi-component plasmas including negative ions at critical densities, *J. Plasma Phys.*, **39**, 71–79.

Verheest, F. (1990) Nonlinear parallel Alfvén waves in cometary plasmas, *Icarus*, **86**, 273–282.

Verheest, F. (1992a) Parallel solitary Alfvén waves in warm multispecies beam-plasma systems. Part 2. Anisotropic pressures, *J. Plasma Phys.*, **47**, 25–37.

Verheest, F. (1992b) Nonlinear dust-acoustic waves in multispecies dusty plasmas, *Planet. Space Sci.*, **40**, 1–6.

Verheest, F. (1993) Are weak dust-acoustic double layers adequately described by modified Korteweg-de Vries equations?, *Phys. Scripta*, **47**, 274–277.

Verheest, F. (1996) Waves and instabilities in dusty space plasmas, *Space Sci. Rev.*, **77**, 267–302.

Verheest, F. and B. Buti (1992) Parallel solitary Alfvén waves in warm multispecies beam-plasma systems. Part 1, *J. Plasma Phys.*, **47**, 15–24.

Verheest, F. and M.A. Hellberg (1997) Bohm sheath criteria and double layers in multispecies plasmas, *J. Plasma Phys.*, **57**, 465–477.

Verheest, F. and M.A. Hellberg (1999) General discussion of nonlinear electrostatic modes in multispecies plasmas, *Physica Scripta*, **T82**, 98-105.

Verheest, F. and G.S. Lakhina (1996) Oblique solitary Alfvén modes in relativistic electron-positron plasmas, *Astrophys. Space Sci.*, **240**, 215–224.

Verheest, F. and P. Meuris (1995) Whistler-like instabilities due to charge fluctuations in dusty plasmas, *Phys. Lett. A*, **198**, 228–232.

Verheest, F. and P. Meuris (1996a) Electromagnetic instabilities in solar wind interaction with dusty cometary plasmas, in: *Solar Wind Eight*. Eds. D. Winterhalter, J.T. Gosling, S.R. Habbal, W.S. Kurth and M. Neugebauer, AIP Press, Woodbury, New York, pp. 343–346.

Verheest, F. and P. Meuris (1996b) Nonlinear electromagnetic modes in plasmas with variable dust charges, *Phys. Lett. A*, **210**, 198–201.

Verheest, F. and P. Meuris (1996c) Korteweg-de Vries equation for oblique modes in magnetized multi-ion plasmas, *J. Phys. Soc. Japan*, **65**, 2522–2525.

Verheest, F. and P. Meuris (1998) Interaction of the solar wind with dusty cometary plasmas, *Physica Scripta*, **T75**, 84–87.

Verheest, F. and P.K. Shukla (1995) Linear and quasilinear Alfvén waves in dusty plasmas with anisotropic pressures, *Physica Scripta*, **T60**, 136–139.

Verheest, F. and P.K. Shukla (1997) Nonlinear waves in multispecies self-gravitating dusty plasmas, *Physica Scripta*, **55**, 83–85.

Verheest, F., M.A. Hellberg, G.J. Gray and R.L. Mace (1996) Electrostatic solitons in multispecies electron-positron plasmas, *Astrophys. Space Sci.*, **239**, 125–139.

Verheest, F., P.K. Shukla, N.N. Rao and P. Meuris (1997a) Dust-acoustic waves in self-gravitating dusty plasmas with fluctuating dust charges, *J. Plasma Phys.*, **58**, 163–170.

Verheest, F., P. Meuris, R.L. Mace and M.A. Hellberg (1997b) Alfvén-Jeans and magnetosonic modes in multispecies self-gravitating dusty plasmas, *Astrophys. Space Sci.*, **254**, 253–267.

Verheest, F., M.A. Hellberg and R.L. Mace (1999a) Self-gravitational magnetosonic modes in dusty plasmas with quasi-inertialess plasma constituents, *Phys. Plasmas*, **6**, 279–284.

Verheest, F., G.S. Lakhina and B.T. Tsurutani (1999b) Intermediate electromagnetic turbulence at comets, *J. Geophys. Res.*, **104**, 24863–24868.

Vladimirov, S.V. (1994b) Propagation of waves in dusty plasmas with variable charges on dust particles, *Phys. Plasmas*, **1**, 2762–2767.

Vladimirov, S.V. and N.F. Cramer (1996) Nonlinear Alfvén waves in magnetized plasmas with heavy impurities or dust, *Phys. Rev. E*, **54**, 6762–6768.

Vladimirov, S.V. and V.N. Tsytovich (1998) Dissipative drift waves in partially ionized plasmas containing high-Z impurities or dust, *Phys. Rev. E*, **58**, 2415–2423.

Vranješ, J. and Čadež, V.M. (1990) Gravitational instability of a homogeneous gas cloud with radiation, *Astrophys. Space Sci.*, **164**, 329–331.

Walch, B., M. Horányi and S. Robertson (1995) Charging of dust grains in plasma with energetic electrons, *Phys. Rev. Lett.*, **75**, 838–841.

Wang, X. and A. Bhattacharjee (1997) Hydrodynamic waves and correlation functions in dusty plasmas, *Phys. Plasmas*, **4**, 3759–3764.

Wang, X. and A. Bhattacharjee (1998) Pair correlations in strongly coupled dusty plasmas, *Phys. Rev. E*, **58**, 4967–4972.

Whipple, E.C. (1985) Potentials of surfaces in space, *Rep. Prog. Phys.*, **44**, 1197–1250.

Whipple, E.C., T.G. Northrop and D.A. Mendis (1985) The electrostatics of a dusty plasma, *J. Geophys. Res.*, **90**, 7405–7413.

Wilson, G.R. (1991) The plasma environment, charge state, and currents of Saturn's C and D rings, *J. Geophys. Res.*, **96**, 9689–9701.

Winske, D. and M. Rosenberg (1998) Nonlinear development of the dust acoustic instability in a collisional dusty plasma, *IEEE Trans. Plasma Sci.*, **26**, 92–96.

Winske, D., C.S. Wu, Y.Y. Li, Z.Z. Mou and S.Y. Guo (1985) Coupling of newborn ions to the solar wind by electromagnetic instabilities and their interaction with the bow shock, *J. Geophys. Res.*, **90**, 2713–2726.

Winske, D., S.P. Gary, M.E. Jones, M. Rosenberg, V.W. Chow and D.A. Mendis (1995) Ion heating in a dusty plasma due to the dust/ion acoustic instability, *Geophys. Res. Lett.*, **22**, 2069–2072.

Wu, C.S. and R.C. Davidson (1972) Electromagnetic instabilities produced by neutral-particle ionization in interplanetary space, *J. Geophys. Res.*, **77**, 5399–5406.

Xu, W.J., N. D'Angelo and R.L. Merlino (1993) Dusty plasmas: The effect of closely packed grains, *J. Geophys. Res.*, **98**, 7843–7847.

Yaroshenko, V.V. (1997) Instability of longitudinal waves in a dusty plasma of a narrow planetary ring, *Plasma Phys. Rep.*, **23**, 433–439.

Yu, M.Y. and R. Bharuthram (1994) Self-similar cylindrical expansion of impurity particles in a plasma, *J. Plasma Phys.*, **52**, 345–352.

Yu, M.Y. and H. Luo (1995) Adiabatic self-similar expansion of dust grains in a plasma, *Phys. Plasmas*, **2**, 591–593.

INDEX